INSULIN ACTION

Developments in Molecular and Cellular Biochemistry

Series Editor: Naranjan S. Dhalla, Ph.D., M.D. (Hon.), FACC

KLUWER ACADEMIC PUBLISHERS – DORDRECHT / BOSTON / LONDON

Insulin Action

Edited by

ASHOK K. SRIVASTAVA

Centre Hospitalier de l'Université de Montreal
Pavillion Hotel Dieu
3840, Rue Saint-Urbain
Montreal, Quebec
H2W 1T8 Canada

and

BARRY I. POSNER

Polypeptide Hormone Laboratory
Strathcona Anatomy & Dentistry Building
3640 University Street, Room W3, 15
Montreal, Quebec
H3A 2B2 Canada

Kluwer Academic Publishers

Dordrecht / Boston / London

Library of Congress Cataloging-in-Publication Data

```
Insulin action / edited by Ashok K. Srivastava, Barry Posner.
      p.   cm. -- (Developments in molecular and cellular
  biochemistry)
    ISBN 0-7923-8113-0
    1. Insulin--Physiological effect--Congresses.  2. Insulin-
  -Receptors--Congresses.   I. Srivastava, Ashok K.  II. Posner, Barry
  I., 1937-   .  III. Series.
    [DNLM: 1. Insulin--physiology.  2. Receptors, Insulin--physiology.
  W1 DE998D 1998 / WK 820 I56 1998]
  QP572.I5I49   1998
  572'.565--dc21
  DNLM/DLC
                                                        97-49618
                                                           CIP
```

ISBN 0-7923-8113-0

Published by Kluwer Academic Publishers,
P.O. Box 17, 3300 AA Dordrecht, The Netherlands.

Sold and distributed in North, Central and South America
by Kluwer Academic Publishers,
101 Philip Drive, Norwell, MA 02061, U.S.A.

In all other countries, sold and distributed
by Kluwer Academic Publishers,
P.O. Box 322, 3300 AH Dordrecht, The Netherlands.

Printed on acid-free paper

Printed in the Netherlands.

Molecular and Cellular Biochemistry:

An International Journal for Chemical Biology in Health and Disease

CONTENTS VOLUME 182, Nos. 1 & 2, May 1998

INSULIN ACTION
Ashok K. Srivastava and Barry I. Posner, guest editors

Molecular and Cellular Biochemistry **182**: 1–2, 1998.

Preface

In 1996 the 75th anniversary of the discovery of insulin was celebrated at the University of Toronto, the scene of that discovery in 1921. This volume was stimulated by the scientific program which was staged at that time and brought together much of the world's best talent to discuss and analyze the most recent developments in our understanding of pancreatic function, insulin secretion, the interaction of insulin with its target tissues, the mechanism of insulin action at the cellular level, and the defects which underlie both Type I (insulin dependent diabetes mellitus, IDDM) and Type II (noninsulin dependent diabetes mellitus, NIDDM) forms of the disease. We have chosen to focus the present volume on work related to insulin action.

Thus a first group of papers delineates early signalling events following the binding of insulin to its receptor including the key substrates (viz. IRS-1 etc.) which are tyrosine phosphorylated by the activated insulin receptor kinase (IRK) which then results in the docking of transducing proteins via SH2 domains (White, Ogawa *et al.*) and the consequent activation of more distal steps in the signalling pathway. These more distal steps include the activation of the ras pathway and an attendant group of ser/thr kinases (Ceresa and Pessin), as well as a number of other ser/thr kinases (J. Avruch). Possible roles for P-ser/P-thr phosphatases are also addressed in a paper by Ragolia and Begum.

A second group of papers addresses factors involved in modulating IRK function. Thus the rapid internalization of the activated IRK into endosomes implicates this compartment as a site at which insulin signalling occurs but also as a site at which IRK function is critically regulated (e.g. by the presence of a unique insulin degrading enzyme) (Di Guglielmo *et al.*). The role of other aspects of compartmentalization (e.g. caveoli) as a mode by which insulin action is regulated is explored by Mastic *et al*. The nature of a unique class of proteins which bind to the tyrosine-phosphorylated IRK and inhibit its signalling capacity, presumably by competing with SH2 transducing proteins required to effect signalling, is described by Liu and Roth. The key role of phosphotryosine phosphatases (PTPs) in regulating IRK activity is also considered (Drake and Posner, Goldstein *et al.*, Byon *et al.*). Work in this area has lead to the recognition that one or more PTP(S) associated primarily with endosomes dramatically restrains autoactivation of the

IRK and maintains it in the basal state. Since inhibition of this enzyme(s) promotes IRK activation and insulin signalling in the complete absence of insulin a basis for developing a novel class of insulin mimetics has emerged (Drake and Posner). Goldstein *et al.* summarize work implicating the transmembrane receptor-type PTP, LAR as playing a key regulatory role while Byon *et al.* discuss the role of PTP-1B as a negative regulator of insulin action.

The effect of insulin and other agents on target tissues is expounded in a group of papers starting with the observations of Fantus and Tsiani on the insulin mimetic effects of vanadate. Of particular interest is their demonstration that vanadate acts via a different mechanism from that of insulin, a feature which might recommend vanadate-based agents in the treatment of insulin resistant states seen in Type II diabetics. The role of insulin in regulating Na^+/K^+ ATPase is explored as well as the possible mechanisms by which this regulation is effected (Sweeney and Klip). The regulation of glycogen synthase, a major molecular target of insulin, is studied by Srivastava and Pandey who examine the role of several insulin-stimulated pathways in effecting glycogen synthase activation. The significance of the insulin-sensitive glucose transporter (GLUT4) of skeletal muscle in modulating whole body glucose homeostasis and insulin sensitivity is highlighted by the work of Charron and Katz using several transgenic mouse lines overexpressing or lacking skeletal GLUT4 molecules.

A final group of studies deals with pathogenic considerations responsible for NIDDM. Zierath *et al.* provide data from human studies implicating hyperglycemia itself as a significant factor which down regulates the insulin signalling cascade and contributes to the insulin resistance seen in this condition. Lamothe *et al.* describe the use of transgenic approaches to investigating those genes whose disruption influences insulin action and/or pancreatic function. Peraldi and Spiegelman review the work implicating tumor necrosis factor α(TNF-α) as an important effector of insulin resistance in obesity states. The work of Goldfine *et al.* in delineating the role of a membrane glycoprotein (PC-1) as an inhibitor of the IRK and an agent contributing to insulin resistance states is summarized. Finally the role of the antidiabetic agent pioglitazone in promoting insulin action in hepatocytes from diabetic animals is summarized by Pugazhenthi and Khandelwal.

We wish to take this opportunity to thank the contributing authors for their cooperation and sustained interest in putting this volume together. We thank Ms. Susanne Bordeleau-Chenier and Ms. Sheryl Jackson for their superb secretarial assistance, and wish to express our appreciation to Dr. Naranjan S. Dhalla, the Editor-in-Chief, *Molecular and Cellular Biochemistry* for facilitating the publication of this focussed issue.

ASHOK K. SRIVASTAVA
Centre Hospitalier de l'Universite de Montreal
Campus Hotel Dieu
3840 rue Saint-Urbain
Montreal, Quebec, H2W 1T8, Canada

BARRY I. POSNER
Polypeptide Laboratory and SMBD-Jewish General Hospital
McGill University, Room W315
Strathcona Medical Building, 3640 University Street
Montreal, Quebec, H3A 2B2, Canada

Molecular and Cellular Biochemistry **182**: 3–11, 1998.
© 1998 *Kluwer Academic Publishers. Printed in the Netherlands.*

The IRS-signalling system: A network of docking proteins that mediate insulin action

Morris F. White

Research Division, Joslin Diabetes Center and the Graduate Program in Biomedical and Biological Sciences, Harvard Medical School, Boston, MA, USA

Abstract

New molecules discovered during the past ten years have created a rational framework to understand signalling transduction by a broad range of growth factors and cytokines, including insulin. Insulin action is initiated through the insulin receptor, a transmembrane glycoprotein with intrinsic protein tyrosine kinase activity. The tyrosine kinase mediates the insulin response through tyrosine phosphorylation of various cellular substrates, in particular the IRS-proteins. During insulin-stimulated tyrosine phosphorylation, the IRS-proteins mediate a broad biological response by binding and activating various enzymes or adapter molecules. Although we are far from a complete understanding of the insulin signalling system and its failure, enough pieces of the puzzle are falling into place that mechanism-based solutions to insulin resistance encountered with type II diabetes may soon be attainable. (Mol Cell Biochem **182**: 3–11, 1998)

Key words: IRS-1 proteins, PI3-kinase, Ras-MAP kinase, SHP2, diabetes

Introduction

The integration of multiple transmembrane signals is especially important during development and maintenance of the nervous system, communication between cells of the immune system, evolution of transformed cells, and metabolic control [1]. Tyrosine phosphorylation plays a key role in many of these processes by directly controlling the activity of receptors or enzymes at early steps in signalling cascades, or by coordinating the assembly of multicomponent signalling complexes around activated receptors or docking proteins [2]. During insulin signalling, IRS-proteins function as insulin receptor-specific docking proteins to engage multiple downstream signalling molecules.

The insulin receptor substrate, IRS-1, was the first docking protein identified and serves as the prototype for this class of molecule [3]. Four related proteins have recently expanded this family, including IRS-2, IRS-3, Gab-1 and p62^{dok} [4–6]. These docking proteins are not related by extensive amino acid sequence identities, but are related functionally as insulin receptor substrates (IRS-proteins). IRS-proteins contain several common structures: an NH_2-terminal PH and/or PTB domain that mediates protein-lipid or protein-protein interactions; multiple COOH-terminal tyrosine residues that create SH2-protein binding site; proline-rich regions to engage SH3 or WW domain; and serine/threonine-rich regions which may regulate overall function through other protein-protein interactions (Fig. 1).

Docking proteins provide several important features for receptor signalling. First, they amplify receptor signals by eliminating the stoichiometric constraints encountered by receptors which directly recruit SH2-proteins to their autophosphorylation sites. They also dissociate the intracellular signalling complex from the endocytic pathways of the activated receptor. This feature may be especially important for insulin-stimulated biological effects such as glucose uptake, which involves signal transmission to membrane compartments that are inaccessible to the insulin receptor [7]. Docking proteins also expand the repertoire of signalling pathways that can be regulated, since several isoforms can be engaged and phosphorylated by a single activated receptor. Moreover, a single docking protein can integrate signals from various receptors by serving as a common substrate to integrate multiple inputs controlling cellular growth and metabolism [8, 9].

Address for offprints: M.F. White, Research Division, Joslin Diabetes Center, 1 Joslin Place, Boston, MA 02215, USA

4

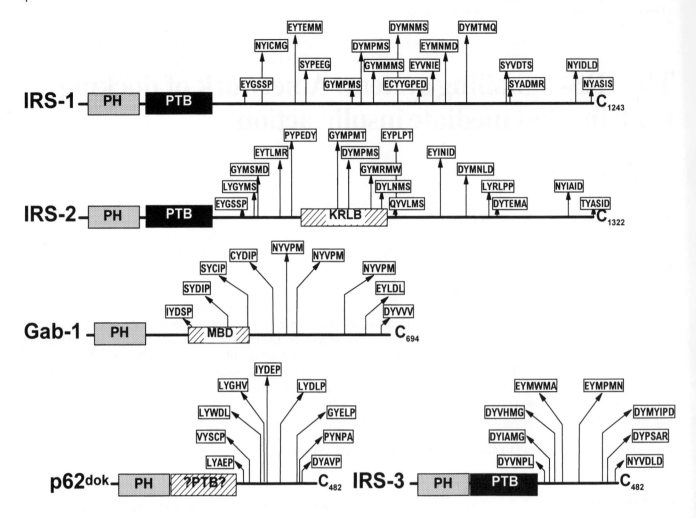

Fig. 1. IRS-protein family members. Linear depiction of IRS-1, 2, Gab-1, p62[dok], and sin showing receptor interaction domains and tyrosine phosphorylation motifs. PH – pleckstrin homology domain; PTB – phosphotyrosine binding domain; KRLB – kinase regulatory loop binding domain; MBD – c-Met binding domain; SH3 – src-homology-3 domain; Pro – proline-rich domain.

Members of the IRS-protein family

The identification of IRS-1 and IRS-2
The finding that insulin stimulates tyrosine phosphorylation of a 185 kDa protein in all cells and tissues lead to the hypothesis that substrate phosphorylation mediates insulin signal transduction [10]. IRS-1 was the first insulin receptor substrate purified and cloned, and serves as a prototype for docking proteins [3]. It is encoded by a single exon on human chromosome 2q36-37 (mouse chromosome 1); IRS-1 has a calculated molecular mass of 132 kDa, but owing to extensive serine phosphorylation it migrates near 185 kDa during SDS-PAGE [3, 11]. Similar but immunologically distinct proteins in myeloid progenitor cells (the IL-4 receptor substrate, 4PS), and in hepatocytes and skeletal muscle from the mice lacking IRS-1, lead to the purification of 4PS and the cloning of IRS-2 [4]. IRS-2 is about 10 kDa larger that IRS-1 and migrates slightly slower during SDS-PAGE. IRS-2 is also encoded by a single exon residing on the short arm of mouse chromosome 8 near the insulin receptor gene (human chromosome 13q341) [12].

A comparison of the amino acid sequences of IRS-1 and IRS-2 reveals several features, including a well conserved pleckstrin homology (PH) domain at the extreme NH$_2$-terminus, followed immediately by a phosphotyrosine binding (PTB) domain that binds to phosphorylated NPXY-motifs (Fig. 1). The COOH-terminal regions of IRS-1 and IRS-2 are rather poorly conserved (35% identity), but contain multiple tyrosine phosphorylation sites in relatively similar positions. IRS-1 and IRS-2 are likely to provide certain redundant functions; however, emerging data suggest that these docking proteins are differentially expressed and may engage different signalling molecules and mediate distinct responses [12, 13].

Small IRS-proteins expand the family

During the past year, 3 small proteins, Gab-, p62[dok] and IRS-3, were cloned as substrates for various tyrosine kinases and activated cytokine receptors. Since these proteins are also substrates for the insulin receptor, we include them as IRS-proteins. These docking proteins are considerably smaller that IRS-1 and IRS-2, but they contain many similar features. One of the most important is the PH domain which serves to link the protein to the insulin receptor. In addition, each protein contains a unique tail of tyrosine phosphorylation sites which bind to various SH2-proteins. However, the PTB domain which is also characteristic of IRS-1 and IRS-2 is completely absent from Gab-1, partially recognizable in pp62[dok], and conserved in IRS-3. These small docking proteins add diversity and crosstalk to several signalling systems including the insulin receptor.

Gab-1 occurs in mammalian tissues and was cloned by expression screening with a Grb-2 probe [5]. A similar protein in drosophila called DOS mediates retinal development downstream of sevenless [14]; however, the biological function of Gab-1 in mammalian tissues is unknown. In common with IRS-1 and IRS-2, Gab-1 contains a relatively conserved NH2-terminal PH domain followed immediately by a short COOH-terminal tail with multiple tyrosine phosphorylation sites, including motifs which bind the PI-3 kinase (Fig. 1). Unlike IRS-1 and IRS-2, there is no PTB domain in Gab-1, which may diminish its coupling to the insulin receptor [5]. Thus insulin signalling through Gab-1 may occur only at high insulin doses or in cells with many insulin receptors, suggesting that Gab-1 mediates the least sensitive insulin responses.

A 62 kDa phosphoprotein, originally called p62[rasGAP], is a common target of several protein-tyrosine kinases, including v-Abl, v-Src, v-Fps, v-Fms, and activated receptors for IGF-1, EGF, csf-1 and insulin [6]. Recently, a substrate for bcr-abl that binds to ras-Gap was purified and cloned [6, 15]. This protein, called p62[dok], reacts with monoclonal antibodies raised against p62[rasGAP], suggesting that it is identical to this insulin receptor substrate [6]. The p62[dok] contains a PH domain at its NH2-terminus that is rather dissimilar to the PH domain in IRS-1, IRS-2 and Gab-1. p62[dok] may contain a functional PTB domain as arginine residues known to bind the phosphorylated NPEY-motif in the insulin receptor appear to be correctly positioned; however, this regions contains little similarity to the PTB domains in IRS-1 and IRS-2 [16]. The COOH-terminus of p62[dok] contains multiple tyrosine phosphorylation sites in motifs that recognize various SH2-proteins, but none of them are expected to bind the PI-3 kinase (Fig. 1). p62[dok] is unique among the IRS-proteins as it does not regulate the PI-3 kinase.

A 60 kDa protein was identified as an insulin receptor substrate several years ago in rat adipocytes, but was difficult to purify and clone until recently [17]. The pp60 clone is called IRS-3, owing to similarities with IRS-1 and IRS-2. A PH and PTB domain with similarity to IRS-1 and IRS-2 are located at the NH2-terminus (Fig. 1). The high conservation in this region is notable and supports the hypothesis that these domains play an important role in the function of the docking proteins. The COOH-tail contains multiple tyrosine phosphorylation sites which occur in motifs recognized to bind the PI-3 kinase, SHP2 and Grb-2 [17]. In adipocytes, pp60 binds more rapidly than IRS-1 or IRS-2 to p85 during insulin stimulation, suggesting that it could be a principal regulator of the PI3-kinase [18]. Moreover, pp60 is the predominant insulin receptor substrate in adipocytes lacking IRS-1. Since pp60 is localized mainly at the plasma membrane it may play an important role in the regulation of the GLUT4 translocation [19].

Multiple mechanisms couple IRS-proteins to the insulin receptor

Separation of the docking proteins (IRS-proteins) from the activated receptors creates the new problem of kinase/substrate coupling. Specific interactions are essential, but these must be transient to allow the receptor to engage multiple molecules to amplify and diversify the signal. These requirements probably exclude SH2 domain interactions, as these tend to be to strong and stoichiometric which could diminish the catalytic turnover. Several solutions appear to be employed by the IRS-proteins. A pleckstrin homology (PH) domain provides one of the interaction domains found at the NH2-terminus of all IRS-proteins; however, the exact mechanism for coupling is unknown. The PH domain appears to mediate specific interactions, as chimeric IRS-1 proteins containing a heterologous PH domain derived from dissimilar proteins do not work; by contrast, chimeric IRS-1 proteins containing the PH domain from IRS-2 or Gab-1 are still sensitive insulin receptor substrates[2]. The identification of the ligand or domains that specifically bind to the PH domains in IRS-proteins is an important direction.

A second means for receptor coupling is provided by the phosphotyrosine binding (PTB) domain. This domain binds specifically but weakly to the phosphorylated NPXY-motifs located in the receptors for insulin, IGF-1 and IL-4 [20]. However, all receptors which engage the IRS-protein do not contain NPXY-motifs, so other binding sequences may be tolerated, otherwise the PTB domain may not be universally employed. In certain cases, other modules may contribute, including the recently identified kinase regulatory loop binding (KRLB) domain in IRS-2 or the c-Met binding domain (MBD) in Gab-1 [21, 22]. Together these interaction domains provide specific mechanisms for receptor substrate coupling.

IRS-proteins engaged multiple signalling proteins

One of the major mechanisms used by IRS-proteins to generate downstream signals is the direct binding to the SH2 domains of various signalling proteins. Several enzymes and adapter proteins have been identified which associate with IRS-1, including PI-3 kinase, SHP2, Fyn, Grb-2, nck, and Crk [23–28]; other partners associate through unknown mechanism which do not depend on tyrosine phosphorylation, including SV40 large T antigen, 14-3-3 and $\alpha_{v\beta3}$ [29–31]. PI-3 kinase is the best studied signalling molecule activated by IRS-1. It plays an important role in the regulation of a broad array of biological responses by various hormones, growth factors and cytokines, including mitogenesis [32, 33], differentiation [34], chemotaxis [35, 36], membrane ruffling [37], and insulin-stimulated glucose transport [38]. Moreover, PI-3 kinase activity is required for neurite extension and inhibition of apoptosis in PC12 cells [33] and cerebella neurons [39], suggesting that it plays an important role in neuronal survival.

PI-3 kinase
PI-3 kinase was originally identified as a dimer composed of a 11 0 kDa catalytic subunit (p110α or p110β) associated with an 85 kDa regulatory subunit (p85α or p85β). During a search for new SH2-proteins that bind to IRS-1, we cloned a smaller regulatory subunit that occurs predominantly in brain and testis, called p55[PIK] (Fig. 2). The COOH-terminal portion of p55[PIK] is similar to p85, including a proline-rich motif, two SH2 domains, and a consensus binding motif for p110 [40]. However, p55[PIK] contains a unique 30-residue NH$_2$-terminus which replaces the Src homology-3 (SH3) domain and the Bcr-homology region found in p85 [41]. Like p85, p55[PIK] associates with tyrosine phosphorylated IRS-1, and this association activates the PI-3 kinase [41]. Two other small regulatory subunits (p55α and p50α) are encoded by alternative splicing of the p85α gene [42–44]. These various regulatory subunits confer considerable variety to PI-3 kinase.

PI-3 kinase plays an important role in many insulin-regulated metabolic processes, including glucose uptake, general and growth-specific protein synthesis, and glycogen synthesis. Binding of the SH2 domain in p85 to phosphorylated YMXM-motifs in IRS-1 activates the associated catalytic domain, and this is maximal when both of the SH2 domains are occupied [45]. Double occupancy is easily accomplished when both YMXM-motifs are located within the same peptide suggesting that the second binding event occurs more readily through an intramolecular reaction [46]. IRS-1 and IRS-2 contain about nine YMXM-motifs, which are ideal docking sites for PI-3 kinase. IRS-3 is also well equipped to engage PI-3 kinase as four YMXM-motifs are located within a 50 residue region (Fig. 1). These closely spaced motifs are ideal to activate the PI-3 kinase.

Based on the results of various inhibitor studies, several enzymes appear to carry the signal initiated by PI-3 kinase activation to its final destinations (Fig. 2). The p70[s6k], PKB, PKC$_\zeta$ and others are thought to be downstream of PI-3 kinase [47–49]. The regulation of PKB is complex (Fig. 3), involving occupancy of its PH domain and serine/threonine phosphorylation [50]. PKB and PKC$_\zeta$ are implicated in various biological responses, including translocation of GLUT4 to the plasma membrane, general and growth-regulated protein synthesis, and glycogen synthesis [50–53].

SHP2
SHP2 is a phosphotyrosine phosphatase with two SH2 domains that is expressed in most mammalian cells [54]. Several growth factor receptors, including the EGFR, the PDGFR, and c-kit bind specifically to the SH2 domains in SHP2 [55–58]; a homologue in *Drosophila*, csw, mediates signals from the PDGF receptor homologue, *torso* [59]. During insulin stimulation, SHP2 binds two tyrosine residues at the extreme COOH-terminus of IRS-1 [26, 60]; a similar pair of sites occur in IRS-2, whereas IRS-3 contains a single potential site at Tyr^{466} [17]. SHP2 also binds to phosphorylated Tyr$_{1146}$ in the regulatory loop of the insulin receptor [61]. In addition, SHP2 associates with a 115 kDa protein during insulin stimulation [60]. This protein is located in the plasma membrane, contains YXX(LNI)-motifs and may be a direct substrate for the insulin receptor and other tyrosine kinase receptors [62]. The association of SHP2 with various docking proteins may serve to localize this phosphatase in various subcellular regions where it modulates distinct signalling pathways.

Several reports suggest that SHP2 mediates downstream signals from the insulin receptor since a catalytically inactive mutant inhibits insulin-stimulated MAP kinase and c-*fos* transcription in intact cells [63–65]. This dominant negative effect is partially reversed by co-expression of v-ras or Grb2, indicating that SHP2 may act upstream of Ras, possibly as an adapter protein [64]. Alternatively, SHP2 may diminish the tyrosine phosphorylation of IRS-1, providing a mechanism to attenuate certain signals. This hypothesis is supported by recent findings that reduced levels of SHP2 in mice increase IRS-1 tyrosine phosphorylation and its associated PI-3 kinase activity during insulin stimulation [66]. Consistent with this, expression in 32D cells of mutant IRS-1 lacking the SHP2 binding motifs increases IRS-1 tyrosine phosphorylation and associated PI-3 kinase activity, resulting in 2 fold enhanced insulin-stimulated [^{32}S]methionine-incorporation [67]. Thus, SHP2 may be essential to balance the converging and opposing signals essential for insulin and cytokine action.

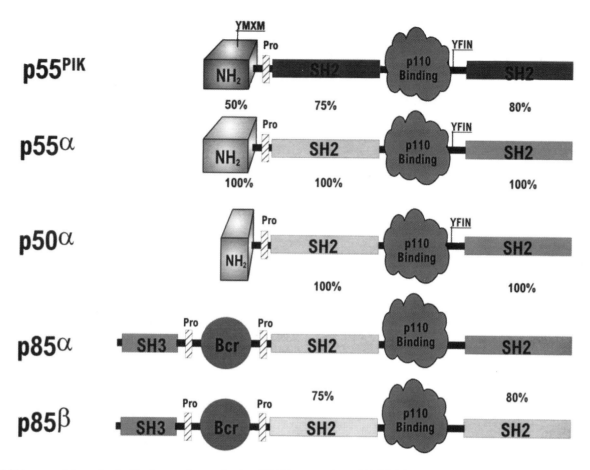

Fig. 2. PI-3 kinase regulatory subunits. The five regulatory subunits of PI-3 kinase are pictured, highlighting the common and unique structural features. Pro – proline-rich region; NH$_2$ – amino terminal regions; Bcr – Bcr-homology region. Percentages refer to the percent identity relative to p85α.

IRS-proteins and the MAP kinase cascade

In addition to PI-3 kinase and SHP2, tyrosine phosphorylated motifs in the IRS-proteins bind to the SH2-domains in several small adapter proteins, including Grb-2, nck and crk [24, 27, 28]. In addition to SH2 domains, these proteins contain multiple SH3 domains that bind various downstream signalling molecules that regulate metabolism, growth and differentiation [2, 68]. Flanking its SH2 domain, Grb-2 contains two SH3-domains that associate constitutively with mSOS, a guanine nucleotide exchange protein that stimulates GDP/GTP exchange on p21^{ras} [24, 69, 70]. Nck is a 47 kDa adapter protein composed of three SH3 domains and a single SH2 domain; crk contains two SH3 domains and a single SH2 domain [2] These adapter proteins bind through their SH3 domains to a variety of signalling proteins, and are targeted to specific subcellular locations by tyrosine phosphorylated membrane receptors or docking proteins.

The recruitment by growth factor receptors of Grb2/mSos to membranes containing p21^{ras} is one of the mechanisms employed to activate the MAP kinase cascade [68]. During insulin stimulation, Grb-2 engages IRS-1, IRS-2, Shc or SHP2, although the preferred interactions depend on the cell background. In skeletal muscle IRS-1 is the dominant Grb-2 binding protein, whereas Grb-2 binds poorly to Shc in this background. Thus mice lacking IRS-1 display an 80% reduction in insulin-stimulated MAP kinase even though Shc phosphorylation is normal [71]. By contrast, Shc plays a major role during insulin stimulation of MAP kinase in cultured cells, and in many cell culture systems Shc is believed to be the major Grb2/Sos activator during insulin stimulation [23, 69].

Although the MAP kinase cascade is a well documented insulin signalling pathway, it is not very sensitive to insulin. First, the Grb-2 binding site (Tyr$_{895}$) in IRS-1 is weakly phosphorylated, requiring relatively high insulin receptor expression; similarly, insulin-stimulated Shc tyrosine phosphorylation requires a strong insulin signal, usually achieved by overexpression of the insulin receptor [23]. In 32D myeloid cells which express few insulin receptors and no murine IRS-proteins, ectopic expression of IRS-1 alone is insufficient to stimulated Grb-2 binding, even though IRS-1 becomes tyrosine phosphorylated and bound to p85

8

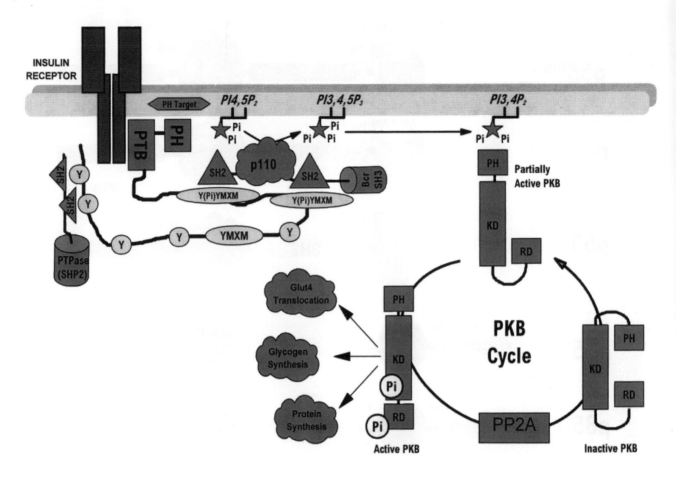

Fig. 3. Model of PKB Activation. Schematic representation of a working model of the signal transduction pathways and molecular mechanism of the cycle of PKB activation. KD – kinase domain; RD – regulatory domain.

[23]. By contrast, overexpression of the insulin receptor alone mediates insulin-stimulated MAP kinase activation without IRS-proteins, apparently through tyrosine phosphorylation of endogenous Shc [23]. Thus low insulin receptor levels are sufficient to mediate IRS-1 phosphorylation, but only on a few sites that activate PI-3 kinase. By contrast, Grb-2 binding to IRS-1 occurs at high receptor levels. Since insulin-stimulated tyrosine phosphorylation of Shc also requires relatively high receptor levels Grb-2/Sos activation is generally insensitive to insulin.

IRS-proteins and diabetes

IRS-proteins regulate many biological processes, including the control of glucose metabolism, protein synthesis, and cell survival, growth and transformation. Although not all insulin signals are mediated by the IRS-proteins, major physiological responses to insulin are probably absent without them. Mice lacking IRS-1 are mildly insulin-resistant con-

firming that IRS-proteins mediate insulin signals in the intact mammals. Without IRS-1, glucose levels are maintained at normal levels during fasting by elevated circulating insulin, and after a glucose challenge the serum glucose is reduced slowly even by an exaggerated insulin release [72, 73]. However, the absence of IRS-1 does not cause NIDDM [74]. Apparently, insulin resistance caused by the absence of IRS-1 is adequately compensated by elevated insulin production and secretion, and signalling through other IRS-proteins.

A few studies have investigated insulin responses in various tissues of the IRS $1^{(-/-)}$ mouse to determine the nature of the compensatory signalling. In liver, insulin signalling is nearly normal even though IRS-2 expression is not elevated; however, IRS-2 tyrosine phosphorylation increases which mediates a typical PI-3 kinase response [75]. Perhaps the increase phosphorylation of IRS-2 occurs because competition from IRS-1 is absent; however, this compensation does not occur in skeletal muscle [71]. Muscle from the IRS-1$^{(-/-)}$ mice retains only a 20% response to insulin, including PI-3 kinase, MAP kinase and p70^{s6k}

glucose uptake, glycogen synthesis and protein synthesis which reflects the low level of IRS-1 phosphorylation that occurs in this tissue [71]. The respectable control of glucose homeostasis must arise from a nearly normal inhibition of hepatic gluconeogenesis (although this has not been measured directly), and slow but reasonable glycogen production in muscle; although insulin weakly stimulates glucose uptake in skeletal muscle, once inside its conversion to glycogen is largely substrate driven.

A heterozygous disruption of the insulin receptor is slightly more severe than a heterozygous disruption of IRS-1, although both animals are generally euglycemic throughout their lives. At birth, the compound heterozygous mice (IR$^{+/-}$/IRS$^{+/-}$) are slightly more resistant to insulin than the individual heterozygotes. During the first 4 months of life, insulin secretion adequately compensates to maintain normal glycemia. However, between 4–6 months of age about half of the mice develop diabetes owing to severe uncompensated insulin resistance [74]. The β-cell mass and serum insulin levels increase in parallel to the insulin resistance, but are still inadequate. The molecular basis for the pathophysiology may eventually provide a better understanding of NIDDM. The compound heterozygous mouse model provides the best evidence that the enzyme: substrate relation between the insulin receptor and IRS-1 is important for carbohydrate metabolism.

Reduced function of the insulin receptor and IRS-1 could contribute to NIDDM in humans. However, mutations in the insulin receptor and IRS-1 are rare in humans, and probably do not contribute significantly to the disease. In a few cases, polymorphisms in IRS-1 have been identified in human IRS- 1, including Ser$_{513}$, Ala$_{972}$, and Arg$_{1221}$ [76, 77]. The Ala$_{972}$ mutation occurs in 10.7% of NIDDM subjects from various ethnic backgrounds, but also at 5.8% in control subjects. Subsequent analysis of this mutation in 32D cells suggests that it partially reduces the ability of IRS-1 to activate PI-3 kinase [76].

Other mechanisms to reduce the function of the insulin receptor or IRS-1 could play a role in NIDDM. Recently, TNFα-induced serine phosphorylation of IRS-1 has been implicated as an important cause of insulin resistance. In 32D cells, inhibition of the insulin receptor kinase by TNFα requires expression of IRS-1 [9]. The mechanism of this effect is unclear, but it may involve an interaction between IRS-1 and elements in the TNFα signalling cascade, such as sphingomyelinase or the JNK kinases [78–80]. Since adipocytes secrete TNFα, a molecular explanation for insulin resistance in obesity and diabetes may involve this pathway [8, 9, 81]. Future experiments in more physiological systems are required to support these hypotheses, and extensions to human is required. If NIDDM is even partially due to reduced IRS-protein expression or phosphorylation, then it may be possible to identify drugs which enhance its expression or reduce its serine phosphorylation to rescue a normal signalling capacity.

Future perspectives

Significant progress during the past 10 years has established a new paradigm for our understanding of insulin signalling. IRS-proteins represent an important multi-functional interface between many receptors and intracellular signalling pathways. Additional members of the IRS-protein family will be identified to further validate the role of docking proteins insulin signalling. With this growing knowledge, rational approaches for the discovery of new drugs that modify various elements of the insulin signalling pathway will be developed. And with persistent effort, the molecular basis of NIDDM will be revealed.

Acknowledgements

This work was supported by DK38712, DK43808 and DK48712.

References

1. Hunter T: Oncoprotein networks. Cell 88: 333–346, 1997
2. Pawson T: Protein modules and signalling networks. Nature 373: 573–580, 1995
3. Sun XJ, Rothenberg PL, Kahn CR, Backer JM, Araki E, Wilden PA, Cahill DA, Goldstein BJ, White MF: The structure of the insulin receptor substrate IRS-1 defines a unique signal transduction protein. Nature 352: 73–77, 1991
4. Sun XJ, Wang LM, Zhang Y, Yenush L, Myers MG Jr, Glasheen EM, Lane WS, Pierce JH, White MF: Role of IRS-2 in insulin and cytokine signalling. Nature 377: 173–177, 1995
5. Holgado-Madruga M, Emlet DR, Moscatello DK, Godwin AK, Wong AJ: A Grb2-associated docking protein in EGF- and insulin-receptor signalling. Nature 379: 560–563, 1996
6. Yamanashi Y, Baltimore D: Identification of the Abl- and rasGAP-associated 62 kDa protein as a docking protein, Dok. Cell: 205–211, 1997
7. Heller-Harrison RA, Morin M, Guilhenne A, Czech MP: Insulin-mediated targeting of phosphatidylinositol 3-kinase to GLUT4-containing vesicles. J Biol Chem: 10200–10204, 1996
8. Hotamisligil GS, Spiegelman BM: Tumor necrosis factor alpha: A key component of the obesity-diabetes link. Diabetes: 1271–1278, 1994
9. Hotamisligil GS, Peraldi P, Budvari A, Ellis RW, White MF, Spiegelman BM: IRS-1 mediated inhibition of insulin receptor tyrosine kinase activity in TNF-α-and obesity-induced insulin resistance. Science 271: 665–668, 1996
10. White MF, Maron R, Kahn CR: Insulin rapidly stimulates tyrosine phosphorylation of a M$_r$ 185,000 protein in intact cells. Nature 318: 183–186, 1985
11. Araki E, Haag BL III, Kahn CR: Cloning of the mouse insulin receptor substrate-1 (IRS-1) gene and complete sequence of mouse IRS-1. Biochim Biophys Acta 1221: 353–356, 1994

12. Sun XJ, Pons S, Wang LM, Zhang Y, Yenush L, Burks D, Myers MG Jr, Glasheen E, Copeland NG, Jenkins NA, Pierce JH, White MF: The IRS-2 gene on murine chromosome 8 encodes a unique signalling adapter for insulin and cytokine action. Mol Endocrinol 11: 251–262, 1997

13. Bruning JC, Winnay J, Cheatham B, Kahn CR: Differential signalling by IRS-1 and IRS-2 in IRS-1 deficient cells. Mol Cell Biol: 1997 (in press)

14. Raabe T, Riesgo-Escovar J, Liu X, Bausenwein BS, Deak P, Maroy P, Hafen E: DOS, a novel pleckstrin homology domain-containing protein required for signal transduction between sevenless and ras 1 in *Drosophila*. Cell 85: 911–920, 1996

15. Carpino N, Wisniewski D, Strife A, Marshak D, Kobayashi R, Stillman B, Clarkson B: p62dok: A constitutively tyrosine-phosphorylated, GAP-associated protein in chronic myelogenous leukemia progenitor cells. Cell 88: 197–204, 1997

16. Eck MJ, Dhe-Paganon S, Trub T, Nolte RT, Shoelson SE: Structure of the IRS-1 PTB domain bound to the juxtamembrane region of the insulin receptor. Cell 85: 695–705, 1996

17. Lavan BE, Lane WS, Lienhard GE: The 60 kDa phosphotyrosine protein in insulin-treated adipocytes is a new member of the insulin receptor substrate family. J Biol Chem 272: 11439–11443, 1997

18. Smith-Hall J, Pons S, Patti ME, Burks DJ, Yenush L, Sun XJ, Kahn CR, White MF: The 60 kDa insulin receptor substrate functions like an IRS-protein (pp60^{IRS3}) in adipocytes. Biochemistry: 1997, (in press)

19. Kelly KL, Ruderman NB: Insulin-stimulated phosphatidylinositol 3-kinase. J Biol Chem 268: 4391–4398, 1993

20. Wang LM, Keegan AD, Li W, Lienhard GE, Pacini S, Gutkind JS, Myers MG Jr, Sun XJ, White MF, Aaronson SA, Paul WE, Pierce JH: Common elements in interleukin 4 and insulin signalling pathways in factor dependent hematopoietic cells. Proc Natl Acad Sci USA 90: 4032–4036, 1993

21. Sawka-Verhelle D, Tartare-Deckert S, White MF, Van Obberghen E: IRS-2 binds to the insulin receptor through its PTB domain and through a newly identified domain comprising amino acids 591–786. J Biol Chem 271: 5980–5983, 1996

22. He W, Craparo A, Zhu Y, O'Neill TJ, Wang LM, Pierce JH, Gustafson TA: Interaction of insulin receptor substrate-2 (IRS-2) with the insulin and insulin-like growth factor 1 receptors. J Biol Chem 271: 11641–11645, 1996

23. Myers MG Jr, Wang LM, Sun XJ, Zhang Y, Yenush L, Schlessinger J, Pierce JH, White MF: The role of IRS-I/GRB2 complexes in insulin signalling. Mol Cell Biol 14: 3577–3587, 1994

24. Skolnik EY, Batzer AG, Li N, Lee CH, Lowenstein EJ, Mohammadi M, Margolis B, Schlessinger J: The function of GRB2 in linking the insulin receptor to ras signalling pathways. Science 260: 1953–1955, 1993

25. Sun XJ, Pons S, Asano T, Myers MG Jr, Glasheen EM, White MF: The fyn tyrosine kinase binds IRS-1 and forms a distinct signalling complex during insulin stimulation. J Biol Chem 271: 10583–10587, 1996

26. Kuhne MR, Pawson T, Lienhard GE, Feng GS: The insulin receptor substrate 1 associates with the SH2-containing phosphotyrosine phosphatase Syp. J Biol Chem 268: 11479–11481, 1993

27. Beitner-Johnson D, Blakesley VA, Shen-Orr Z, Jimenez M, Stannard B, Wang LM, Pierce JH, LeRoith D: The proto-oncogene product c-Crk associates with insulin receptor substrate-1 and 4PS. J Biol Chem 271: 9287–9290, 1996

28. Lee CH, Li W, Nishimura R, Zhou M, Batzer AG, Myers MG Jr, White MF, Schlessinger J, Skolnik EY: Nck associates with the SH2 domain docking proteins IRS-1 in insulin stimulated cells. Proc Natl Acad Sci USA 90: 11713–11717, 1993

29. Vuori K, Ruoslahti E: Association of insulin receptor substrate-1 with integrins. Science 266: 1576–1578, 1994

30. Fei ZL, D'Ambrosio C, Li S, Surmacz E, Baserga R: Association of insulin receptor substrate 1 with simian virus 40 large T antigen. Mol Cell Biol 15: 4232–4239, 1995

31. Craparo A, Freund R, Gustafson TA: 14-3-3 epsilon interacts with the insulin-like growth factor 1 receptor and insulin receptor substrate I in phosphotyrosine-independent manner. J Biol Chem 272: 11663–11670, 1997

32. Valius M, Kazlauskas A: Phospholipase C-gamma 1 and phosphatidylinositol 3 kinase are the downstream mediators of the PDGF receptor's mitogenic signal. Cell 73: 321–334, 1993

33. Yao R, Cooper GM: Requirement for phosphatidylinositol-3 kinase in the prevention of apoptosis by nerve growth factor. Science 267: 2003–2006, 1995

34. Kimura K, Hattori S, Kabuyama Y, Shizawa Y, Takayanagi J, Nakamura S, Toki S, Matsuda Y, Onodera K, Fukui Y: Neurite outgrowth of PC12 cells is suppressed by wortmannin, a specific inhibitor of phosphatidylinositol 3-kinase. J Biol Chem 269: 18961–18967, 1994

35. Kundra V, Escobedo JA, Kazlauskas A, Kim HK, Rhee SG, Williams LT, Zetter BR: Regulation of chemotaxis by the platelet-derived growth factor receptor-beta. Nature 367: 474–476, 1994

36. Okada T, Sakuma L, Fukui Y, Hazeki O, Ui M: Blockage of chemotactic peptide-induced stimulation of neutrophils by wortmannin as a selective inhibitor of phosphatidylinositol 3-kinase. J Biochem 269: 3563–3567, 1994

37. Wennstrom S, Hawkins P, Cooke F, Hara K, Yonezawa K, Kasuga M, Jackson T, Claessonwelsh L, Stephens L: Activation of phosphoinositide 3-kinase is rectivation of phosphoinositide 3-kinase is required for PDGF-stimulated membrane ruffling. Curr Biol 4: 385–393, 1994

38. Okada T, Kawano Y, Sakakibara T, Hazeki O, Ui M: Essential role of phosphatidylinositol 3-kinase in insulin-induced glucose transport and antilipolysis in rat adipocytes. J Biochem 269: 3568–3573, 1994

39. Dudek H, Datta SR, Franke TF, Bimbaum MJ, Yao R, Cooper GM, Segal RA, Kaplan DR, Greenberg ME: Regulation of neuronal survival by the serine-threonine protein kinase Akt. Science 275: 661–665, 1997

40. Dhand R, Hara K, Hiles I, Bax B, Gout 1, Panayotou G, Fry MJ, Yonezawa K, Kasuga M, Waterfield MD: PI 3-kinase: Structural and functional analysis of intersubunit interactions. EMBO J 13: 511–521

41. Pons S, Asano T, Glasheen EM, Miralpeix M, Zhang Y, Fisher TL, Myers MG Jr, Sun XJ, White MF: The structure and function of p55^{PIK} reveals a new regulatory subunit for the phosphatidylinositol-3 kinase. Mol Cell Biol 15: 4453–4465, 1995

42. Antonetti DA, Algenstaedt P, Kahn CR: Insulin receptor substrate 1 binds two novel splice variants of the regulatory subunit of phosphatidylinositol 3-kinase in muscle and brain. Mol Cell Biol 16: 2195–2203, 1996

43. Inukai K, Anai M, van Breda E, Hosaka T, Katagiri H, Funaki M, Fukushima Y, Ogihara T, Yazaki Y, Kikuchi M, Oka Y, Asano T: A novel 55 kDa regulatory subunit for phosphatidylinositol 3-kinase structurally similar to p55^{PIK} is generated by alternative splicing of the p85α gene. J Biol Chem 271: 5317–5320, 1996

44. Fruman DA, Cantley LC, Carpenter CL: Structural organization and alternative splicing of the murine phosphoinositide 3-kinase p85α gene. Genomics 37: 113–121, 1996

45. Backer JM, Schroeder GG, Kahn CR, Myers MG Jr, Wilden PA, Cahill DA, White MF: Insulin stimulation of phosphatidylinositol 3-kinase activity maps to insulin receptor regions required for endogenous substrate phosphorylation. J Biol Chem 267: 1367–1374, 1992

46. Rordorf-Nikolic T, Van Horn DJ, Chen D, White MF, Backer JM: Regulation of phosphatidylinositol 3-kinase by tyrosyl phosphoproteins. Full activation requires occupancy of both SH2 domains in the 85 kDa regulatory subunit. J Biol Chem 270: 3662–3666, 1995

47. Myers MG Jr, Zhang Y, Aldaz GAI, Grammer TC, Glasheen EM, Yenush L, Wang LM, Sun XJ, Blenis J, Pierce JH, White MF: YMXM motifs and signalling by an insulin receptor substrate 1 molecule without tyrosine phosphorylation sites. Mol Cell Biol 16: 4147–4155, 1996

48. Franke TF, Yang S, Chan TO, Datta K, Kazlauskas A, Morrison DK, Kaplan DR, Tsichlis PN: The protein kinase encoded by the Akt proto-oncogene is a target of the PDGF-activated phosphatidylinositol 3-kinase. Cell 81: 727–736, 1995

49. Diaz-Meco MT, Lozano J, Municio MM, Berra E, Frutos S, Sanz L, Moscat J: Evidence for the *in vitro* and *in vivo* interaction of Ras with protein kinase C zeta. J Biol Chem 269: 31706–31710, 1994

50. Alessi DR, Andjelkovic M, Caudwell B, Cron P, Morrice N, Cohen P, Hemmings BA: Mechanism of activation of protein kinase B by insulin and IGF-1. EMBO J 15: 6541–6551, 1996

51. Cross DAE, Alessi DR, Cohen P, Andjelkovich M, Hemmings BA: Inhibition of glycogen synthase kinase-3 by insulin mediated protein kinase B. Nature 378: 785–787, 1996

52. Bandyopadhyay G, Standaert ML, Zhao LM, Yu B, Avignon A, Galloway L, Kamam P, Moscat J, Farese RV: Activation of protein kinase C (α, β, and zeta) by insulin in 3T3/L I cells. J Biol Chem 272: 2551–2558, 1997

53. Hemmings BA: Akt signalling Linked membrane events to life and death decisions. Science 275: 628–630, 1997

54. Freeman RM, Jr., Plutzky J, Neel BG: Identification of a human src homology 2-containing protein-tyrosine-phosphatase: A putative homolog of *Drosphila* corkscrew. Proc Natl Acad Sci USA 89: 11239–11243, 1992

55. Feng GS, Hui C-C, Pawson T. SH-2 containing phosphotyrosine phosphatase as a target of protein-tyrosine kinase. Science 259: 1607–1614, 1993

56. Lechleider RJ, Freeman RM Jr, Neel BG: Tyrosyl phosphorylation and growth factor receptor association of the human corkscrew homologue, SHPTP2. J Biol Chem 268: 13434–13438, 1993

57. Lechleider RJ, Sugimoto S, Bennett AM, Kashishian AS, Cooper JA, Shoelson SE, Walsh CT, Neel BG: Activation of the SH2-containing phosphotyrosine phosphatase SH-PTP2 by its binding site, phosphotyrosine 1009, on the human platelet-derived growth factor receptor B. J Biol Chem 268: 21478–21481, 1993

58. Tauchi T, Feng GS, Marshall MS, Shen R, Mantel C, Pawson T, Broxmeyer HE: The ubiquitously expressed Syp phosphatase interacts with c-kit and Grb2 in hematopoietic cells. J Biol Chem 269: 25206–25211, 1994

59. Perkins LA, Larsen I, Perrimon N: *corkscrew* encodes a putative protein tyrosine phosphatase that functions to transduce the terminal signal from the receptor tyrosine kinase torso. Cell 12: 225–236, 1992

60. Eck MJ, Pluskey S, Trub T, Harrison SC, Shoelson SE: Spatial constraints on the recognition of phosphoproteins by the tandem SH2 domains of the phosphatase SH-PTP2. Nature 379: 277–280, 1996

61. Kharitonenkov A, Schnekenburger J, Chen Z, Knyazev P, Ali S, Zwick E, White MF, Ullrich A: Adapter function of PTP1D in insulin receptor/IRS-1 interaction. J Biol Chem 270: 29189–29193, 1995

62. Yamao T, Matozaki T, Amano K, Matsuda Y, Takahashi N, Ochi F, Fujioka Y, Kasuga M: Mouse and human SHPS-I: Molecular cloning of cDNAs and chromosomal localization of genes. Biochem Biophys Res Commun 231: 61–67, 1997

63. Yamauchi K, Milarski KL, Saltiel AR, Pessin JE: Protein-tyrosine--phosphatase SBPTP2 is a required positive effector for insulin downstream signalling. Proc Natl Acad Sci USA 92: 664–668, 1995

64. Noguchi T, Matozaki T, Horita K, Fujioka Y, Kasuga M: Role of SH-PTP2, a protein-tyrosine phosphatase with Src homology 2 domains, in insulin-stimulated ras activation. Mol Cell Biol 14: 6674–6682, 1994

65. Sasaoka T, Rose DW, Jhun BH, Saltiel AR, Draznin B, Olefsky JM: Evidence for a functional role of Shc proteins in mitogenic signalling induced by insulin, insulin-like growth factor-1, and epidermal growth factor. J Biol Chem 269: 13689–13694, 1994

66. Arrandale JM, Gore-Willse A, Rocks S, Ren JM, Zhu J, Davis A, Livingston JN, Rabin DU: Insulin signalling in mice expressing reduced levels of Syp. J Biol Chem 271: 21353–21358, 1996

67. Mendez R, Myers MG Jr, White MF, Rhoads RE: Stimulation of protein synthesis, eukaryotic translation initiation factor 4E phosphorylation, and PHAS-1 phosphorylation by insulin requires insulin receptor substrate-1 and phosphotidylinositol-3-kinase. Mol Cell Biol 16: 2857–2864, 1996

68. Schlessinger J: How receptor tyrosine kinases activate Ras. Trends Biochem Sci 18: 273–275, 1993

69. Skolnik EY, Lee CH, Batzer AG, Vicentini LM, Zhou M, Daly RJ, Myers MG Jr, Backer JM, Ullrich A, White MF, Schlessinger J: The SH2/SH3 domain-containing protein GRB2 interacts with tyrosine-phosphorylated IRS-1 and Shc: Implications for insulin control of ras signalling. EMBO J 12: 1929–1236, 1993

70. Gale NW, Kaplan S, Lowenstein EJ, Schlessinger J, Bar-Sagi D: Grb2 mediates the EGF-dependent activation of guanine nucleotide exchange on Ras. Nature 363: 88–92, 1993

71. Yamauchi T, Tobe K, Tamemoto H, Ueki K, Kaburagi Y, Yamamoto-Handa R, Takahadhi Y, Yoshizawa F, Aizawa S, Akanuma Y, Sonenberg N, Yazaki Y, Kadowaki T: Insulin signalling and insulin actions in the muscles and livers of insulin-resistant, insulin receptor substrate 1-deficient mice. Mol Cell Biol 16: 3074–3084, 1996

72. Araki E, Lipes MA, Patti ME, Bruning JC, Haag BL, Ill, Johnson RS, Kahn CR: Alternative pathway of insulin signalling in mice with targeted disruption of the IRS-1 gene. Nature 372: 186–190, 1994

73. Tamemoto H, Kadowaki T, Tobe K, Yagi T, Sakura H, Hayakawa T, Terauchi Y, Ueki K, Kaburagi Y, Satoh S, Sekihara H, Yoshioka S, Horikoshi H, Furuta Y, Ikawa Y, Kasuga M, Yazaki Y, Aizawa S: Insulin resistance and growth retardation in mice lacking insulin receptor substrate-1. Nature 372: 182–186, 1994

74. Bruning JC, Winnay J, Bonner-Weir S, Taylor SI, Accili D, Kahn CR: Development of a novel polygenic model of NIDDM in mice heterozygous for *IR* and *IRS-1* null alleles. Cell 88: 561–572, 1997

75. Patti ME, Sun XJ, Bruning JC, Araki E, Lipes MA, White MF, Kahn CR: IRS-2/4PS is an alternative substrate of the insulin receptor in IRS-1 deficient transgenic mice. Diabetes 44:(abstr) 31A, 1995

76. Almind K, Inoue G, Pedersen O, Kahn CR: A common amino acid polymorphism in insulin receptor substrate-1 causes impaired insulin signalling. Evidence from transfection studies. J Clin Invest 97: 2569–2575, 1996

77. Clausen JO, Hansen T, Bjorbaek C, Echwald SM, Urhammer SA, Rasmussen S, Andersen CB, Hansen L, Almind K, Winther K, Haraldsdottir J, Borch-Johnsen K, Pedersen O: Insulin resistance: Interactions between obesity and a common variant of insulin receptor substrate-1. Lancet 346: 397–402, 1995

78. Verheij M, Bose R, Lin XH, Yao B, Jarvis WD, Grant S, Birrer MJ, Szabo E, Son LI, Kyriakis JM, Haimovitz-Friedman A, Fuks Z, Kolesnick RN: Requirement for ceramide-initiated SAPK/JNK signalling in stress-induced apoptosis. Nature 380: 75–79, 1996

79. Saklatvala J, Davis W, Guesdon F. Interleukin 1 (IL1) and tumour necrosis factor (TNF) signal transduction. Phil Trans R Soc Lond Biol Sci 351: 151–157, 1996

80. Hirai S, Izawa M, Osada S, Spyrou G, Ohno S: Activation of the JNK pathway by distantly related protein kinases, MEKK and MUK. Oncogene 12: 641–650, 1996

81. Hotamisligil GS, Amer P, Caro JF, Atkinson RL, Spiegelman BM: Increased adipose tissue expression of tumor necrosis factor-α in human obesity and insulin resistance. J Clin Invest 95: 2409–2415, 1995

Molecular and Cellular Biochemistry **182**: 13–22, 1998.

Role of binding proteins to IRS-1 in insulin signalling

Wataru Ogawa, Takashi Matozaki and Masato Kasuga

Second Department of Internal Medicine, Kobe University School of Medicine, Kobe, Japan

Abstract

Insulin elicits its divergent metabolic and mitogenic effects by binding to its specific receptor, which belongs to the family of receptor tyrosine kinases. The activated insulin receptor phosphorylates the intracellular substrate IRS-1, which then binds various signalling molecules that contain SRC homology 2 domains, thereby propagating the insulin signal. Among these IRS-1-binding proteins, the Grb2-Sos complex and the protein tyrosine phosphatase SHP-2 transmit mitogenic signals through the activation of Ras, and phosphoinositide 3-kinase is implicated in the major metabolic actions of insulin. Although substantial evidence indicates the importance of IRS-1 in insulin signal transduction, the generation of IRS-1-deficient mice has revealed the existence of redundant signalling pathways. (Mol Cell Biochem **182**: 13–22, 1998)

Key words: insulin signalling, insulin receptor substrate-1 (IRS-1), IRS-1 binding proteins, phophatidylinositol 3-kinase, Ras-MAP kinase

Introduction

The principle role of insulin is to control the plasma glucose concentration by stimulating glucose transport into muscle and adipose cells, as well as by reducing glucose output from the liver [1]. These effects of insulin occur through activation of effectors such as glucose transporters and glycogen synthase, or through regulation of the amount of specific protein participants in metabolic pathways [2]. In addition, insulin is a major hormonal regulator of lipid metabolism, inhibiting lypolysis and stimulating fatty acid synthesis [3]. It also stimulates protein synthesis by affecting multiple steps including amino acid transport and the initiation of translation [4]. Furthermore, insulin promotes cell growth and differentiation of specific cells and tissues.

Insulin exerts this wide variety of biological effects through interaction with its specific receptor, which belongs to a large family of receptor tyrosine kinases [5]. In response to ligand binding, most tyrosine kinase receptors undergo autophosphorylation and thereby create high-affinity binding sites for various signalling molecules that contain SRC homology (SH) 2 domains [6]. However, the insulin receptor and closely related members of this receptor family, rather than binding SH2 domain-containing signalling molecules directly, phosphorylate a specific intracellular substrate and this substrate, termed insulin receptor substrate 1 (IRS-1), that, in turn, recruits SH2 containing proteins [7, 8]. This review focused on our current understanding of the insulin receptor-IRS-1 signalling system.

Structure and function of IRS-1

IRS-1 was originally identified as a 185-kDa tyrosine phosphorylated protein (pp185) in insulin-treated hepatoma cells [9]. This protein was subsequently shown to undergo rapid and marked tyrosine phosphorylation in response to insulin or insulin-like growth factor 1 in a variety of cells and tissues [7]. IRS-1 was first purified from rat liver [10] and its corresponding cDNA was subsequently cloned [11]. The deduced amino acid sequence of IRS-1 revealed that it contains 22 potential tyrosine phosphorylation sites, at least 8 of which become phosphorylated after insulin stimulation in intact cells [12]. These phosphorylated tyrosine residues form binding sites for signalling proteins that possess SH2 domains. A combination of site-directed

Address for offprints: W. Ogawa, Second Department of Internal Medicine, Kobe University School of Medicine, Kobe 650, Japan

14

mutagenesis of the tyrosine phosphorylation sites in IRS-1 and *in vitro* competitive binding experiments with tyrosine-phosphorylated peptides revealed that the 85 kDa regulatory subunit of phosphoinositide (PI) 3-kinase binds to Y_{608}MPM and Y_{939}MNM, Grb2 binds to Y_{895}VNI, and SHP-2 binds to Y_{1172}IDL and Y_{1222}ASI [13–16] (Fig. 1).

In its far NH$_2$-terminal region, IRS-1 contains a pleckstrin homology (PH) domain, a poorly conserved region of about 120 amino acids that was first identified as an internal repeats in pleckstrin (major substrate of protein kinase C in platelets) and which is present in a variety of signalling molecules [17–19]. It has been suggested that PH domains mediate molecular interactions by analogy of other common functional domains, such as SH2 or SH3 domains. Indeed, several molecules has been shown to bind directly to these domains [20–22], although a common ligand has not been identified. Although the precise function of the IRS-1 PH domain is not known, tyrosine phosphorylation of a mutant IRS-1 that lacks the PH domain was reduced in intact cells [23], indicating that the PH domain contributes to the interaction of IRS-1 with the insulin receptor.

Immediately downstream of the PH domain, lies a second domain that is thought to be important in the interaction of IRS-1 with the insulin receptor. This domain, termed the phosphotyrosine-binding (PTB) domain was shown to interact directly with the autophosphorylated insulin receptor in the yeast two-hybrid system [24]. *In vitro* binding experiments showed that the NPXpY motif in the juxtamembrane region of the β subunit of the insulin receptor mediates this interaction [25]. A similar domain was identified in Shc, a substrate of various receptor tyrosine kinases, including the insulin receptor, and the SHC PTB domain also binds to the NPXpY motif [26, 27].

Consistent with these observations, mutations in the juxtamembrane region of the insulin receptor attenuate its ability to catalyze the tyrosine phosphorylation of either IRS-1 or SHC [28, 29]. Although the amino acid sequence similarity between the PTB domains of IRS-1 and SHC is low, they show similar three-dimensional structures [30, 31]. A similar structure is also adopted by PH domains [32], suggesting that the PTB domain is an another type of PH domain. Although apparently inconsistent with the data obtained with the yeast two-hybrid system and *in vitro* binding experiments, a deletion mutant of IRS-1 that lacks its PTB domain becomes tyrosine phosphorylated and transmits insulin-induced signals to a similar extent as wild-type IRS-1 [23]. Further *in vivo* studies are thus required to clarify physiological functions of the PTB and PH domains of IRS-1.

PI 3-kinase and the metabolic effects of insulin

The first downstream molecule that was shown to associate with IRS-1 is PI 3-kinase [11]. This enzyme phosphorylates the D-3 position of the inositol ring of phosphoinositides, is composed of 110 kDa catalytic (p110) and 85 kDa regulatory (p85) subunits, and produces phosphatidylinositol 4, 5-bisphosphate [PI-(4, 5) P_2] and PI-(3, 4, 5) P_3 in cells [33]. The cDNAs for two subtypes of both p110 (p110α and p110β) and p85 (p85α and p85β) have been cloned from mammalian libraries [33]. The p85 subunit contains one SH3 domain and two SH2 domains, through the latter of which it associates with tyrosine-phosphorylated IRS-1 [34, 35] (Fig. 2). It also contains a bcr homology region and proline-rich sequences, although the roles of these regions in insulin

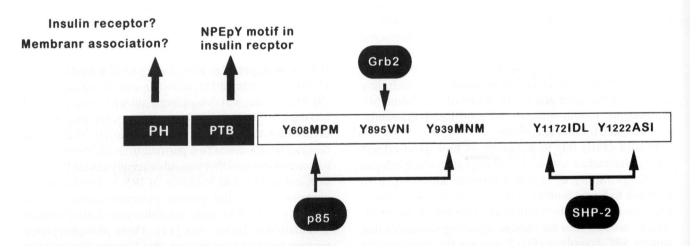

Fig. 1. Structural feature of IRS-1. IRS-1 contains PH and PTB domains on its NH$_2$ terminal region. These domains are thought to play an important role in interaction with the insulin receptor. IRS-1 becomes phosphorylated on its multiple tyrosine residues in response to insulin. Among these tyrosine phosphorylation sites, Y_{608}MPM and Y_{939}MNM bind to the 85 kDa regulatory subunit of phosphoinositide (PI) 3-kinase, Y_{895}VNI binds to Grb2, and Y_{1172}IDL and Y_{1222}ASI bind to SHP-2.

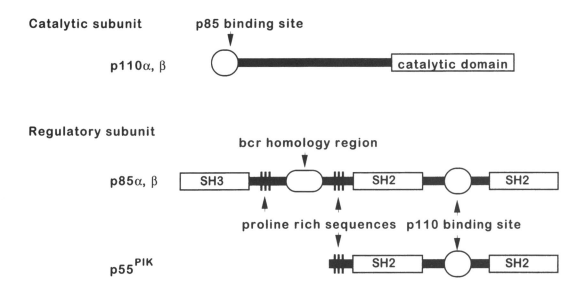

Fig. 2. Structural features of PI 3-kinase catalytic and regulatory subunits. p110 contains the binding site for regulatory subunits on its NH$_2$ terminal region and catalytic domain on its COOH terminal region. p85 possesses two SH2 and one SH3 domains and binds to p110 via the region between the two SH2 domains. The roles of the bcr homology region and the proline rich sequences are not known. p55^{PIK} dose not contain either SH3 or bcr homology region.

signal transduction are not known. The catalytic activity of p110 is enhanced by binding of p85 [36], which occurs through the NH$_2$ terminal region of p110 and the region between the two SH2 domain of p85 [37, 38]. Moreover, the association of p85 with tyrosine-phosphorylated IRS-1 is thought to potentiate further the catalytic activity of p110 [39]. Recently, an additional regulatory subunit of PI 3-kinase with a molecular size of ~55 kDa was identified [40]. This protein, termed p55^{PIK}, also contains two SH2 domains and a binding site for p110, but it does not possess SH3 or bcr homology domains (Fig. 2). Functional differences between the regulatory subunits of PI 3-kinase are still under investigation.

The roles of PI 3-kinase in insulin signal transduction have been investigated by two main approaches. We have con-

structed a mutant p85 that lacks the binding site for p110 [41]. When overexpressed, this protein, termed Δp85, inhibits the insulin-induced increase in PI 3-kinase activity that co-immunoprecipitates with IRS-1 as well as cellular PI-(3, 4, 5) P$_3$ production in CHO cells [41], indicating that it acts as a dominant negative mutant. Insulin-induced glucose transport as well as translocation of GLUT1, the primary glucose transporter in fibroblasts, were also markedly attenuated by Δp85 [41]. Furthermore, overexpression or microinjection of this mutant protein inhibited translocation of GLUT4 [42, 43] (Fig. 3) and glucose transport (H Sakaue, W Ogawa, M Kasuga, unpublished data) in adipocytes. These observations indicate that signals transmitted through PI 3-kinase regulate glucose transport.

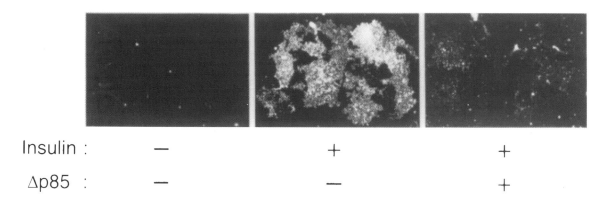

Insulin : — + +

Δp85 : — — +

Fig. 3. Δp85 inhibits insulin-induced translocation of GLUT4 in cultured adipocytes. 3T3-L1 adipocytes were treated with or without 0.1 µm insulin, then the plasma membrane fragments were prepared for immunofluorescence microscopy with antibodies to GLUT4 [43]. Insulin dramatically increases the fluorescence intensity in the plasma membrane (right and middle panel). This effect is abolished by microinjection of recombinant Δp85 (left panel).

The second approach to exploring the role of PI 3-kinase has been to examine the effects of two structurally unrelated, cell-permeable reagents that inhibit the catalytic activity of p110. Wortmannin, a fungal metabolise, and the synthetic inhibitor LY294002, were each shown to inhibit both the lipid kinase activity of p110 and the total cellular production of PI-(3, 4, 5) P_3 induced by insulin [44]. Both reagents also inhibit insulin-induced glucose uptake and translocation of glucose transporters [45, 46]. Various other biological activities of insulin are also inhibited by these reagents. Wortmannin- or LY294002-sensitive effects of insulin include antilipolysis [45], phosphorylation and activation of cAMP phosphodiesterase [47], stimulation of fatty acid synthesis [48], activation of acetyl CoA carboxylase [48] and activation of glycogen synthase [48–51], inhibition of glycogen synthase kinase 3β (GSK3β) activity [48, 51], stimulation of protein synthesis [52], phosphorylation of eukaryotic translation initiation factor 4E (eIF4E) and eIF4E binding protein (4E-BP1)/phosphorylated heat- and acid-stable protein regulated by insulin (PHAS-I) [52–54], stimulation of the transcription of specific genes such as that encoding phosphoenolpyruvate carboxykinase (PEPCK) [55], regulation of the cytoskeleton [56, 57] and activation of the serine-threonine kinases, p70 S6 kinase [53, 58–60], and Akt [61, 62].

A disadvantage of the pharmacological approach is its inability of discriminate between p110 and other wortmannin-sensitive enzymes. At least two other mammalian PI 3-kinases, phosphatidylinositol-specific PI 3-kinase [63] and G protein-activated PI 3-kinase (also known as p110γ) [64], are inhibited by wortmannin. Moreover, mTOR/RAFT, a putative target of rapamycin, also possess a wortmannin-sensitive protein kinase activity [54]. It is not known whether any of these additional wortmannin-sensitive proteins play a role in insulin signalling; however, some biological activities of insulin are sensitive to wortmannin but not to Δp85 [50, 60], suggesting that an unidentified wortmannin target may participate in insulin signal transduction.

Downstream effectors of PI 3-kinase

Regulation of the cytoskeleton is an important biological effect of insulin. Insulin promotes membrane ruffling, also know as the formation of lammelipodia, in a manner that is sensitive to both wortmannin and Δp85 [56, 65]. Membrane ruffling is regulated by the small GTP-binding protein Rac; a dominant negative Rac mutant abolishes the effect of insulin on this process, whereas a constitutively active Rac mutant induces membrane ruffling, in a manner that is not inhibited by PI 3-kinase inhibitors [57, 65]. Activation of Rac (conversion from the GDP-bound state to GTP-bound state) is promoted by platelet-derived growth factor [66] or by

insulin (K Kotani, W Ogawa, M Kasuga, unpublished data) in a wortmannin-sensitive manner. Although the mechanism of its activation is not fully understood, these observations indicate that Rac is a downstream effector of PI 3-kinase. This conclusion is consistent with recent data showing that the serine-threonine kinase PAK, a putative target of Rac, is activated by insulin in L6 myoblasts and that this activation is inhibited by wortmannin [67]. However, a dominant negative form of Rac did not affect glucose transport [68] indicating that insulin signalling pathways may bifurcate after PI 3-kinase, with different branches regulating glucose metabolism and the cytoskeleton.

The PH domain-containing serine-threonine kinase Akt, also known as PKB or RAC-PK, is activated by various growth factor including insulin [61, 69, 70]. Activation of Akt is inhibited by either wortmannin or Δp85 [61, 69, 70], indicating that Akt is also a downstream effector of PI 3-kinase. The mechanisms by which PI 3-kinase might activate Akt, however, is not clear. Although Franke *et al.* [69] showed that D3-phosphorylated phosphoinositides interact with and activate Akt *in vitro* [69], the latter effect was not observed by James *et al.* [71]. Serine-threonine phosphorylation of Akt appears to be the primary mechanism by which its enzymatic activity is regulated [62, 72], which, together with the observation that a kinase deficient Akt mutant is still phosphorylated in response to insulin in intact cells [61, 72] may indicate the existence of an Akt kinase that functions downstream of PI 3-kinase.

The kinase activity of GSK3β is inhibited by insulin in a wortmannin-sensitive manner [48, 51]. The observation that Akt phosphorylates and inhibits the activity of GSK3β *in vitro* [73] suggests that GSK3β is a physiological target of Akt. Furthermore, oncogenic or membrane-targeted mutants of Akt activate p70 S6 kinase in the absence of extracellular stimuli [70, 72], suggesting that p70 S6 kinase also may be regulated by Akt. Unfortunately, a simple kinase-deficient mutant of Akt dose not appear to function in a dominant negative manner, at least for signals to p70 S6 kinase [72]. The generation of an alternative dominant negative mutant of Akt may shed light on signalling downstream of this protein.

Grb2 and the Ras-MAP kinase cascade

Grb2 is an adapter protein that consists predominantly of one SH2 and two SH3 domains, and is apparently devoid of enzymatic activity [74]. Through one of its SH3 domains, Grb2 forms a complex with Sos, a guanine nucleotide exchange factor for Ras. Insulin induces the association of the Grb2-Sos complex with IRS-1 as well as the subsequently conversion of Ras from GDP-bound to GTP-bound form [75, 76] and activation of the mitogen-activated protein

(MAP) kinase cascade [77]. Activated Ras causes phosphorylation and activation of c-Raf, which subsequently phosphorylates and activates MAP kinase kinase (or MEK); the latter, a dual-specificity protein kinase that catalyze the phosphorylation of tyrosine and serine residues, then phosphorylates and activates MAP kinases [77].

Studies with dominant negative mutants of Ras or Sos have revealed that the Pas-MAP kinase cascade is important in the stimulatory effect of insulin on cell growth and DNA synthesis [78, 79]. Furthermore, introduction of IRS-1 antisense RNA or antibodies to IRS-1 into cells, or a point mutation in the Grb2 binding site of IRS-1 attenuate the effect of insulin on DNA synthesis [15, 80, 81], indicating that the IRS-1-Grb2 complex plays a major role in the mitogenic signalling pathway of insulin. In contrast, metabolic actions of insulin, such as stimulation of glucose uptake and activation of glycogen synthase were not affected by dominant negative mutant of Ras or Sos or by a synthetic

inhibitor of MEK [49, 50, 79, 82], indicating that the Ras-MAP kinase pathway does not contribute to the major metabolic actions of insulin. Grb2 also binds to dynamin and the Grb2-dynamin complex associates with IRS-1 in response to insulin [83]. Because the Grb2-dynamin complex is implicated in epidermal growth factor-induced receptor endocytosis [84], its association with IRS-1 may indicate a similar role in insulin receptor endocytosis.

Shc also binds to the Grb2-Sos complex. Shc undergoes tyrosine phosphorylation in response to various growth factors, and the SH2 domain of Grb2 has been shown to bind to phosphorylated Tyr[317] of Shc [85]. Microinjection of specific antibodies to or fragments of Shc inhibited DNA synthesis induced by insulin [85, 86], indicating that Shc also mediates insulin mitogenic signalling probably through the Ras-Map kinase cascade. The relative contributions of Shc and IRS-1 in regulating the Ras-Map kinase cascade are not clear and may depend on cell and tissue type.

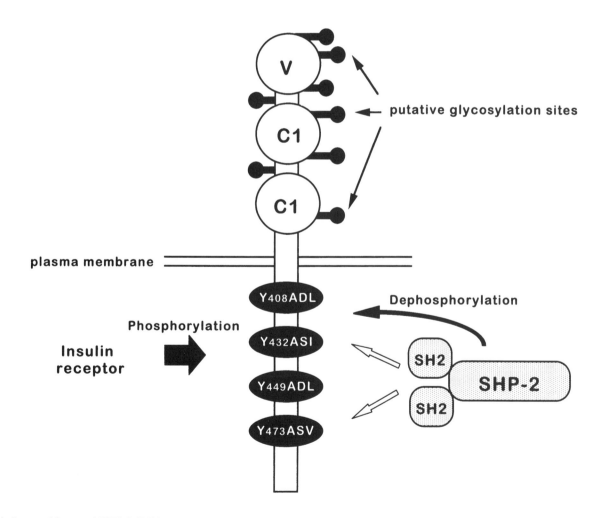

Fig. 4. Structural feature of SHPS-1. SHPS-1 possesses three immunoglobulin-like domains (one V region and two C1 regions), and multiple glycosylation sites in its extracellular portion. It contains four YXX(L/V/I) tyrosine phosphorylation and putative SH2 domain-binding motifs.

18

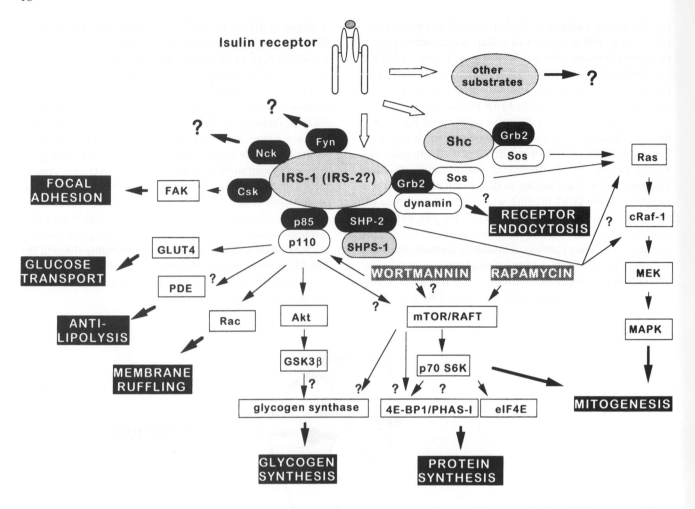

Fig. 5. A model of insulin signalling pathway. Insulin receptor phosphorylates tyrosine residues of IRS-1, -2, Shc and other intracellular substrates. These substrates recruit various signalling molecules containing SH2 domains, through which insulin's divergent metabolic and mitogenic signal is transmitted. Less well characterized substrates might mediate some of important signals of insulin.

SHP-2 and other IRS-1 binding proteins

SHP-2 (formerly known as SH-PTP2, PTP1D, or Syp), SH2 domain-containing tyrosine phosphatase associates with IRS-1 in response to insulin [87, 88]. Synthetic phosphotyrosine-containing peptides corresponding to tyrosine phosphorylation motifs of IRS-1 increase phosphatase activity of SHP-2 *in vitro* [16], suggesting that binding of IRS-1 modulates SHP-2 activity. Although signals transmitted through SHP-2 are not well characterized, microinjection of the SH2 domains of SHP-2 or antibodies to SHP-2 inhibited insulin-induced DNA synthesis in intact cells [89]. Furthermore, overexpression of a catalytically inactive SHP-2 attenuated signalling through the Ras-MAP kinase cascade [90, 91] indicating that SHP-2 contribute to insulin mitogenic signalling through this pathway. These data are consistent with the observation that Corkscrew, a *Drosophila* homolog of SHP-2 apparent to be important

in the activation of Ras-MAP kinase cascade during fly development [92]. The step at which SHP-2 affects Ras-MAP kinase cascade in mammalian cells is controversial; Noguchi *et al.* [90] showed that overexpression of a catalytically inactive SHP-2 inhibited Ras activation, whereas Sawada *et al.* [93] showed that Ras activation was not affected but that Raf and MEK were inhibited [93].

One potential substrate of SHP-2 is a tyrosine-phosphorylated protein of ~115 kDa (pp115), the tyrosine phosphorylation of which was markedly enhanced by overexpression of catalytically inactive SHP-2 [94, 95]. Recently, pp115 was purified and its corresponding cDNA was isolated [96]. This protein, termed SHP substrate 1 (SHPS-1), is a membrane glycoprotein that contains three immunoglobulin-like domains in its extracellular region and four YXX(L/V/I) tyrosine phosphorylation and putative SH2 domain-binding motifs in its intracellular region (Fig. 4). Insulin induces tyrosine phosphorylation of SHPS-1 and subsequent association with SHP-2 [95, 96] or SHP-1 (Y Fujioka, T Matozaki, M Kasuga, unpublished data), a closely related isoform of SHP-2. Although SHPS-1 requires further characterization, this protein may contribute to signalling mediated by SHP-2.

Carboxyl-terminal SRC kinase (Csk) is a cytoplasmic tyrosine kinase that phosphorylates and inactivates SRC. Csk also phosphorylates other SRC-type tyrosine kinases. Insulin stimulation promotes association of Csk with IRS-1 [97]. Overexpression of Csk in fibroblast potentiated insulin-induced dephosphorylation of the focal adhesion kinase pp125^{FAK}, and a kinase-deficient mutant of Csk inhibited this effect of insulin [97], suggesting that Csk may participate in the regulation of focal adhesion proteins by insulin.

Fyn, another cytoplasmic tyrosine kinase, and Nck, an adapter protein that shows structural similarity to Grb2, each associate with IRS-1 [8, 98]. Moreover, molecules such as SV40 large T antigen [99] and integrins [100] that do not contain SH2 domains also form complexes with IRS-1. The physiological significance of these interactions is currently unknown.

Is IRS-1 essential for all actions of insulin?

Given the extensive biochemical and cell biological data described so far (Fig. 5), it is somewhat surprising that IRS-1 deficient mice survive and do not show severe defects in glucose homeostasis. Such mice are 50% smaller than wild-type animals, although they show impaired glucose tolerance, they are not diabetic [101, 102]. A complicating factor in interpreting these observations is the existence of IRS-2 in IRS-1 deficient mice [103]. This protein, originally designated 4PS for interleukin-4 receptor phosphoprotein substrate, shares sequence similarity with IRS-1, and is capable of mediating insulin signalling [104]. IRS-2 therefore likely compensates, at least in part, for disrupted signalling in IRS-1 deficient mice, although the relative contribution of IRS-1 and IRS-2 to insulin signalling under physiological conditions are not known. Disruption of the IRS-2 gene and the generation of double-knockout mice, lacking both IRS-1 and IRS-2, may clarify this issue.

In addition to IRS-1 and 2, several other substrates of the insulin receptor have been identified [105–108]. One of these proteins, pp60, is apparently present only in adipocytes and hepatic cells, both of which are physiological targets of insulin, and associates with PI 3-kinase [105, 106]. Thus, this and other less well characterized substrates of the insulin receptor also may play important roles in the biological actions of insulin.

References

1. Birnbaum MJ: The insulin-sensitive glucose transporter. Int Rev Cytol 137: 239–297, 1993
2. O'Brien RM, Granner DK: Regulation of gene expression by insulin. Biochem J 278: 609–619, 1991
3. Sparks JD, Sparks CE: Insulin regulation of triacylglycerol-rich lipoprotein synthesis and secretion. Biochim Biophys Acta 1251: 9–32, 1994
4. Kimball SR, Vary TC, Jefferson LS: Regulation of protein synthesis by insulin. Annu Rev Physiol 56: 321–348, 1994
5. Lee J, Pilch PF: The insulin receptor: Structure, function, and signalling. Am J Physiol 266: C319–C334, 1994
6. Schlessinger J: SH2/SH3 signalling proteins. Curr Opin Genet Dev 4: 25–30, 1994
7. Myers MG Jr, Sun XJ, White MF: The IRS-1 signalling system. Trends Biochem Sci 19: 289–294, 1994
8. Myers MG Jr, White MF: Insulin signal transduction and the IRS proteins. Annu Rev Pharmacol Toxicol 36: 615–658, 1996
9. White MF, Maron R, Khan CR: Insulin rapidly stimulates tyrosine phosphorylation of a M_r 185,000 protein in intact cells. Nature 318: 183–186, 1985
10. Rothenberg PL, Lane WS, Karasik A, Backer J, White MF, Khan CR: Purification and partial sequence analysis of pp185, the major cellular substrate of the insulin receptor tyrosine kinase. J Biol Chem 266: 8302–8311, 1991
11. Sun XJ, Rothenberg P, Khan CR, Backer JM, Araki E, Wilden PA, Cahill DA, Goldstein BJ, Khan CR: Structure of the insulin receptor substrate IRS-1 defines a unique signal transduction protein. Nature 352: 73–77, 1991
12. White MF, Kahn CR: The insulin signalling system. J Biol Chem 269: 1–4, 1994
13. Sun XJ, Crimmins DL, Myers MG Jr, Miralpeix M, White MF: Pleiotropic insulin signals are engaged by multisite phosphorylation of IRS-1. Mol Cell Biol 13: 7418–7428, 1993
14. Skolinik KY, Lee C-H, Batzer A, Vicentini LM, Zhou M, Daly R, Myers MG Jr, Backer JM, Ullrich A, White MF, Schlessinger J: The SH2/ SH3 domains containing protein GRB2 interacts with tyrosine-phosphorylated IRS-1 and Shc: Implications for insulin control of ras signalling. EMBO J 12: 1929–1936, 1993
15. Myers MG Jr, Wang LM, Sun XJ, Zhang YT, Yenush L, Schlessinger J, Pierce JH, White MF: The role of IRS-1/GRB2 complexes in insulin signalling. Mol Cell Biol 14: 3577–3587, 1994
16. Sugimoto S, Wandness TJ, Shoelsen SE, Neel BG, Walsh CT: Activation of the SH2-containing protein tyrosine phosphatase, SH-PTP2, by phosphotyrosine containing peptides derived from insulin receptor substrate-1. J Biol Chem 269: 13614–13622, 1994
17. Haslam RJ, Koide HB, Hemmings BA: Pleckstrin domain homology. Nature 363: 309–310, 1993
18. Mayer BJ, Ren R, Clark KL, Baltimore D: A putative modular domain present in diverse signalling proteins. Cell 73: 629–630, 1993
19. Musacchio A, Gibson T, Rice P, Thompson J, Saraste M: The PH domain: A common piece in the structural patchwork of signalling proteins. Trends Biochem Sci 18: 342–348, 1993
20. Touhara K, Inglese J, Pitcher JA, Shaw G, Lefkowitz RJ: Binding of G protein beta gamma-subunits to pleckstrin homology domains. J Biol Chem 269: 10217–10220, 1994
21. Yao L, Kawakami Y, Kawakami T: The pleckstrin homology domain of Bruton tyrosine kinase interacts with protein kinase C. Proc Natl Acad Sci USA 91: 9175–9179, 1994
22. Harlan JE, Hajduk PJ, Yoon HS, Fesik SW: Pleckstrin homology domains bind to phosphatidylinositol-4,5-bisphosphate. Nature 371: 168–170, 1994

23. Yenush L, Makati KJ, Smith-Hall J, Ishibashi O, Myers MG Jr, White MF: The pleckstrin homology domain is the principle link between the insulin receptor and IRS-1. J Biol Chem 271: 24300–24306, 1996

24. Craparo A, O'Neill TJ, Gustafson TA: Non-SH2 domains within insulin receptor substrate 1 and Shc mediate their phosphotyrosine dependent interaction with the NPEY motif of the insulin like growth factor 1 receptor. J Biol Chem 270: 15639–15643, 1995

25. Wolf G, Trüb T, Ottinger E, Groninga L, Lynch A, White MF, Miyazaki M, Lee J, Shoelsen SE: PTB domains of IRS-1 and Shc have distinct but overlapping binding specificity. J Biol Chem 270: 27402–27410, 1995

26. Kavanaugh WM, Truck CW, Williams LT: PTB domain binding to signalling proteins through a sequence motif containing phosphotyrosine. Science 268: 1177–1179, 1995

27. Gustafson TA, He W, Caparo A, Schaub CD, O'Neill TJ: Phosphotyrosine-dependent interaction of Shc and IRS-1 with the NPEY motif of the insulin receptor via a novel non-SH2 domain. Mol Cell Biol 15: 2500–2508, 1995

28. Backer JM, Schroeder GG, Cahill DA, Ullrich A, Siddle K, White MF: Cytoplasmic juxtamembrane region of the insulin receptor: A critical role in ATP binding, endogenous substrate phosphorylation, and insulin-stimulated bioeffects in CHO cells. Biochemistry 30: 6366–6372, 1991

29. Yonezawa K, Ando A, Kaburagi Y, Yamamoto-Honda R, Kitamura T, Hara K, Nakafuku M, Okabayashi Y, Kadowaki T, Kaziro Y, Kasuga M: Signal transduction pathways from insulin receptors to Ras. Analysis by mutant insulin receptors. J Biol Chem 269: 4634–4640, 1994

30. Eck MJ, Dhe-Paganon S, Trüb T, Notle RT, Shoelson SE: Structure of the IRS-1 PTB domain bound to the juxtamembrane region of the insulin receptor. Cell 85: 695–705, 1996

31. Zhou M-M, Ravichandran KS, Olenjniczak ET, Petros AM, Meadows RP, Harlan JE, Wade WS, Burakoff SJ, Fesik SW: Structure and ligand recognition of the phosphotyrosine binding domain of Shc. Nature 378: 584–592, 1995

32. Lemmon MA, Ferguson KM, Schlessinger J: PH domain: Diverse sequences with a common fold recruit signalling molecules to the cell surface. Cell 85: 621–624, 1996

33. Carpenter LC, Cantley LC: Phosphoinositide kinases. Curr Opin Cell Biol 8: 153–158, 1996

34. Yonezawa K, Ueda H, Hara K, Nishida K, Ando A, Chavanieu A, Matsuba H, Shii K, Yokono K, Fukui Y, Calas B, Grigorescu F, Dhand R, Gout I, Otsu M, Waterfield M, Kasuga M: Insulin-dependent formation of a complex containing an 85 kDa subunit of phosphatidylinositol 3-kinase and tyrosine-phosphorylated insulin receptor substrate 1. J Biol Chem 267: 25958–25966, 1992

35. Backer JM, Myers MG Jr, Shoelsen SE, Chin DJ, Sun XJ, Mirapleix M, Hu P, Margolis B, Skolnik KY, Schlessinger J, White MF: Phosphatidylinositol 3'-kinase is activated by association with IRS-1 during insulin stimulation. EMBO J 9: 3469–3479, 1992

36. Hu Q, Klippel A, Muslin AJ, Fantl WJ, Williams LT: Ras-dependent induction of cellular responses by constitutive active phosphatidylinositol 3-kinase. Science 268: 100–102, 1995

37. Dhand R, Hara K, Hiles I, Bax B, Gout I, Pnayotou G, Fry MJ, Yonezawa K, Kasuga M, Waterfield MD: PI 3-kinase: Structural and functional analysis of intersubunit interaction. EMBO J 13: 511–521, 1994

38. Klippel A, Escobedo JA, Hirano M, Williams LT: The interaction of small domains between the subunits of phosphatidylinositol 3-kinase determines enzyme activity. Mol Cell Biol 14: 2675–2685, 1994

39. Myers MG Jr, Backer JM, Sun XJ, Shoelsen SE, Hu P, Schlessinger J, Yoakim M, Schaffhausen B, White MF: IRS-1 activates the phosphatidylinositol 3'-kinase by associating with the src homology 2 domain of p85. Proc Natl Acad Sci USA 89: 10350–10354, 1992

40. Pons S, Asano T, Glasheen E, Miralpeix M, Zhang Y, Fisher TL, Myers MG Jr, Sun XJ, White MF: The structure and function of p55^{PIK} reveal a new regulatory subunit for phosphatidylinositol 3-kinase. Mol Cell Biol 15: 4453–4465, 1995

41. Hara K, Yonezawa K, Sakaue H, Ando A, Kotani K, Kitamura T, Kitamura Y., Ueda H, Stephens L, Jackson T R, Hawkins PT, Dhand R, Clark AE, Holman GD, Waterfield MD, Kasuga M: 1-Phosphatidylinositol 3-kinase activity is required for insulin stimulated glucose transport but not for Ras activation in CHO cells. Proc Natl Acad Sci USA 91: 7415–7419, 1994

42. Quon MJ, Chen H, Ing BR, Liu M-L, Zarnowski MJ, Yonezawa K, Kasuga M, Cushman SW, Taylor SI: Roles of 1-phosphatidylinositol 3-kinase and ras in regulating translocation of GLUT4 in transfected rat adipose cells. Mol Cell Biol 15: 5403–5411, 1995

43. Kotani K, Carozzi AJ, Sakaue H, Hara K, Robinson LJ, Clark SF, Yonezawa K, James DE, Kasuga M: Requirement for phosphoinositide 3-kinase in insulin-stimulated GLUT4 translocation in 3T3-L1 adipocytes. Biochem Biophys Res Commun 209: 343–348, 1995

44. Ui M, Okada T, Hazeki K, Hazeki O: Wortmannin as a unique probe for an intracellular signalling protein, phosphoinositide 3-kinase. Trends Biochem Sci 20: 303–307, 1994

45. Okada T, Kawano Y, Sakakibara T, Hazeki O, Ui M. Essential role of phosphatidylinositol 3-kinase in insulin-induced glucose transport and antilipolysis in rat adipocytes. J Biol Chem 269: 3568–3573, 1994

46. Clarke JF, Young PW, Yonezawa K, Kasuga M, Holman GD: Inhibition of the translocation of GLUT1 and GLUT4 in 3T3-L1 cells by the phosphatidylinositol 3-kinase inhibitor, wortmannin. Biochem J 300: 631–635, 1994

47. Rhan T, Ridderstråle M, Tronqvist H, Maganielllo V, Fredrikson G, Belfrage P, Degerman E: Essential role of phosphatidylinositol 3-kinase in insulin-induced activation and phosphorylation of the cGMP-inhibited cAMP phosphodiesterase in rat adipocytes. FEBS Lett 350: 314–318, 1994

48. Moule SK, Edgell NJ, Welsh GI, Diggle TA, Foulstone EJ, Heesom KJ, Proud CG, Denton RM: Multiple signalling pathways involved in the stimulation of fatty acid and glycogen synthesis by insulin in rat epididymal fat cells. Biochem J 311: 595–601, 1995

49. Yamamoto-Honda R, Tobe K, Kaburagi Y, Ueki K, Asai S, Yachi M, Shirouzu M, Akanuma Y, Yokoyama S, Yazaki Y, Kadowaki T: Upstream mechanism of glycogen synthase activation by insulin and insulin-like growth factor-1. Glycogen synthase activation is antagonized by wortmannin or LY294002 but not by rapamycin or by inhibiting p21^{ras}. J Biol Chem 270: 2729–2734, 1995

50. Sakaue H, Hara K, Noguchi T, Matozaki T, Kotani K, Ogawa W, Yonezawa K, Waterfield MD, Kasuga M: Ras-independent and wortmannin-sensitive activation of glycogen synthase by insulin in Chinese hamster ovary cells. J Biol Chem 270: 11304–11309, 1995

51. Welsh GI, Foulstone EJ, Young SW, Tavaré JM, Proud CG: Wortmannin inhibits the effects of insulin and serum on the activities of glycogen synthase kinase-3 and mitogen-activated protein kinase. Biochem J 303: 15–20, 1994

52. Mendez R, Myers MG Jr, White MF, Rhoads R: Stimulation of protein synthesis, eukaryotic translation initiation factor 4E phosphorylation, and PHAS-I phosphorylation by insulin require insulin receptor substrate 1 and phosphatidylinositol 3-kinase. Mol Cell Biol 16: 2857–2864, 1996

53. von Mateuffel SR, Gingras A-C, Ming X-F, Sonnenberg N: 4E-BP1 phosphorylation is mediated by the FRAP-p70^{s6k} pathway and is independent of mitogen-activated protein kinase. Proc Natl Acad Sci USA 93: 4076–4080, 1996

54. Brunn GJ, Williams J, Sabers C, Widerrecht G, Lawrence JC Jr, Abraham RT: Direct inhibition of the signalling function of the mammalian target of rapamycin by the phosphoinositide 3-kinase inhibitors, wortmannin and LY294002. EMBO J 15: 5265–5267, 1996

55. Sutherland C, O'Brien RM, Granner DK: Phosphatidylinositol 3-kinase, but not p70/p85 ribosomal S6 protein kinase, is required for the regulation of phosphoenolpyruvate carboxykinase (PEPCK) gene expression by insulin. J Biol Chem 270: 15501–15506, 1995

56. Kotani K, Yonezawa K, Hara K, Ueda H, Kitamura Y, Sakaue H, Ando A, Chavanieu A, Calas B, Grigorecu F, Nishiyama M, Waterfield MD, Kasuga M: Involvement of phosphoinositide 3-kinase in insulin- or IGF-1-induced membrane ruffling. EMBO J 13: 2313–2321, 1994

57. Hall A: Small GTP-binding proteins and the regulation of the actin cytoskeleton. Annu Rev Cell Biol 10: 31–53, 1994

58. Chung J, Grammer TC, Lemon KP, Kazlauskas A, Blenis J: PDGF- and insulin-dependent pp70s6k activation mediated by phosphatidylinositol-3-OH kinase. Nature 370: 71–75, 1994

59. Cheatham B, Vlahos CJ, Cheatham L, Wang L, Blenis J, Khan RC: Phosphatidylinositol 3-kinase activation is required for insulin stimulation of pp70 S6 kinase, DNA synthesis, and glucose transporter translocation. Mol Cell Biol 14: 4902–4911, 1994

60. Hara K, Yonezawa K, Sakaue H, Kotani K, Kotani K, Kojima A, Waterfield M D, Kasuga M: Normal activation of p70 S6 kinase by insulin in cells overexpressing a dominant negative 85 kD subunit of phosphoinositide 3-kinase. Biochem Biophys Res Commun 208: 735–741, 1995

61. Khon AD, Kovacina KS, Roth RA: Insulin stimulates the kinase activity of RAC-PK, a pleckstrin homology domain containing ser/ thr kinase. EMBO J 14: 4288–4295, 1995

62. Alessi DR, Andjelkovic M, Caudwell B, Cron P, Morrice N, Cohen P, Hemmings BA: Mechanism of activation of protein kinase B by insulin and IGF-1. EMBO J 15: 6541–6551, 1996

63. Volinia S, Dhand R, Vanhaesebroeck B, MacDougall LK, Stein R, Zvelebil MJ, Domin J, Panaretou C, Waterfield MD: A human phosphatidylinositol 3-kinase complex related to the yeast Vps34p-Vps15p protein sorting system. EMBO J 14: 3339–3348, 1995

64. Stoyanov B, Volinia S, Hanck T, Rubio I, Loubtchenkov M, Malek D, Stoyanov S, Vanhaesebroeck B, Dhand R, Nurberg B, Gierschik P, Seedorf K, Hsuan JJ, Waterfield MD, Wetzker R: Cloning and characterization of a G-protein-activated human phosphoinositide-3 kinase. Science 269: 690–693, 1995

65. Kotani K, Hara K, Kotani K, Yonezawa K, Kasuga M: Phosphoinositide 3-kinase as an upstream regulator of the small GTP-binding protein rac in the insulin signalling of membrane ruffling. Biochem Biophys Res Commun 208: 985–990, 1995

66. Hawkins PT, Eguinoa A, Qui R-G, Stoke D, Cooke FT, Walters R, Wennström S, Claesson-Welsh L, Evans T, Symons M, Stephens L: PDGF stimulates an increase in GTP-Rac via activation of phospho-inositide 3-kinase. Curr Biol 5: 393–403, 1995

67. Tsakiridis T, Taha C, Grinstein S, Klip A: Insulin activates a p21-activated kinase in muscle cells via phosphatidylinositol 3-kinase. J Biol Chem 271: 19663–19667, 1996

68. Marcusohn J, Isakoff SJ, Rose E, Symons M, Skolnik KY: The GTP-binding protein Rac does not couple PI 3-kinase to insulin-stimulated glucose transport in adipocytes. Curr Biol 5: 1296–1302, 1995

69. Franke TF, Yang S-I, Chan TO, Datta K, Kazlauskas A, Morrison DK, Kaplan DR, Tsichlis PN: The protein kinase encoded by the Akt protooncogene is a target of the PDGF-activated phosphatidylinositol 3-kinase. Cell 81: 727–736, 1995

70. Burgring BMT, Coffer PJ: Protein kinase B (c-Akt) in phosphatidyl-inositol-3-OH kinase signal transduction. Nature 376: 599–602, 1995

71. James SR, Downes CP, Gigg R, Grove SJA, Holmes AB, Alessi DR: Specific binding of the AKT-1 protein-kinase to phosphatidylinositol 3, 4, 5-triphosphate without subsequent activation. Biochem J 315: 709–713, 1996

72. Kohn AD, Takeuchi F, Roth RA: Akt, a pleckstrin homology domain containing kinase, is activated primarily by phosphorylation. J Biol Chem 271: 21920–21926, 1996

73. Cross DAK, Alessi DR, Cohen P, Andjelkovich M, Hemmings BA: Inhibition of glycogen synthase kinase-3 by insulin mediated by protein kinase B. Nature 378: 785–789, 1995

74. Lowenstein EJ, Daly RJ, Batzer AG, Li W, Margolis B, Lammers R, Ullrich A, Skolnik KY, Bar-Sagi D, Schlessinger J: The SH2 and SH3 domain-containing protein GRB2 links receptor tyrosine kinases to ras signalling. Cell 70: 431–442, 1992

75. Baltensperger K, Kozma LM, Cherniack AD, Klaulund JK, Chawala A, Banerjee U, Czech MP: Binding of the Ras activator son of sevenless to insulin receptor substrate-1 signalling complexes. Science 260: 1950–1952, 1993

76. Skolnik KY, Batzer A, Li N, Lee C-H, Lowenstein E, Mohammadi M, Margolis B, Schlessinger J: The function of GRB2 in linking the insulin receptor to ras signalling pathways. Science 260: 1953–1955, 1993

77. Roberts TM: A signal chain of events. Nature 360: 534–535, 1992

78. Moller W, Bos JL: The role of ras proteins in insulin signal transduction. Horm Metab Res 24: 214–218, 1992

79. Sakaue M, Bowtell D, Kasuga M: A dominant-negative mutant of mSOS1 inhibits insulin-induced Ras activation and reveals Ras-dependent and -independent insulin signalling pathways. Mol Cell Biol 15: 379–388, 1995

80. Waters SB, Yamauchi K, Pessin JE: Functional expression of insulin receptor substrate-1 is required for insulin-stimulated mitogenic signalling. J Biol Chem 268: 22231–22234, 1993

81. Rose DW, Saltiel AR, Majumdar M, Decker SJ, Olefsky JM: Insulin receptor substrate 1 is required for insulin-mediated mitogenic signal transduction. Proc Natl Acad Sci USA 91: 797–801, 1994

82. Lazar DF, Wiese RJ, Brady MJ, Mastick CC, Waters SB, Yamauchi K, Pessin JE, Cuatrecasas P, Saltiel AR: Mitogen-activated protein kinase kinase inhibition does not block the stimulation of glucose utilization by insulin J Biol Chem 270: 20801–20807, 1995

83. Ando A, Yonezawa K, Gout I, Nakata T, Ueda H, Hara K, Kitamura Y, Noda Y, Takenawa T, Hirokawa N, Kasuga M: A complex of GRB2-dynamin binds to tyrosine phosphorylated insulin receptor substrate-1 after insulin treatment. EMBO J 13: 3033–3038, 1994

84. Wang ZX, Moran MF: Requirement for the adapter protein GRB2 in EGF receptor endocytosis. Science 272: 1935–1939, 1996

85. Sasaoka T, Ishihara H, Sawa T, Ishiki M, Morioka H, Imamura T, Usui I, Takata Y, Kobayashi M: Functional importance of amino-terminal domain of Shc for interaction with insulin and epidermal growth factor receptors in phosphorylation-independent manner. J Biol Chem 271: 20082–20087, 1996

86. Sasaoka T, Drazrin B, Leitner JW, Langlois WJ, Olefsky JM: Shc is the predominant signalling molecule coupling insulin receptors to activation of guanine nucleotide releasing factor and p21ras-GTP formation. J Biol Chem 269: 10734–10738, 1994

87. Kuhné MR, Pawson T, Lienhard GE, Feng G-S: The insulin receptor substrate 1 associates with the SH2-containing phosphotyrosine phosphatase Syp J Biol Chem 268: 11479–11481, 1993

88. Matozaki T, Kasuga M: Roles of protein tyrosine phosphatases in growth factor signalling. Cell Signal 8: 13–19, 1996

89. Xiao S, Roses DW, Sasaoka T, Maegawa H, Bruke TR Jr, Roller PP, Shoelsen SE, Olefsky JM: Syp is a positive mediator of growth factor-stimulated mitogenic signal transduction. J Biol Chem 269: 21244–21248, 1994

90. Noguchi T, Matozaki T, Horita K, Fujioka Y, Kasuga M: Role of SH-PTP2, a protein-tyrosine phosphatase with src homology-2 domains, in insulin-stimulated Ras activation. Mol Cell Biol 14: 6674–6682, 1994

91. Yamauchi K, Milarski KL, Saltiel AR, Pessin JE: Protein-tyrosine phosphatase SH-PTP2 is a required positive effector for insulin downstream signalling. Proc Natl Acad Sci USA 92: 664–668, 1995

92. Herbst RP, Carroll PM, Allard JD, Schilling J, Raabe T, Simon MA: Daughter of sevenless is a substrate of the phosphotyrosine phosphatase corkscrew and functions during sevenless signalling. Cell 85: 899–909, 1996

22

93. Sawada T, Kim L, Saltiel AR: Expression of a catalytically inert Syp blocks activation of MAP kinase pathway downstream of p21^{ras}. Biochem Biophys Res Commun 214: 737–743, 1995

94. Milarski KL, Saltiel AR: Expression of catalytically inactive Syp phosphatase in 3T3 cells blocks stimulation of mitogen-activated protein kinase by insulin. J Biol Chem 269: 21239–21243, 1994

95. Noguchi T, Matozaki T, Fujioka Y, Yamao T, Tsuda M, Takada T, Kasuga M: Characterization of a 115 kDa protein that binds to SH-PTP2, a protein-tyrosine phosphatase with Src homology 2 domains, in Chinese hamster ovary cells. J Biol Chem 271: 27652–27658, 1996

96. Fujioka Y, Matozaki T, Noguchi T, Iwamatsu A, Yamao T, Takahashi N, Tsuda M, Takada T, Kasuga M: A novel membrane glycoprotein, SHPS-1, that binds the SH-2-domain-containing protein tyrosine phosphatase SHP-2 in response to mitogens and cell adhesion. Mol Cell Biol 16: 6887–6889, 1996

97. Tobe K, Sabe H, Yamamoto T, Yamauchi T, Asai S, Kaburagi Y, Tamemoto H, Ueki K, Kimura H, Akanuma Y, Yazaki Y, Hanafusa H, Kadowaki T: Csk enhances insulin-stimulated dephosphorylation of focal adhesion proteins. Mol Cell Biol 16: 4765–4772, 1996

98. Lee C-H, Li W, Nishimura R, Zhou M, Batzer AG, Myers MG Jr, White MF, Schelessinger J, Skolnik KY: Nck associates with the SH2 domain-docking protein IRS-1 in insulin-stimulated cells. Proc Natl Acad Sci USA 90: 11713–11717, 1993

99. Fei ZL, D'Ambrosio C, Li S, Surmacz E, Baserga R: Association of insulin receptor substrate 1 with simian virus 40 large T antigen. Mol Cell Biol 15: 4232–4239, 1995

100. Vouli K, Rouslahti E: Association of insulin receptor substrate-1 with integrins. Science 266: 1576–1578, 1994

101. Tamemoto H, Kadowaki T, Tobe K, Yagi T, Sakura H, Hayakawa T, Terauchi Y, Ueki K, Kaburagi Y, Satho S, Sekihara H, Yoshioka S, Horikoshi H, Furuta Y, Ikawa Y, Kasuga M, Yazaki Y, Aizawa S: Insulin resistance and growth retardation in mice lacking insulin receptor substrate-1. Nature 372: 182–186, 1994

102. Araki E, Lipes MA, Patti M-E, Brüning JC, Haag B III, Johnson RS, Khan CR: Alternative pathway of insulin signalling in mice with targeted disruption of the IRS-1 gene. Nature 372: 186–190, 1994

103. Patti ME, Sun XJ, Bruening JC, Araki E, Lipes MA, White MF, Kahn CR: 4PS/insulin receptor substrate (IRS)-2 is the alternative substrate of the insulin receptor in IRS-1 deficient mice. J Biol Chem 270: 24670–24673, 1995

104. Sun XJ, Wang L-M, Zhang Y, Yenush L, Myers MG Jr, Glasheen E, Lane WS, Pierce JH, White MF: Role of IRS-2 in insulin and cytokine signalling. Nature 377: 173–177, 1995

105. Lavan BE, Lienhard GE: The insulin elicited 60 kDa phosphotyrosine protein in rat adipocytes is associated with phosphatidylinositol 3-kinase. J Biol Chem 268: 5921–5928, 1993

106. Hosomi Y, Shii K, Ogawa W, Matsuba H, Yoshida M, Okada Y, Yokono K, Kasuga M, Baba S, Roth RA: Characterization of a 60 kilodalton substrate of the insulin receptor kinase. J Biol Chem 269: 11498–11502, 1994

107. Holgado-Madruga M, Emlet DR, Moscatello DK, Godwin AK, Wong AJ: A Grb2-associated docking protein in EGF- and insulin-receptor signalling. Nature 379: 560–564, 1996

108. Yeh TC, Ogawa W, Danielsen AG, Roth RA: Characterization and cloning of a 58/53 kDa substrate of the insulin receptor tyrosine kinase. J Biol Chem 271: 2921–2928, 1996

Molecular and Cellular Biochemistry **182**: 23–29, 1998.
© 1998 *Kluwer Academic Publishers. Printed in the Netherlands.*

Insulin regulation of the Ras activation/inactivation cycle

Brian P. Ceresa and Jeffrey E. Pessin
Department of Physiology and Biophysics, The University of Iowa, Iowa City, USA

Abstract

In addition to mediating a number of metabolic functions, insulin also uses mitogenic pathways to maintain cellular homeostasis. Many of these mitogenic responses are mediated by signals through the small molecular weight guanine nucleotide binding protein, Ras. In the last decade, great progress has been made in understanding the molecular mechanisms which regulate the insulin mediated conversion of Ras from its inactive, GDP-bound state, to the activated GTP-bound form. More recently, it has been appreciated that insulin also regulates the inactivation of this pathway, namely by uncoupling the protein complexes whose formation is required for Ras activation. This review addresses molecular mechanism which both positively and negatively regulate this mitogenic signalling pathway. (Mol Cell Biochem **182**: 23–29, 1998)

Key words: insulin, Ras, SOS, IRS-1, Shc, MAP kinase

Introduction

Binding of insulin to its cognate receptor elicits a number of metabolic and mitogenic responses (reviewed in [1, 2]). Without argument, the principal function of insulin is to regulate the uptake, synthesis, and storage of cellular energy, however, it is quite evident that through mitogenic signalling insulin plays a substantial role in cellular homeostasis by coordinating protein synthesis, cell growth and division. One of the most well-characterized mitogenic pathways which insulin regulates is the Ras/Mitogen Activating Protein (MAP) kinase pathway. The Ras pathway is activated by a number of growth factors, hormones, and cytokines of which the detailed signal transduction pathways have only recently become apparent. Despite the ability of multiple ligands to activate this pathway, each extracellular signal seems to regulate the Ras pathway in its own distinct manner. Nevertheless, the cellular consequences of activation of this pathway includes cell growth, transformation, proliferation, activation, and, in some cases, growth inhibition [3]. In fact, activating mutants of Ras are found in nearly 10% of all human tumors and are associated with greater than 80% of all colon and pancreatic cancers [4].

In the early 1990's, when the observation was made that insulin could modulate the Ras pathway it sparked a flurry of investigation of the physiological significance of this event and molecular mechanism regulating its activation [5, 6]. To date, although there are still many unanswered questions, we now have a detailed understanding of the protein interactions that lead from receptor tyrosine kinases to Ras activation. These findings have converged with other studies examining downstream pathways and have provided a molecular framework directly connecting plasma membrane receptor signals to nuclear transcriptional events. In this article, we will utilize the insulin receptor tyrosine kinase as a model system linking proximal tyrosine phosphorylation to the activation of Ras and its subsequent coupling to the MAP kinase pathway.

Ras activation

Ras is one member of a large family of small molecular weight GTP binding proteins. These proteins differ from the heterotrimeric GTP binding proteins in that they are typically 20–25 kDa in size, monomeric and have relatively low intrinsic rates of guanylnucleotide exchange and GTP hydrolytic (GTPase) activities. Currently there are greater than 50 Ras-related low molecular weight GTP binding proteins which have been categorized into several subfamilies

Address for offprints: B.P. Ceresa, Department of Physiology and Biophysics, The University of Iowa, Iowa City, IA 52242, USA

24

based upon sequence relationships and biological functions. These low molecular GTP binding proteins play multiple roles in the regulation of growth, cell fate determination, macromolecular biosynthesis, motility, organization of the cytoskeleton, nuclear transport and vesicular trafficking. Defects in the regulation of these pathways can have severe consequences and can result in multiple abnormalities including tumorigenesis and immunological disorders.

The p21 Ras family consists of four related GTP binding proteins termed H-Ras, K-Ras, N-Ras and R-Ras which play important roles in mediating differentiation of a variety of specialized cell types and in the proliferative response of others [4, 7, 8]. Although many receptor tyrosine kinases can stimulate Ras through a similar molecular paradigm, the specific coupling of the insulin receptor kinase to Ras activation is illustrated in Fig. 1. The insulin receptor is comprised of two extracellular α subunits and two trans-membrane β subunits linked by disulfide bonds to form an $\alpha_2\beta_2$ complex [9]. Binding of insulin to the α subunits of the receptor induces intramolecular autophosphorylation of the β subunits resulting in the phosphorylation of multiple tyrosine residues and activation of the intracellular kinase domains of the β subunit [10]. Once activated, the insulin receptor tyrosine phosphorylates multiple substrates of which insulin receptor substrates 1 and 2 (IRS1 and IRS2) and Shc have been the best characterized [11–13]. Multiple

signalling roles have been ascribed to the IRS proteins, however, Shc appears to primarily function in the Ras activation pathway [14, 15].

Shc was originally isolated and the cDNA cloned based upon its homology with the SH2 domain sequences from the human c-fos gene [16]. The Shc gene codes for three related mRNAs (3.8, 3.4 and 2.8 kb) that translate into three proteins of 46, 52 and 66 kDa. The 46 and 52 kDa isoforms result from the use of two alternative translation initiation sites within a single 3.4 kb transcript [16]. Antibodies directed against the 46/52 Shc protein also cross react with a 66 kDa species. This isoform is translated from at least one of the other two transcripts which arises due to alternative splicing of a single Shc gene [17]. Structurally, all three isoforms are similar in that they contain an amino terminal phosphotyrosine binding (PTB) domain, a central collagen-homology (CH) domain, and a carboxyl terminal Src homology 2 (SH2) domain [16]. The 52 kDa Shc isoform is the major substrate for the insulin receptor kinase whereas EGF induces tyrosine phosphorylation of both the 46 and 52 kDa Shc isoforms [12, 18, 19]. At present the functional role for the 66 kDa Shc isoform not well understood but recent data suggests that it may limit the extent of Ras activation [20].

Nevertheless, phosphorylation of Shc on tyrosine 317 results in the formation of a docking site for the SH2

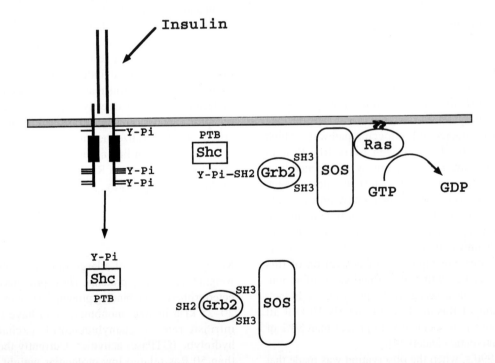

Fig. 1. A schematic model for the major pathway by which insulin results in Ras activation. Insulin binding to the insulin receptor (IR), results in the activation of the intracellular β subunit tyrosine protein kinase activity. The insulin receptor can then phosphorylate multiple intracellular substrates including IRS1 and Shc. The tyrosine phosphorylation of Shc results in the formation of a Grb2 SH2 binding site. The pre-assembled Grb2-SOS complex can then associate with Shc and through an unknown mechanism becomes targeted and/or activated to the plasma membrane location of Ras.

domain of the adapter protein Grb2 [21]. Grb2 is relatively small molecular weight protein (23 kDa) containing one SH2 domain which is flanked by two SH3 domains[5, 22]. In contrast to PTB and SH2 domains which direct binding interactions with tyrosine phosphorylated residues, SH3 domains have specificity for proline rich sequences [21, 23]. Although Grb2 has been found to associate with a number of effector proteins, the proline rich carboxyl terminal domain of SOS is one major target [23–25]. The mammalian SOS protein, named for its homology to the Son-ofSeveness gene product from *Drosophila melangoster*, is a Ras guanyl-nucleotide exchange factor that catalyzes the replacement of GDP bound to Ras with GTP [26, 27]. As with all GTP binding proteins, the conversion from the GDP-bound state to GTP results in a conformation change that results in activation and the ability to couple with downstream effector molecules.

Based upon these data, a molecular model has been proposed that directly links receptor tyrosine kinase signalling to Ras activation (Fig. 1). Following insulin stimulation, the activated insulin receptor kinase phosphorylates Shc, generating a recognition site for the Grb2 SH2 domain. In the unstimulated or basal state, the SH3 domains of Grb2 associate with SOS. Upon insulin stimulation, tyrosine phosphorylated Shc interacts with the pre-existing Grb2-SOS complex resulting in the formation of a ternary Shc-Grb2-SOS complex. The Shc PTB domain and/or the SH2 domain can then either interact with the tyrosine phosphorylated insulin receptor or an as yet unknown plasma membrane docking protein. In either case, the membrane localization and/or activation of the Shc-Grb2-SOS complex appears necessary in order to direct SOS to the plasma membrane location of Ras. In this manner, a series of protein-protein interactions mediated by tyrosine phosphorylation signals results in the rapid conversion of Ras from the inactive GDP-bound to the active GTP-bound state.

The role of IRS in Ras activation

As described above, the primary pathway for insulin activation of Ras is dependent upon Shc, however, Ras activation can occur through IRS1 under certain circumstances. IRS1 is a 185 kDa cytosolic protein that is another the major substrate for the insulin receptor kinase [13]. Following tyrosine phosphorylation, many of IRS1's phosphotyrosine residues serve as docking sites for important signalling molecules including phosphatidylinositol 3-kinase, SHPTP2, Nck, and Grb2 [28, 29]. While both Shc and IRS1 can compete for the cellular pool of Grb2 and each can effectively led to Ras activation [30–32], IRS1 phosphorylation is not sufficient to activate this pathway. For example, the cytokine interleukin-4 (IL4) stimulates IRS1 phosphorylation through

activation of the JAK family of protein tyrosine kinases [33, 34], however, IL4 stimulation does not result in Shc phosphorylation and can not activate the Ras pathway [35]. This occurs despite the tyrosine phosphorylation of IRS1 and its association with the SH2 domains of multiple effector proteins including Grb2. In addition, co-immunoprecipitation studies quantitating the extent of Shc-Ras versus IRS1-Ras signalling demonstrated 7 fold more guanine nucleotide exchange activity in Shc immunoprecipitates than IRS1 immunoprecipitates from insulin-treated Rat 1 fibroblasts [14]. Furthermore, point mutations in the insulin receptor (tyrosines 1162 and 1163 to phenylalanine) which prevent phosphorylation of IRS1 and IRS1-Grb2 formation, have no effect of the tyrosine phosphorylation of 52 kDa Shc, with little attenuation of Ras activation [15]. These data have been generally interpreted to indicate that although insulin is capable of activating Ras through both Shc and IRS1, the predominant pathway occurs through an insulin receptor-Shc interaction rather than an IRS1-Grb2 interaction.

Ras-dependent downstream signalling pathway

Studies from numerous laboratories using a combination of genetic, molecular and cell biological approaches in various systems have been able to develop a coherent series of protein-protein interactions that connect Ras activation to a serine/threonine kinase cascade leading to downstream biological responsiveness [4, 7, 8]. Nearly simultaneously, several laboratories observed that the serine/threonine kinase c-Raf 1 specifically associates with Ras when in the GTP-bound state and dissociates when Ras is converted into the inactive GDP-bound state [36–41]. This finding provided the missing link between Raf activation its ability to activate a dual functional (tyrosine/threonine kinase) termed MEK [42–44]. MEK is an atypical protein kinase that once activated expresses both threonine and tyrosine kinase activity and phosphorylates the ERK family of MAP kinases on a TEY motif [45, 46]. This phosphorylation event activates the ERK kinase and allows for its translocation into the nucleus providing a mechanism by which both cytosolic proteins and nuclear transcription factors can function as substrates [47–54]. A schematic representation of this pathway is present in Fig. 2.

Although it is clear that the Ras/Raf/MEK/ERK pathway plays an important role in the growth factor regulation of mitogenesis, there remains considerable controversy for its role in insulin action, particularly since insulin is predominantly a metabolic hormone. It was initially suggested that Ras activation resulted in the stimulation of insulin's metabolic effect to enhance glucose uptake in adipocytes and cardiac muscle [55, 56]. However, multiple growth factors can stimulate Ras activation and the downstream

26

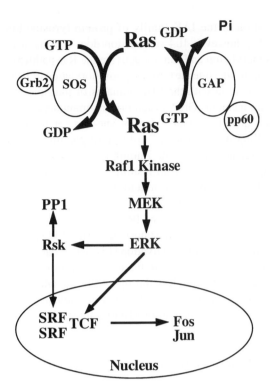

Fig. 2. A schematic illustration of the Ras/Raf/MEK/ERK kinase cascade. Once Ras is activated by the exchange of GDP for GTP, the GTP-bound Ras associates with c-Raf 1 resulting in the activation of the Raf protein kinase activity. Activated Raf can then phosphorylate MEK on two serine residues resulting the activation of MEK. In turn, MEK phosphorylates ERK on a tyrosine and threonine residue that activates and inducing the translocation of ERK from the cytoplasm into the nucleus. Several cytoplasmic and nuclear proteins are substrates for activated ERK including MAPKAP, pp90Rsk, TCF and CREB.

ERK kinase cascade but exhibit none of insulin's metabolic effects [57]. This has been more rigorously explored by testing the role of the Ras pathway in insulin's metabolic functions through constitutive activation and inhibition of Ras. In 3T3-L1 adipocytes, microinjection of constitutively active Ras had no effect on the translocation of the major insulin sensitive glucose transporter, GLUT4 [58]. Conversely, when dominant interfering Ras, neutralizing Ras antibodies, or lovastatin is used to inhibit Ras function, there is no change in either GLUT4 expression, translocation, or in insulin-stimulated glucose uptake [58, 59]. Similarly, inhibition of MEK or ERK activity does not effect insulin stimulated glucose transport, glycogen synthesis or lipogenesis [60–63]. However, in L6 cells, expression of an inducible form of dominant interfering Ras resulted in decreased insulin mediated GLUT3 expression [64]. Thus, the inability of Ras to modulate the metabolic functions of insulin, particularly glucose transporter expression or activation, suggests that the Ras pathway does not plays a significant role in these pathways.

Regulation of guanine nucleotide exchange

As alluded to in Fig. 1, insulin increases Ras activity by increasing the rate of guanine nucleotide exchange. However, it is equally plausible that insulin could also regulate Ras activation by inhibition of GTP hydrolysis through negatively regulating GTPase Activating Protein (GAP) activity. However, there is no experimental evidence that insulin has any effect on Ras GTP hydrolysis. For example, insulin treatment of Rat-1 fibroblasts stimulates [^{32}P]α-GTP binding to Ras with little change in the rate of hydrolysis of the gamma phosphate from [^{32}P]γ-GTP bound to Ras [65, 66]. These data indicate that insulin increases the rate of GDP dissociation from Ras. Since other studies have demonstrated that guanine nucleotide exchange is mediated by a direct interaction between SOS and Ras, there is strong evidence indicating that insulin regulates Ras GDP/GTP exchange activity rather than GTP hydrolysis [26, 27, 67].

Similar to most but not all growth factors, the insulin activation of Ras is transient. Following insulin stimulation, cells exhibit a maximal amount of bound GTP within 2–4 min which returns to near basal levels within 15–30 min [19, 68]. To achieve such a rapid inactivation there must be a cellular mechanism in place that effectively terminates guanine nucleotide exchange activity. One proposed mechanism is the uncoupling of the Grb2-SOS complex via serine/threonine phosphorylation of SOS. Indirect support for this module was based upon the observation that mutants which prevent Grb2-SOS interaction markedly impair the Ras activation pathway [62]. In addition, stable overexpression of Grb2 in L6 myoblasts increased the number of Grb2-SOS complexes present and resulted in enhanced activation of the Ras pathway [69].

More recently, several laboratories have demonstrated that insulin results in the serine/threonine phosphorylation of SOS [68, 70, 71]. This phosphorylation event coincided with the disassociation of the Grb2-SOS complex and desensitization of the Ras pathway as measured by the amount of GTP versus GDP bound to immunoprecipitated Ras [68, 71, 72]. These data suggest a model in which enhanced SOS activity is maintained until the Grb2-SOS complex dissociates causing either decreased nucleotide exchange activity or prevents the targeting of SOS to the plasma membrane location of Ras (Fig. 3).

The molecular mechanism accounting for the serine/threonine phosphorylation of SOS has been a matter of considerable debate. Confounding the problem is that SOS can serve as a substrate for several kinases including ERK both *in vitro* and *in vivo*. In Chinese hamster ovary cells expressing the human insulin receptor (CHO/IR) inhibition of MEK with the MEK specific inhibitor, PD98059, or a dominant-interfering MEK mutant prevented serine/threonine phosphorylation of SOS. However, since MEK also

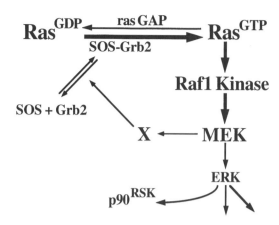

Fig. 3. Hypothesized mechanism accounting for the MEK-dependent serine/threonine phosphorylation of SOS and dissociation of the Grb2-SOS complex. Following activation of the Ras/Raf/MEK/ERK pathway, a MEK-dependent kinase(s) phosphorylate the carboxyl terminal domain of SOS. This phosphorylation events results in a decreased affinity of the Grb2 SH3 domains for SOS and thereby induces the dissociation of the Grb2-SOS complex. The uncoupling of this complex either results in the inactivation and/or loss of SOS targeting to the plasma membrane location of Ras. Since under these conditions there is no longer any stimulatory activity, the constitutive GTPase Activating Protein (GAP) activity can then induce the hydrolysis of GTP-bound Ras back to the GDP state.

phosphorylates and activates MAP kinase, it is difficult to assess the individual contribution of each kinase [71]. To clarify this issue, the SOS phosphorylating activity was resolved from ERK activity by expression of the MAP kinase specific phosphatase, MKP1. Furthermore, physical separation of ERK activity from another MEK-dependent kinase activity was achieved by FPLC anion exchange chromatography [73]. These data directly demonstrated the presence of a SOS kinase activity in CHO cells which is distinct from the ERK activity.

However, this apparent feedback pathway for Ras inactivation is not universal, as other mechanisms for Ras desensitization appear to be present [74]. For example, EGF stimulation results the activation of Ras through the formation of a Shc-Grb2-SOS complex, there is no dissociation of the Grb2-SOS complex during the Ras inactivation phase [19, 73]. Alternatively, evidence has been presented that Ras inactivation follows EGF receptor dephosphorylation and internalization, Shc dephosphorylation and/or uncoupling of Shc from the Grb2-SOS complex [70, 74–76]. The inability of EGF to induce dissociation of the Grb2-SOS complex is quite surprising as EGF is a potent stimulator of the Ras/Raf/MEK/ERK pathway and serine/threonine phosphorylation of SOS. The apparent differences between EGF and insulin on the Grb2-SOS complex may reflect subtle differences in the Ras activation mechanisms. In the case of EGF, the tyrosine phosphorylated EGF receptor provides a high affinity membrane docking site for the Shc-Grb2-SOS

complex which does not occur following insulin stimulation. Furthermore, mutations of the EGF receptor which still allow Shc phosphorylation but prevent EGF receptor targeting are now capable of inducing Grb2-SOS dissociation [77]. This may be, in part, due to differences between the sites of serine/threonine phosphorylation by insulin versus EGF [76, 78, 79].

Summary

Like all small molecular weight G-proteins, Ras serves as a molecular switch whose on/off status is regulated by the binding of GTP and GDP to its guanine nucleotide binding site. Conversion of Ras from its inactive, GDP-bound state, requires dissociation of GDP from the guanine nucleotide binding site and binding of the more cellular abundant GTP. This is accomplished by the use of receptor tyrosine kinases that induce the formation and membrane targeting of a signalling complex composed of Shc, Grb2 and SOS. The inactivation of Ras does not appear to be dependent on changes in GTPase activity or targeting but appears to rely on mechanisms that inhibit the SOS activation pathway. The elucidation of the complex interplay between insulin and other growth factors in regulating Ras will ultimately provide new information with regard to the cellular control of signalling specificity and the control of biological responsiveness.

Acknowledgement

Dr. B.P. Ceresa was the recipient of a postdoctoral fellowship award from the Juvenile Diabetes Foundation International.

References

1. Saltiel AR: Diverse signalling pathways in the cellular actions of insulin. Am J Physiol 270: E375–E385, 1996
2. Cheatham B, Kahn CR: Insulin action and the insulin signalling network. Endo Rev 16: 117–142, 1995
3. Satoh T, Nakafuku M, Kaziro Y: Function of Ras as a molecular switch in signal transduction. J Biol Chem 267: 24149–24152, 1992
4. Barbacid M: Ras genes. Ann Rev Biochem 56: 779–827, 1987
5. Medema RH, Wubbolts R, Bos JL: Two dominant inhibitory mutants of p21ras interfere with insulin induced gene expression. Mol Cell Biol 11: 5963–5967, 1991
6. Burgering BM, Medema RH, Maassen JA, Wetering ML, van der Eb AJ, McCormick F, Bos J L: Insulin mediated gene expression mediated by p21ras activation. EMBO J 10: 1103–1109, 1991
7. Downward J: Regulatory mechanisms for Ras proteins. BioEssays 14: 177–184, 1992
8. Marshall MS: Ras target proteins in eukaryotic cells. FASEB J 9: 1311–1318, 1995

28

9. Boni-Schnetzler M, Kaligian A, DelVecchio R, Pilch P: Ligand-dependent intersubunit association within the insulin receptor complex activates its intrinsic activity. J Biol Chem 263: 6822–6828, 1988

10. Czech M: The nature and regulation of the insulin receptor: Structure and function. Ann Rev Physiol 47: 357–381, 1985

11. Sun XJ, Wang LM, Zhang Y, Yenush L, Myers MG Jr, Glasheen E, Lane WS, Pierce JH, White MF: Role of IRS-2 in insulin and cytokine signalling. Nature 377: 173–177, 1995

12. Pronk GJ, McGlade I, Pelicci G, Pawson T, Bos JL: Insulin-induced phosphorylation of the 46 and 52 kDA Shc Proteins. J Biol Chem 268: 5748–5753, 1993

13. White MF, Maron R, Kahn CR: Insulin rapidly stimulates tyrosine phosphorylation of a M_r-185,000 protein in intact cells. Nature 318: 183–186, 1985

14. Sasaoka T, Draznin B, Leitner JW, Langlois WJ, Olefsky JM: Shc is the predominant signalling molecule coupling insulin receptors to activation of guanine nucleotide releasing factor and p21ras-GTP formation. J Biol Chem 269: 10734–10738, 1994

15. Ouwens DM, van der Zon GCM, Pronk GJ, Bos JL, Moller W, Cheatham B, Kahn CR, Maassen JAT: A mutant insulin receptor induces formation of a Shc-growth factor receptor bound protein 2 (Grb2) complex and p21ras-GTP without detectable interaction of insulin receptor substrate 1 (IRS1) with Grb2. J Biol Chem 269: 33116–33123, 1994

16. Pelicci G, Lanfrancone L, Grignani F, McGlade I, Cavallo F, Forni G, Nicoletti I, Grignani F, Pawson T, Pelicci PG: A novel transforming protein (SHC) with an SH2 domain is implicated in mitogenic signal transduction. Cell 70: 93–104, 1992

17. Migliaccio E, Mele S, Salcini AE, Pelicci G, Lai KV, Superti-Furga G, Pawson T, Fiore PPD, Pelicci PG: Opposite effects of the p52shc/p46shc and p66shc splicing isoforms on the EGF receptor-MAP kinase-fos signalling pathway. EMBO J 16: 706–716, 1997

18. Okada S, Yamauchi K, Pessin JE: Shc isoform-specific tyrosine phosphorylation by the insulin and epidermal growth factor receptors. J Biol Chem 270: 20737–20741, 1995

19. Waters SB, Chen D, Kao AW, Okada S, Holt KH, Pessin JE: Insulin and epidermal growth factor receptors regulate distinct pools of Grb2-SOS in the control of Ras activation. J Biol Chem 271: 18224–18230, 1996

20. Kao AW, Waters SB, Okada S, Pessin JE: Insulin stimulates the phosphorylation of the 66 and 52 kDa Shc isoforms by distinct mechanisms. Endocrinology 138: 2474–2480, 1997

21. Skolnik EY, Lee C-H, Batzer A, Vicentini LM, Zhou M, Daly R, Meyers MJJ, Backer JM, Ullrich A, White MF, Schlessinger J: The SH2/Sh3 domain-containing protein GRB2 interacts with tyrosine-phosphorylated IRS1 and Shc: Implications for insulin control of Ras signalling. EMBO J 12: 1929–1936, 1993

22. Lowenstein EJ, Daly RJ, Batzer AG, Li W, Margolis B, Lammers R, Ullrich A, Skolnik EY, Bar-Sagi D, Schlessinger J: The SH2 and SH3 domain-containing protein Grb2 links receptor tyrosine kinases to Ras signalling. Cell 70: 431–442, 1992

23. Rozakis-Adcock M, Fernley R, Wade J, Pawson T, Bowtell D: The SH2 and SH3 domains of mammalian Grb2 couple the EGF receptor to the Ras activator mSOS1. Nature 363: 83–85, 1993

24. Egan SE, Giddings BW, Brooks MW, Buday L, Sizeland AM, Weinberg RA: Association of Sos Ras exchange protein with Grb2 is implicated in tyrosine kinase signal transduction and transformation. Nature 363: 45–51, 1993

25. Li N, Batzer A, Daly R, Yajnik V, Skolnik E, Chardin P, Bar-Sagi D, Margolis B, Schlessinger J: Guanine-nucleotide-releasing factor hSos1 binds to Grb2 and links receptor tyrosine kinases to Ras signalling. Nature 363: 85–88, 1993

26. Simon MA, Bowtell DDL, Dodson GS, Laverty TR, Rubin GM: Ras 1 and a putative guanine nucleotide exchange factor perform crucial steps in signalling by the sevenless proteins tyrosine kinase. Cell 87: 701–718, 1991

27. Bowtell D, Fu P, Simon M, Senior P: Identification of murine homologues of the *Drosophila* son of sevenless gene: Potential activators of Ras. Proc Natl Acad Sci USA 89: 6511–6515, 1992

28. Myers MG Jr, White MF: Insulin signal transduction and the IRS proteins. Ann Rev Pharm Toxicol 36: 615–658, 1996

29. Waters SB, Pessin JE: Insulin receptor substrate 1 and 2 (IRS1 and IRS2): What a tangled web we weave. Trends Cell Biol 6: 1–4, 1996

30. Myers MG Jr, Wang L-M, Sun XJ, Zhang Y, Yenush L, Schlessinger J, Pierce JH, White MF: Role of IRS-1-GRB-2 complexes in insulin signalling. Mol Cell Biol 14: 3577–3587, 1994

31. Yamauchi K, Pessin JE: Insulin receptor substrate-1 (IRS1) and Shc compete for a limited pool of Grb2 in mediating insulin downstream signalling. J Biol Chem 269: 31107–31114, 1994

32. Yonezawa K, Ando A, Kaburagi Y, Yamamoto-Honda R, Kitamura T, Hara K, Nakafuku M, Okabayashi Y, Kadowaki T, Kaziro Y, Kasuga M: Signal transduction pathways from insulin receptors to Ras. J Biol Chem 269: 4634–4640, 1994

33. Keegan AD, Nelms K, White M, Wang LM, Pierce JH, Paul WE: An IL-4 receptor region containing an insulin receptor motif is important for IL-4-mediated IRS-1 phosphorylation and cell growth. Cell 76: 811–820, 1994

34. Yin T, Tsang ML, Yang YC: JAK1 kinase forms complexes with interleukin-4 receptor and 4PS/insulin receptor substrate-l-like protein and is activated by interleukin-4 and interleukin-9 in T lymphocytes. J Biol Chem 269: 26614–26617, 1994

35. Pruett W, Yuan Y, Rose E, Batzer AG, Harada N, Skolnik EY: Association between Grb2/SOS and insulin receptor substrate 1 is not sufficient for activation of extracellular signal-regulated kinases by interleukin-4: Implications for Ras activation by insulin. Mol Cell Biol 15: 1778–1785, 1995

36. Koide H, Satoh T, Nakafuku M, Kaziro Y: GTP- dependent association of Raf-1 with Ha-Ras: Identification of Raf as a target downstream of Ras in mammalian cells. Proc Natl Acad Sci USA 90: 8683–8686, 1993

37. Moodie SA, Willumsen BM, Weber MJ, Wolfman A: Complexes of Ras GTP with Raf-1 and mitogen-activated protein kinase kinase. Science 260: 1658–1661, 1993

38. Zhang XF, Settleman J, Kyriakis JM, Takeuchisuzuki E, Elledge SJ, Marshall MS, Bruder JT, Rapp UR, Avruch J: Normal and oncogenic p21(ras) proteins bind to the amino-terminal regulatory domain of c-Raf-1. Nature 364: 308–313, 1993

39. Warne PH, Viciana PR, Downward J: Direct interaction of Ras and the amino-terminal region of Raf-1 *in vitro*. Nature 364: 352–354, 1993

40. Van Aelst L, Barr M, Marcus S, Polverino A, Wigler M: Complex formation between RAS and RAF and other protein kinases. Proc Natl Acad Sci USA 90: 6213–6217, 1993

41. Vojtek AB, Hollenberg SM, Cooper JA: Mammalian Ras interacts directly with the serine/threonine kinase Raf. Cell 74: 205–214, 1993

42. Zheng CF, Guan KL: Cloning and characterization of two distinct human extracellular signal-regulated kinase activator kinases, MEK1 and MEK2. J Biol Chem 268: 11435–11439, 1993

43. Huang WD, Alessandrini A, Crews CM, Erikson RL: Raf-1 forms a stable complex with Mek1 and activates Mek1 by serine phos-phorylation. Proc Natl Acad Sci USA 90: 10947–10951, 1993

44. Crews CM, Alessandrini A, Erikson RL: The primary structure of MEK, a protein kinase that phosphorylates the ERK gene product. Science 258: 478–480, 1992

45. Boulton TG, Yancopoulos GD, Gregory JS, Slaughter C, Moomaw C, Hsu J, Cobb MH: An insulin stimulated protein kinase similar to yeast kinases involved in cell cycle control. Science 249: 64–65, 1990

46. Anderson D, Koch CA, Grey L, Ellis C, Morgan MF, Pawson T: Binding of SH2 domains of phospholipase Cyl, GAP, and Src to activated growth factor receptors. Science 250: 979–982, 1990

47. Haystead TAJ, Haystead CMM, Hu C, Lin TA, Lawrence JC: Phosphorylation of PHAS-I by mitogen-activated protein (MAP) kinase. Identification of a site phosphorylated by MAP kinase *in vitro* and in response to insulin in rat adipocytes. J Biol Chem 269: 23185–23191, 1994

48. Chuang CF, Ng SY: Functional divergence of the MAP kinase pathway. ERK1 and ERK2 activate specific transcription factors. FEBS Lett 346: 229–234, 1994

49. Gille H, Sharrocks AD, Shaw PE: Phosphorylation of transcription factor p62TCF by MAP kinase stimulates ternary complex formation at c-fos promoter. Nature 358: 414–417, 1992

50. Gille H, Korteniann M, Thomae O, Moomaw C, Slaughter C, Cobb MH, Shaw PE: ERK phosphorylation potentiates Elk-l-mediated ternary complex formation and transactivation. EMBO J 14: 951–962, 1995

51. Janknecht R, Ernst WH, Pingoud V, Nordheim A: Activation of ternary complex factor Elk-1 by MAP kinase. EMBO J 12: 5097–5104, 1993

52. Lin TA, Kong X, Haystead TAJ, Pause A, Belsham G, Sonenberg N, Lawrence JC: PHAS-I as a link between mitogen-activated protein kinase and translation initiation. Science 266: 653–656, 1994

53. Nakajima T, Kinoshita S, Sasagawa T, Sasaki K, Naruto M, Kishimoto T, Akira S: Phosphorylation at threonine-235 by a Ras-dependent mitogen-activated protein kinase cascade is essential for transcription factor NF-IL6. Proc Natl Acad Sci USA 90: 2207–2211, 1993

54. Whitmarsh AJ, Shore P, Sharrocks AD, Davis RJ: Integration of MAP kinase signal transduction pathways at the serum response element. Science 269: 403–407, 1995

55. Manchester J, Kong X, Lowry LH, Lawrence JCJ: Ras signalling in the activation of glucose transport by insulin. Proc Natl Acad Sci USA 91: 4644–4648, 1994

56. Kozma L, Baltensperger K, Klarlund J, Parras A, Santos E, Czech MP: The Ras signalling pathway mimics insulin action on glucose transporter translocation. Proc Natl Acad Sci USA 90: 4460–4464, 1993

57. Berghe NVD, Ouwens DM, Maassen JA, Mackelenbergh MGH, Sips HCM, Krans HMJ: Activation of the Ras/mitogen-activated protein kinase signalling pathway alone is not sufficient to induce glucose uptake in 3T3-L1 adipocytes. Mol Cell Biol 14: 2372–2377, 1994

58. Hausdorff SF, Frangioni JV, Birnbaum MJ: Role of p21ras in insulin-stimulated glucose transport in 3T3-L1 adipocytes. J Biol Chem 269: 21391–21394, 1994

59. Reusch JE-B, Bhuripanyo P, Carel K, Leitner JW, Hsieh P, DePaolo D, Draznin B: Differential requirement for p21ras activation in the metabolic signalling by insulin. J Biol Chem 270: 2036–2040, 1995

60. Weise RJ, Mastick CC, Lazar DF, Saltiel AR: Activation of mitogen-activated protein kinase and phosphatidylinositol 3'-kinase is not sufficient for the hormonal stimulation of glucose uptake, lipogenesis, or glycogen synthesis in 3T3-L1 adipocytes. J Biol Chem 270: 3442–3446, 1995

61. Figar DC, Birnbaum MJ: Characterization of the mitogen-activated protein kinase/90 kilodalton ribosomal protein S6 kinase signalling pathway in 3T3-L1 adipocytes and its role in insulin-stimulated glucose transport. Endocrinology 134: 728–735, 1994

62. Sakaue M, Bowtell D, Kasuga M: A dominant-negative mutant of mSOS1 inhibits insulin-induced Ras activation and reveals Ras-dependent and -independent insulin signalling pathways. Mol Cell Biology 15: 379–388, 1995

63. Gabbay RA, Sutherland C, Gnudi L, Kahn BB, O'Brien RM, Granner DK, Flier JS: Insulin regulation of phosphorenolpyruvate carboxykinase gene expression does not require activation of the Ras/mitogen-activated protein kinase signalling pathway. J Biol Chem 271: 1890–1897, 1996

64. Taha C, Mitsumoto Y, Liu Z, Skolnik EY, Klip A: The insulin-dependent biosynthesis of GLUT1 and GLUT3 glucose transporters in L6 muscle cells is mediated by distinct pathways. J Biol Chem 270: 24678–24681, 1995

65. Medema RH, Vries-Smits AMMD, Zon GCMVD, Maassen JA, Bos HL: Ras activation by insulin and epidermal growth factor through enhanced exchange of guanine nucleotides on p21 ras. Mol Cell Biol 13: 155–162, 1993

66. Drazin B, Chang L, Leitner JW, Takata Y, Olefsky J: Insulin activates p21 Ras and guanine nucleotide releasing factor in cells expressing wild type and mutant insulin receptors. J Biol Chem 268: 19998–20001, 1993

67. Shou C, Farnsworth CL, Neel BG, Feig LA: Molecular cloning of cDNAs encoding a guanine-nucleotide-releasing factor for Ras p21. Nature 358: 351–354, 1992

68. Langlois WJ, Sasaoka T, Saltiel AR, Olefsky JM: Negative feedback regulation and desensitization of insulin and epidermal growth factor-stimulated p21ras activation. J Biol Chem 270: 25320–25323, 1995

69. Skolnik EY, Batzer A, Li N, Lee C-H, Lowenstein E, Mohammadi M, Margolis B, Schlessinger J: The function of GRB2 in linking the insulin receptor to Ras signalling pathways. Science 260: 1953–1955, 1993

70. Waters SB, Yamauchi K, Pessin JE: Insulin stimulated disassociation of the SOS-Grb2 complex. Mol Cell Biol 15: 2791–2799, 1995

71. Waters SB, Holt KH, Ross SE, Syu L-J, Guan K-L, Saltiel AR, Koretzky GA, Pessin JE: Desensitization of Ras activation by a feedback disassociation of the SOS-Grb2 complex. J Biol Chem 270: 20883–20886, 1995

72. Cherniack AD, Klarlund JK, Conway BR, Czech MP: Disassembly of son-of-sevenless proteins from Grb2 during p21 Ras desensitization by insulin. J Biol Chem 269: 1485–1488, 1995

73. Holt KH, Kasson BG, Pessin JE: Insulin stimulation of MEK-dependent but ERK-independent SOS protein kinase. Mol Cell Biol 16: 577–583, 1996

74. Klarlund JK, Cherniack AD, Czech MP: Divergent mechanisms for homologous desensitization of p21Ras by insulin and growth factors. J Biol Chem 270: 23421–23428, 1995

75. Rozakis-Adcock M, Geer PVD, Mbamalu G, Pawson T: MAP kinase phosphorylation of mSos1 promotes dissociation of mSos1-Shc and mSos1-EGF receptor complexes. Oncogene 11: 1417–1427, 1995

76. Buday L, Warne PH, Downward J: Downregulation of the Ras activation pathway by MAP kinase phosphorylation of SOS. Oncogene 11: 1327–1331, 1995

77. Holt KH, Waters SB, Yamauchi K, Decker SJ, Saltiel AR, Motto DG, Koretzky GA, Pessin JE: Epidermal growth factor receptor targeting prevents uncoupling of the Grb2-SOS complex. J Biol Chem 271: 8300–8306, 1995

78. Zhao H, Okada S, Pessin JE, Koretzky G: 1997, (Manuscript in preparation)

79. Corbalan-Garcia S, Yang S-S, Degenhardt KR, Bar-Sagi D: Identification of the mitogen-activated protein kinase phosphorylation sites on human SOS1 that regulate interaction with Grb2. Mol Cell Biol 16: 5674–5682, 1996

Molecular and Cellular Biochemistry **182**: 31–48, 1998.
© 1998 *Kluwer Academic Publishers. Printed in the Netherlands.*

Insulin signal transduction through protein kinase cascades

Joseph Avruch

Diabetes Unit, Medical Services and the Department of Molecular Biology, Massachusetts General Hospital, and the Department of Medicine, Harvard Medical School, Boston, MA, USA

Abstract

This review summarizes the evolution of ideas concerning insulin signal transduction, the current information on protein ser/thr kinase cascades as signalling intermediates, and their status as participants in insulin regulation of energy metabolism. Best characterized is the Ras-MAPK pathway, whose input is crucial to cell fate decisions, but relatively dispensable in metabolic regulation. By contrast the effectors downstream of PI-3 kinase, although less well elucidated, include elements indispensable for the insulin regulation of glucose transport, glycogen and cAMP metabolism. Considerable information has accrued on PKB/cAkt, a protein kinase that interacts directly with Ptd Ins 3′OH phosphorylated lipids, as well as some of the elements further downstream, such as glycogen synthase kinase-3 and the p70 S6 kinase. Finally, some information implicates other erk pathways (e.g. such as the SAPK/JNK pathway) and Nck/cdc42-regulated PAKs (homologs of the yeast Ste 20) as participants in the cellular response to insulin. Thus insulin recruits a broad array of protein (ser/thr) kinases in its target cells to effectuate its characteristic anabolic and anticatabolic programs. (Mol Cell Biochem **182**: 31–48, 1998)

Key words: insulin action, protein serine/threonine kinase, Ras-Raf, MAP kinase, ribosomal S6 protein kinase (RSKs), phosphatidyl inositol-3 kinase

Introduction

Insulin is the major hormone of energy storage; a detailed understanding has been available since the 1970's as to how insulin regulates inter-organ fuel economy and redistributes substrate flows within its target cells [1]. In general terms, this is accomplished by alterations in the activity of trans-membrane transporters (especially in skeletal muscle), intracellular enzyme activities and the expression of a variety of genes that encode the metabolic enzymes and transporters whose activities serve to optimize the uptake and storage of circulating fuels. These changes in cell physiology are all initiated when insulin binds to and activates its cell surface receptor, and considerable effort has been directed toward understanding how the activated insulin receptor signals the target cells to respond in the characteristic way.

Current understanding of the signal transduction pathways that underlie insulin's major physiologic actions is still incomplete; nevertheless, remarkable advances have occurred in the last several years. It is now clear that activation of the insulin receptor tyrosine kinase, acting through the tyrosine phosphorylation of characteristic substrates such as IRS family of proteins, creates binding sites that enable the recruitment and activation of multiple, independent intracellular signal generators, that include e.g. lipid kinases such as PI-3 kinase and regulators of small GTPases, such as Ras [2]. These secondary signalling elements also operate by creating new binding sites (e.g. Ras-GTP and Ptd Ins 3-P lipids) which recruit cytosolic effectors to their site of activation and/or action (e.g. the surface membrane). In turn, a major effector mechanism utilized by these signal generators for the transmission, amplification, further diversification of these signals to all components and compartments of the cell is through the engagement of cascades of (ser/thr) protein kinases [3]. These multiple, insulin responsive protein kinase cascades can be seen as parallel but functionally interacting effector arms that regulate not only the cellular apparatus responsible for insulin's final metabolic actions, but also cross regulate the signal transduction pathways engaged by insulin's physiologic

Address for offprints: J. Avruch, Diabetes Research Laboratory, Department of Molecular Biology, Massachusetts General Hospital, Wellman 8, 50 Blossom Street, Boston, MA 02114, USA

antagonists, especially glucagon and β-adrenergic catecholamines. Inasmuch as the insulin receptor is a transmembrane protein tyrosine kinase, how can one rationalize the need for such a complex system simply to achieve the recruitment of protein (ser/thr) kinases such as Raf and PKB to the inner surface of the plasma membrane? Why not just begin with tyrosine phosphorylation of crucial intracellular targets, or with an insulin-activated receptor (ser/thr) kinase? One explanation is suggested by the remarkable conservation of the protein (ser/thr) kinase cascades between lower eukaryotes and mammals; evidently, the intracellular protein (ser/thr) kinases are far more ancient regulatory elements than are either insulin or tyrosine kinases. It is plausible that the protein (ser/thr) kinases, already established in unicellular organisms as versatile elements for transducing signals from the cell surface to the cell interior that regulate cell growth and differentiation, were co-opted into the service of the receptor tyrosine kinases. This review will summarize the evolution of our present concepts of insulin action, current information on the protein (ser/thr) kinase cascades and their status as mediators of insulin signal transduction.

Insulin action: early ideas

The discovery of insulin in 1921, the demonstration of its ability to virtually correct the global disturbances in fuel and energy metabolism that occur in diabetes and its immediate, life saving application to patients remains one of the most dramatic stories in medical history [4]. The understanding of the biochemical apparatus mediating fuel metabolism was then in its infancy, and the discovery of such a physiologically potent and medically important regulator of metabolism captured the attention of many investigators; the effort to understand the processes that underlie the response to insulin set the agenda for a large segment of biochemical research over the next 40–50 years. Moreover, well into the 1970's, insulin was one of the very few highly purified polypeptide hormones available in large amounts for study.

Hypotheses concerning the biochemical basis for insulin's biologic actions were strongly influenced by each succeeding major discovery in metabolic regulation and hormone action. Up through the early 1950's, as the enzymatic machinery underlying energy metabolism was being elucidated, the prevailing ideas concerning insulin action envisioned the hormone acting much like a vitamin, i.e. as a cofactor for, or regulator of an intracellular enzyme, such as hexokinase. The discovery that glucose entry into muscle was the rate limiting step in glucose metabolism under most circumstances *in vivo*, and that insulin activated glucose transport in muscle and adipose tissue raised the possibility that insulin's myriad effects might all follow from the facilitation of glucose entry. It soon became evident that insulin could alter many aspects of plasma membrane function, e.g. amino acid and cation transport, membrane potential, etc. Insulin was also shown to alter intracellular processes e.g. such as protein synthesis or the activity of specific enzymes, such as glycogen synthase, pyruvate dehydrogenase, or hormone sensitive lipase, irrespective of the availability of the extracellular precursor substrate, i.e. glucose or amino acids [1]. The ability of insulin, acting on intact cells, to alter the activity of intracellular enzymes under conditions where changes in the concentrations of the substrates and known allosteric regulators did not occur [5], together with experiments that suggested that insulin could induce these changes without entering the cell [6], pointed to the need to identify a mechanism by which signals generated at the cell surface could be conveyed to targets in the cell interior.

Insulin action: the cAMP era

The discovery of cAMP, the formulation of the second messenger concept and the subsequent identification of the cAMP dependent protein kinase as the intracellular receptor and effector for essentially all of cAMPs actions [8], were developments that greatly influenced the concepts of insulin action. These discoveries provided the first model for transmembrane signalling by polypeptide hormones, and established the importance of regulatory (ser/thr) phosphorylation in cell regulation and hormone action. The reciprocal control of fuel metabolism by insulin and epinephrine/glucagon (acting through cAMP), and the ability of insulin to both lower glucagon/epinephrine-stimulated levels of cAMP (in liver and adipose tissue) and oppose cAMP action, led to the view that, with some exceptions (e.g. activation of glucose transport) the metabolic responses to insulin could be largely explained by insulin's effects on cAMP generation and action [9]. Moreover, like glucagon and epinephrine, insulin also induced rapid changes in the activity of intracellular enzymes that persisted after cell disruption and partial purification, consistent with the effects of covalent protein modification. Many of the physiologically important changes in enzyme activity induced by insulin (e.g. the activation of glycogen synthase, or the inhibition of pyruvate kinase or hormone sensitive triacylglycerol lipase) were opposite to those caused by cAMP-induced phosphorylation, and were shown to be due specifically to insulin-induced dephosphorylation [5]. The intersections and similarities in the responses to insulin and cAMP led to the plausible expectation that insulin also would initiate some or all of its actions by generating a soluble mediator among whose functions would be the control of enzyme phosphorylation in a manner opposing cAMP.

Insulin action: the receptor tyrosine kinase

The discovery that the insulin receptor is a member of the tyrosine kinase family of receptors [2] provided a strong indication that the fundamental signalling mechanisms employed by insulin would be entirely distinct from those employed by the C protein-linked receptors coupled to adenyl cyclase. Subsequent work on the EGF and PDGF receptor kinase subfamilies uncovered the primary signalling mechanism utilized by tyrosine kinases [10]. Thus in contrast to ser/thr phosphorylation, which usually serves to alter the functional performance of the protein substrate, the usual consequence of tyrosine phosphorylation is to create on the substrate a new binding site, whose specificity is defined by the amino acids immediately surrounding the phospho-tyrosine [11]. In this way, multiple proteins that contain domains capable of binding phosphotyrosine (either SH2 or PTB domains) are recruited to the newly created phospho-tyrosine sites [12–15]. Such recruits include proteins that encode catalytic domains such as the protein tyrosine phosphatase SYP, as well as noncatalytic adapter proteins, such as Nck, or the p85 subunit of PI-3 kinase or the Grb2 adapter of the Ras-specific guanyl nucleotide exchange protein, mSOS. Consequent to recruitment these signalling elements undergo activation, either directly because of the binding *per se* or after additional phosphorylation. In addition, these recruits are brought in apposition to their immediate substrate and/or target. The phosphotyrosine-induced recruit-ment of these multiple signal generators diversifies the initial signal and sets up independent, parallel signalling pathways flowing into the cell interior. Establishing this mechanism in the case of insulin awaited the elucidation of the IRS proteins [2]. In contrast to the EGFR and PDGFR, whose tyrosine autophosphorylation sites serve an indispensable docking function, insulin receptor auto-phosphorylation serves primarily to activate the receptor kinase activity toward exogenous proteins [16], i.e. the insulin receptor substrate (IRS) proteins; it is the IRS proteins rather than insulin receptor itself that provide the necessary docking function.

In summary, present evidence indicates that the initial steps of signal transmission downstream from receptor tyrosine kinases occurs through a 'solid-state' mechanism, involving sequential protein-protein interactions. No evidence presently available provides compelling support for the participation of low molecular weight 'messengers' in the initial steps of insulin signalling. The existence of physiologically significant low molecular weight mediators e.g. generated downstream of the IRS-recruitment step, nevertheless remains a hypothetical possibility [5]. More-over, it is now recognized that ability of insulin to inhibit cAMP or Ca^{2+} accumulation or action, however important in metabolic regulation, are not 'primary' consequences of insulin receptor activation, in the sense that adenyl cyclase activation is a 'primary' consequence of β adrenergic receptor activation, but represent cross-talk between unique insulin-directed signalling pathways and the outflows initiated by seven-transmembrane receptors.

Insulin action: the protein (ser/thr) kinase cascades

One of the first clues that insulin activated signal trans-duction pathways distinct and independent of the cAMP kinase regulated pathway, was the finding that in addition to eliciting the expected inhibition of β-adrenergic or glucagon directed protein phosphorylation, insulin unexpectedly also caused a cAMP-independent increase in the ser/thr phos-phorylation of a subset of proteins in bona fide target cells [17]. These observations were unexpected insofar as no action of insulin then known was explained by an *increase* in protein (ser/thr) phosphorylation. Nevertheless, insulin-stimulated (ser/thr) phosphorylations occurred at physiologic con-centrations of hormone, within seconds to minutes after receptor activation; as visualized by autoradiography of cellular polypeptides ^{32}P-labelled *in situ*, increased (ser/thr) phosphorylation involved many more cellular polypeptides than did the somewhat earlier tyrosine phosphorylations. More surprising, the number of proteins that exhibited a net increase in (ser/thr) phosphorylation in response to insulin far outnumbered those undergoing a net (ser/thr) dephos-phorylation. Later studies showed that polypeptide growth factors such as EGF and PDGF also increased the (ser/thr) phosphorylation of a set of polypeptide substrates that overlapped substantially with those responding to insulin.

The rapid onset and reversibility of these insulin/growth factor-stimulated (ser/thr) phosphorylations strongly suggested that a regulatory function for these modifications would eventually be uncovered. The first insulin-stimulated phosphoproteins to be identified were the lipogenic enzymes, ATP-citrate lyase and Acetyl CoA Carboxylase, and the 40S ribosomal protein S6; the regulatory significance of the insulin-stimulated phosphorylation of these proteins is still unknown. Nevertheless, it was anticipated that the protein (ser/thr) kinases which catalyzed these insulin-stimulated phosphorylations probably served as signalling intermediates downstream of the insulin-and growth factor-activated receptor tyrosine kinases. By this view, study of the biochemical mechanisms responsible for activation of these (ser/thr) kinases offered an attractive approach to uncover intermediate steps in insulin/growth factor signal transduction. The detection and characterization of insulin/growth factor- stimulated protein (ser/thr) kinases accelerated substantially once the rather novel extraction conditions required to capture these (ser/thr) kinases in an activated state were identified. Beginning with the ribosomal

S6 kinases from *Xenopus* oocytes and rat liver, and the MAP kinases (also known as erk-1 and erk-2), an increasing number of insulin/growth factor-activated were recognized and purified [18]. By 1992 it was evident that insulin and polypeptide growth factors such as EGF and PDGF rapidly induced the stable activation of a relatively large number of protein (ser/thr) kinases, and concurrently caused other protein (ser/thr) kinases to undergo inactivation. The challenge at that point was twofold: first, to understand the biochemical steps which the activated insulin receptor tyrosine kinase, through the recruitment of secondary signalling elements such as Ras, PI-3 kinase and perhaps others, regulated these numerous protein (ser/thr) kinases. In addition, the role of each of these kinases in elaborating the characteristic metabolic responses to insulin remained to be established.

The Ras-MAP kinase cascade

A prototype format for the organization of these insulin/growth factor-activated (ser/thr) protein kinases emerged with the elaboration of the Ras-MAPK pathway. The ubiquitous and indispensable contribution of this signal transduction pathway in mitogenesis and cell differentiation has evoked considerable investigation. The individual elements, operation of the pathway and the downstream targets in a variety of cellular systems have each been extensively reviewed and the reader should consult these sources for detailed discussion [18–23]. The role of the Ras-MAPK cascade in the metabolic responses to insulin has also been the subject of considerable inquiry, primarily through the use of a variety of recombinant and/or chemical inhibitors [3, 24]. A fairly coherent body of experimental results points to the general conclusion that neither Ras itself nor the specific Ras effector limb consisting of the Raf-MEK-MAPK-Rsk cassette plays a significant role in the insulin activation of glucose transport and glycogen synthase, or in the inhibition of cAMP accumulation or action. Evidence as to the contribution of Ras-MAPK to the insulin regulation of the expression of genes encoding metabolic enzymes is insufficient at present to allow a general conclusion. In contrast, Ras is critical to insulin's ability to promote the differentiation of 3T3 L1 cells to adipocytes. The latter is not unexpected, as the control of cellular differentiation appears to be the most consistent function of the Ras protooncogene based on studies in cultured mammalian cells, as well as from genetic evidence in invertebrates.

Ras

Ras is the founding member of the very large family of small GTPases [25]. These proteins bind and hydrolyze GTP and the rate of GTP/GDP exchange as well as GTP hydrolysis is highly regulated. The Ras proteins (H-Ras, K-Ras 4 a/b, N-Ras) are modified by prenylation at cysteine 186, four residues from the carboxyterminus. Ha-Ras and N-Ras are farnenylated (C15), whereas K-Ras is geranylgeranylated (C20). Prenylation is followed by cleavage of the three carboxyterminal amino acids and methylation of the terminal carboxyl moeity. These modifications greatly increase the hydrophobicity of Ras and are necessary for Ras to insert into the inner leaflet of the plasma membrane. K-Ras contains a polylysine segment just upstream of the prenylation site which together with prenylation is sufficient to enable tight association to the inner surface of the plasma membrane. H-Ras, which lacks a polybasic segment, undergoes a further modification, palmitoylation at Cys 181 and 184, which is required for tight membrane binding [26].

Ras activation results from binding GTP, and receptor tyrosine kinases promote Ras activation through enhanced exchange of GTP for GDP, rather than by inhibition of GTP hydrolysis [27, 28]. Receptor tyrosine kinase stimulation of guanyl nucleotide exchange is mediated primarily by the complex of Shc-or IRS-Grb2 with the Ras-specific exchange factor, mSOS. GTP charging of Ras results in a conformational change in two Ras loops, dubbed switch 1 (amino acids 31–38) and switch 2 (60–72). The switch 1 loop corresponds closely to the Ras 'effector' domain, defined as a segment where mutations greatly inhibit the transforming function of oncogenic (i.e. constitutively active mutant) Ras, without modifying the membrane localization or intrinsic GTPase activity of wildtype, c-Ras [29]. This GTP-dependent reconfiguration of the Ras effector loop creates a high affinity binding site for the (ser/thr) protein kinase known as c-Raf-1 [30]. A large body of genetic and biochemical evidence support the view that c-Raf-1 is a crucial effector of Ras action on mitogenesis and cell differention [20, 31]. It should be emphasized however that important Ras effectors other than the Raf kinases undoubtedly exist; the best characterized candidate Ras effectors other than Raf are the catalytic subunits of each of the PI-3 kinases and a family of guanyl nucleotide exchange factors for the small GTPase known as Ral [32]. In addition perhaps 5–10 proteins of unknown biochemical function have been shown to exhibit a GTP-dependent binding to Ras through the effector loop, *in vitro* and/or *in situ*. These Ras-associating proteins appear to exhibit a subtle but identifiable sequence motif which might be viewed as a canonical Ras-association domain. The nature of the signal outflows governed by these candidate Ras effectors is not known.

Raf

Like Ras, the Raf-1 kinase was first discovered in the form of a mutant retroviral transforming agent (v-raf) [33]. The normal cellular homolog, cRaf-1 is a 648 amino acid, insulin/ mitogen-activated protein (ser/thr) kinase that is nontransforming even when overexpressed. The cRaf-1 catalytic domain is situated in the carboxyterminal half of the protein, and an aminoterminally truncated, carboxy-terminal Raf fragment containing mostly catalytic domain is constitutively active and capable of transforming rodent fibroblasts. The aminoterminal half of cRaf-1 contains a zinc finger structure, and based on the effects of its deletion, the Raf aminoterminal noncatalytic segment can be inferred to perform two functions: it must inhibit the catalytic domain, and also serve as the receptor for activating inputs [21]. The Raf aminoterminal segment mediates Ras binding, and actually contains two independent, contiguous Ras-binding domains. The primary Ras binding domain is situated between c-Raf amino acids 50–150 [34, 35], which as an isolated fragment, binds with nM affinity to the effector loop of Ras-GTP, 100–1000 times more avidly than to Ras-GDP [36]. This is the primary interaction that recruits Raf from its basal, cytosolic location to membrane-bound Ras-GTP. This initial recruitment however is not sufficient to allow a stable interaction productive of Raf activation. Raf kinase activation requires Ras-Raf interaction through a second, independent site, wherein the c-Raf cysteine rich, zinc finger segment (amino acids 139–184) binds in a GTP-independent manner to a Ras epitope distinct from the effector loop, that is only available on prenylated Ras [37, 38]. Whether the prenyl moiety actually participates directly in the binding interaction or serves only to properly configure this second Ras epitope, remains to be determined. In either case, this dual interaction recruits Raf to the membrane and enables the further Raf modifications required for kinase activation to proceed. Several groups reported that grafting the Ras prenylation motif onto the Raf carboxyterminus (to create Raf-CAAX) is sufficient to direct Raf to the inner leaflet of the plasma membrane, and provides some activation of the Raf kinase [39, 40]. Recent work however establishes that despite its constitutive membrane localization, a continued interaction of Raf-CAAX with Ras is necessary for Raf-CAAX activation [41].

The specific biochemical steps that convert Raf to a stable active, Ras-independent state remain incompletely understood. Raf activation requires the binding of Raf to the dimeric 14.3.3 proteins [42, 43], which bind to Raf through several (at least two) sites defined by the consensus sequence RSXS(P)XP [44], wherein the serine is constitutively phosphorylated [45]. The 14.3.3 proteins bind many other cellular proteins, and given the obligatory requirement for a 14.3.3 dimer in Raf activation [42], one or more among these other 14.3.3-associated proteins may be crucial to the Raf activation process; some evidence also suggests that Raf homodimerization may be important to activation [46, 47]. It is likely that Raf activation requires further Raf phosphorylation and this 'final' phosphorylation can be catalyzed by any of several protein kinases, depending on the initiating stimulus. The protein (ser/thr) kinase known as KSR was first identified in *Drosophila* and *C. elegans* through a screen for suppressor mutants of activated Ras alleles [48, 49]. Mammalian KSR has recently been identified as a ceramide-activated kinase that is capable of phosphorylating and activating cRaf-1 *in vitro* [50]. Thus KSR may participate in the activation of Raf by TNFα and related ligands that generate ceramide *in situ*. Other candidate Raf-activating kinases include PKCα [51] and Src family nonreceptor tyrosine kinases [52]. Raf activation by receptor tyrosine kinases however is not inhibited by TPA-induced PKC downregulation, and is accompanied only by increased Raf (ser/thr) phosphorylation [53]; thus the identity of the protein (ser/thr) kinases utilized by the insulin receptor and other RTKs to effect Ras-dependent cRaf-1 activation is not yet known.

In several cell backgrounds cAMP inhibits activation of the MAPK cascade, at the step of Raf activation. Two mechanisms are proposed; kinase A can phosphorylate cRaf-1 *in vitro* at Ser 43, adjacent to the Ras-binding domain, which results in an inhibition of Ras-Raf binding *in vitro* [54]. cAMP also enhances the GTP charging of the small GTPase, Rap-1 [55], which has an effector loop identical to Ras. Rap-1-GTP also binds cRaf-1 *in vitro* [30], but does not enable cRaf-1 activation in situ In fact overexpression of Rap-1 can cause phenotypic reversion of v-Ras transformed cells, presumably be sequestering Ras effectors (? such as cRaf-1) in an inactive state [56]. Although both mechanisms proposed for cAMP-induced inhibition of cRaf-1 are plausible, direct evidence in support of either is lacking. Moreover, it should be noted that in PC12 cells, cAMP activates rather than inhibits of MAPK, an effect attributable to the ability of Rap-1-GTP to activate the BRaf isoform, which like cRaf-1, is also a MEK activator [57].

In summary, Ras-GTP activation of Raf appears to involve the signal-dependent generation of a membrane binding site (i.e. the GTP-reconfigured Ras effector loop), followed by the recruitment of the Raf protein kinase to the membrane. A sustained interaction of Raf with the membrane target, Ras-GTP then occurs, through multiple, physically separate but functionally interacting sites. In the case of cRaf-1, this is followed by a further Raf modification probably by another membrane-associated protein kinase. In contrast, BRaf, the isoform predominant in nervous tissue, can be activated directly *in vitro* by addition of Ras-or Rap-1 GTP [58, 59]. This type of membrane recruitment/ activation process is highly analogous to the scheme first elucidated for the diacylglycerol/phospholipid-induced recruitment and activation of protein kinase C, and also to the PI-3 kinase-induced activation of Akt/PKB (Fig. 1), as will be discussed below.

MEK/MAPK/Rsk

Activated Raf phosphorylates and activates the protein kinases known as MAP kinase-kinase or MEK [60]; other candidate Raf substrates have been identified (e.g. cdc 25A, [61]), but are less well characterized. Other MEK-activating kinases, such as the c-mos protein kinase and several mammalian homologues of the *S. cerevisiae* MEK kinase, Stell (known as MEK kinases) can also phosphorylate and activate MEK *in vitro* and by cotransfection; nevertheless, a substantial body of genetic evidence in *Drosophila* and *C. elegans* points to a Raf as the indispensable MEK-activator operating downstream of receptor tyrosine kinases and Ras-GTP [19]. MEK activation results from the phosphorylation of two serines, residues 218 and 297 situated in the subdomain VIII of the MEK catalytic domain, just upstream of the conserved A/SPE motif [22]. Once activated, MEK phosphorylates the p44 and p42 MAP kinase (erk1 and erk2 respectively), at a thr and tyr in the motif -*TEY*- also situated in catalytic subdomain VIII [23]; the two MAPKs are the only MEK substrates so far identified.

MAP kinase, in contrast to Raf and MEK, phosphorylates numerous cellular targets, with a broad substrate specificity reminiscent of the cAMP-dependent protein kinase. Moreover, the reach of the MAPKs over cellular function extends beyond it's immediate targets, because the physiologic substrates of the MAPKs include many transcription factors, such as elk-1, and several other subfamilies of multifunctional protein (ser/thr) kinases, including the Rsk enzymes [64], and the recently described Mnk subfamily [65]. These MAPK targets, in turn control the abundance or activity of a wide array of proteins/ enzymes. The determinants of MAPK substrate specificity have been partially elucidated, and appear to be representative of the entire large family of ERKs. All of the ERKs examined thus far are 'proline-directed' kinases, in that they exhibit an absolute requirement for a proline residue immediately after the ser/thr phosphorylation site; this motif is actually a negative determinant for kinase A and kinase C [66]. In addition, other amino acids immediately surrounding the SP/TP motif have a modest, but definite influence; thus erk1 and erk2 greatly prefer SP/TP in a context of basic amino acids [66]. Nevertheless, many proteins contain ST/TP couplets in a basic context, but are not phosphorylated *in situ* by the ERKS. Accumulating evidence indicates that a major specificity determinant, equal in importance to the proline requirement, is the presence on physiologic substrates of a high affinity ERK-binding site, that is removed in primary sequence from the SP/TP residues. Thus ERKs phosphorylate protein substrates with much lower K_ms than the corresponding synthetic peptides encompassing the phosphorylation sites. In addition a properly folded protein

MEMBRANE RECRUITMENT OF PROTEIN KINASES

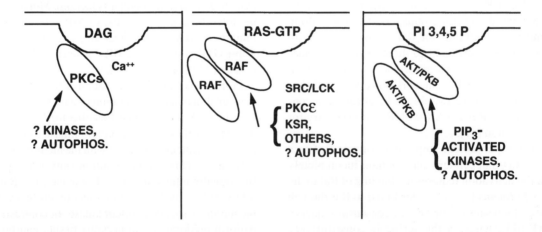

Fig. 1. Receptor tyrosine kinase activation generates a new, plasma membrane-associated binding site for the recruitment and activation of protein kinases. The original examples were provided by PKCs; receptor (although not the insulin receptor) activated phospholipase Cs mediate Ptd Ins 4, 5P or Ptd choline hydrolysis, to yield diacylglycerol (left panel), a high affinity ligand for classical and unconventional protein kinase C isoforms. Diacylglycerol binding directly activates PKC, although an intermediate phosphorylation may be required for some isoforms. Insulin receptor activation results in the GTP-charging of Ras which creates a high affinity ligand for the protein kinases of the Raf family. Once bound to Ras at the membrane, Raf undergoes further modification by one of several (Ser/Thr) or (Tyr) kinases, depending on the initiating stimulus. Insulin receptor activation also activates PI-3 kinase, which generates a variety of 3-OH phosphorylated Ptd Ins derivatives, each of which acts as a binding partner for a variety of protein kinases, including some PKC isoforms. Pictured is PKB/cAkt, which once bound, undergoes activation by multisite phosphorylation. See text for details.

substrate is often critical for high affinity ERK-catalyzed phosphorylation. A MAPK binding site has been identified on the Rsk protein kinase [67], however the ERK-binding site best characterized thusfar is that on the c-jun protein for the ERK subfamily known as the SAPkinases (or jun N-terminal kinases, JNKs). The SAPK binding site on c-jun is situated between c-jun residues 30–45, and SAPK-catalyzed c-jun phosphorylation occurs on serine 63 and 73. Deletion of c-jun residues 30–45 abolishes SAPK binding and SAPK-catalyzed c-jun phosphorylation *in vitro* and *in situ* [68, 69].

Rsk was the very first insulin/mitogen activated protein (ser/thr) kinase to be purified [70] and cloned [71], and the first physiologic substrate of MAPK to be identified [64]. The demonstration that an insulin activated MAPK could directly phosphorylate and activate Rsk (specifically, the *Xenopus* S6 kinase H) provided the first direct evidence for the existence of insulin regulated protein (ser/thr) kinase cascades. Despite these firsts, many aspects of Rsk regulation as well as the identity of the true substrates and physiologic role of the Rsks remains incompletely understood. The three Rsk isoforms each encode two complete protein (ser/thr) kinase catalytic domains [71, 72]; the isolated aminoterminal domains exhibit activity *in vitro*, whereas the function of the carboxyterminal catalytic domain, which appears to contain the major site of MAPK phosphorylation [73], (at least for Rsk2), is uncertain [74, 75]. Interestingly, Rsk3, which is expressed predominantly in lung and skeletal muscle, is activated through phosphorylation by insulin and many growth factors when transiently expressed [72], however unlike Rsk1 and Rsk2, Rsk3 cannot be activated by MAPK *in vitro* [72]. Interestingly, one report describes *in vitro* activation of Rsk3 by SAPK [76] (see below).

Mutations in Rsk2 have been identified in a minority of patients with the Coffin-Lowry syndrome of psychomotor retardation and skeletal malformations [77]. As regards substrates, although Rsk appears to serve as the dominant 40S ribosomal S6 kinase in *Xenopus* oocytes, this is not true in other cells, where the p70 S6 kinase is the dominant 40S kinase, as shown by the parallel inhibition by rapamycin of S6 phosphorylation and p70 S6 kinase, but not Rsk activity [78]. Rsk appears to be a kinase with broad specificity preferring sites in the context RXRXX*S* [79]. Rsk can phosphorylate c-fos, SRF and Nur77, and CREB *in vitro*, but little or no evidence supports a role for Rsk in the regulation of these proteins *in situ* [80]. The intriguing suggestion that Rsk2 participates in the insulin activation of skeletal muscle glycogen synthase by mediating the insulin-stimulated, site-specific phosphorylation of the glycogen binding subunit of protein phosphatase 1, which is proposed to result in a selective increase in skeletal muscle glycogen synthase phosphatase activity [81], remains unconfirmed. Moreover, although Rsk can phosphorylate and inactivate GSK-3 *in vitro* [82], an inhibitor of MEK activation (PD098059) does not block insulin inhibition of GSK-3 in L6 myoblasts [83], despite the ability of the MEK inhibitor to block the EGF-induced inhibition of GSK-3 in fibroblasts [84]. It appears that in L6 myoblasts, the insulin induced inhibition of GSK3 is mediated by the PI-3 kinase-PKB pathway (see below).

In summary, the Ras-MAPK cascade plays a central role in cell fate determination in response to insulin [85, 86], but little role in the acute actions of insulin on the activity of the enzymes governing carbohydrate and lipid metabolism. Considering that many of the targets of this pathway are transcription factors, significant contributions to the insulin regulation of gene expression are likely to be uncovered.

The PI-3 kinase-PKB-GSK-3 pathway

The availability of a variety of inhibitors of the PI-3 kinase has enabled a broad survey of the role of PI-3 kinase in insulin signalling. The available inhibitors include relatively selective cell permeant agents such as wortmannin (especially at < 0.1 μM) [90], and the Lilly compound LY294002 (at ≤10 μM) [91] as well as transfectable recombinant mutant versions of the p85 adaptor protein of the PI-3 kinase [92], which act as dominant inhibitors of the PI-3 kinase catalytic subunits, p110α and β. Using these agents, inhibition of PI-3 kinase has been shown to prevent insulin activation of glucose transport and Glut4 translocation to the surface membrane, the activation of glycogen synthase, insulin inhibition of lipolysis and the insulin activation of cAMP phospho-diesterase [93], among other effects [3]. Thus, in contrast to Ras, PI-3 kinase is clearly a major conduit for the signals that regulate the metabolic responses to insulin. Interestingly, agonists, such as EGF and PDGF, which also activate PI-3 kinase activity in 3T3-L1 adipocytes do not activate many of the wortmannin-sensitive responses characteristic of insulin action, such as activation of glucose transport and glycogen synthase [93, 94]. Conceivably, insulin may activate, in addition to PI-3 kinases, another signal which is not effectively recruited by EGF or PDGF, and which must collaborate with PI-3 kinase to effect responses such as Glut4 translocation and GS activation. Alternatively, inasmuch as the signal generated by PI-3 kinase is a membrane bound Ptd Ins 3P phospholipid, rather than a soluble hydrolysis product analogous to inositol tris phosphate, it is possible that the insulin-stimulated PI-3 kinase activity, which is bound to an IRS polypeptide rather than to the insulin receptor, may be directed to a cellular compartment not available to the EGFR or PDGFR. What-ever the explanation for the apparently unique ability of the insulin receptor to recruit these characteristic metabolic responses, the demonstration that such responses require the activation of PI-3 kinase propelled the effort to identify the downstream effectors of this lipid signal.

As with the GTP charging of Ras, PI-3 kinase catalyzed phosphorylation of the D3 position of Ptd Ins creates a new membrane binding site for proteins that are capable of binding directly to this novel family of phosphoinositides. Many, but not all, proteins which bind Ptd Ins 3P lipids contain pleckstrin homology (PH) domains. Some PH domains bind Ptd Ins 3, 4, 5 P_3 but not Ptd Ins 3, 4, P_2, whereas PKB is activated by Ptd Ins 3, 4 P_2 but not Ptd Ins 3, 4, 5 P_3 [95]. PH domains are very diverse in sequence, and it is not clear whether all such domains bind to D3-phosphorylated Ptd Ins lipids. Nevertheless, as with Ras-GTP, multiple candidate Ptd Ins 3P effectors are being identified, and prominent among these are protein (ser/thr) kinases such as PKB and certain PKCs. In addition, a number of protein kinases, such as GSK-3 and p70 S6 kinase, have their activity modified by wortmannin, but do not bind directly to Ptd Ins 3P lipids; such behavior implies that these kinases lie downstream of Ptd Ins 3P-sensitive regulators. The protein (ser/thr) kinases that interact directly with Ptd Ins 3P lipids best characterized thus far are the novel PKC isoforms, PKCε and PKCη [96] and (perhaps) PKCζ [96, 97], and the protooncogene PKB/cAkt [98]. The role of the PKCs in insulin action has been comprehensively reviewed [99], and this information will not be recapitulated. The evidence for Ptd Ins 3P regulation of the PKCs is largely based on the ability of these lipids (added in the presence of Ptd Ser or Ptd Et) to directly to activate these PKC isoforms *in vitro*, in preference to Ptd Ins 4 and/or 5P; little information is available on the regulation of these novel PKCs *in situ*. Thus although much evidence indicates that a role for the classical Ca^{2+}/DAG-regulated PKCs (α, β, γ) as mediators of insulin action in very unlikely [100], the participation of the Ptd Ins 3P-responsive PKC isoforms in insulin action merits further evaluation; this effort will be greatly facilitated by the identification of specific activators or inhibitors.

Evidence has been presented that Ptd Ins 3, 4, 5 P_3 (and Ptd Ins 4, 5 P_2) can bind to certain SH2 domains, in a manner competitive with phosphotyrosine-containing proteins [101]. This phenomenon has been proposed to reflect PI-3 kinase induced downregulation of RTK signalling.

PKB/cAkt

Isolated independently by three groups, as PKB [102], RAC [103] or cAkt [104], this protein (ser/thr) kinase is the cellular homolog of the vAkt oncogene [105]. The enzyme is a 55 KDa polypeptide that contains a carboxyterminal catalytic domain most similar to those of the PKCs, the N-terminal Rsk catalytic domain and the catalytic domain of the p70 S6 kinase. The PKB aminoterminal noncatalytic region encodes a plecktrin homology domain. Insulin and growth factors rapidly activate PKB through a pathway that is blocked by wortmannin and transfectable PI-3 kinase inhibitors [83, 106, 107]. PDGFR mutants which lack the tyrosine residues that provide the binding sites for the SH2 domains of the p85 adapter subunit of PI-3 kinase fail to activate PI-3 kinase or PKB although activation of MAPK remains intact [108]. Coexpression with a constitutively active PI-3 kinase *in situ* results in a robust activation of PKB [109–111]. Moreover, PKB binds directly to Ptd Ins 3P lipids through its aminoterminal noncatalytic domain, preferring Ptd Ins 3, $4P_2$ to Ptd Ins 3. 4, $5P_3$; mutation at critical amino acids within the pleckstrin homology domain abolishes binding to Ptd Ins 3, $4P_2$ [111 112]. On binding Ptd Ins 3, 4P *in vitro*, PKB homodimerizes and is partially activated. Interestingly, the insulin activation of recombinant PKB in cells that overexpress recombinant insulin receptor is not abolished by deletion of the PKB PH domain [107] whereas deletion or mutation of this domain greatly diminishes RTK induced PKB activation [106, 109, 113, 114]. Activation *in situ* however requires phosphorylation of PKB at two sites, serine 308 in the catalytic subdomain VIII corresponding to the sites of activating phosphorylation of MEK, MAPK (and p70 S6 kinase, see below) and ser473 outside the catalytic domain, near the C-terminus [113, 115, 116]. The latter site is homologous in location to indispensable sites of phosphorylation on various PKCs [117] and on the p70 S6 kinase [118]. The activating PKB phosphorylations are not autophosphorylations, but are catalyzed by other protein kinases, and it appears that separate protein kinases act at ser 308 and ser 473 [119]. The latter site can be efficiently phosphorylated *in vitro* by MAPKAP-kinase 2 [116]. PKB can be activated *in situ* by stress, through a pathway independent of PI-3 kinase [114], however, no evidence as yet specifically implicates the p38/MAPKAP kinase 2 pathway in this response. Serine 308 is phosphorylated by another, as yet unidentified, protein (ser/thr) kinase which itself is completely dependent on added Ptd Ins 3P lipids for activity *in vitro* [119].

Thus the biochemical mechanism underlying the insulin activation of PKB is highly analogous to that operating for Raf activation; the first step is the recruitment of the effector kinase to the membrane by the creation of a new binding site (Ptd Ins 3, $4P_2$ for PKB; Ras-GTP for Raf, homodimerization of the kinase followed by phosphorylation by membrane bound kinase(s); in the case of PKB, at least one of these activating kinases is itself a Ptd Ins. 3, 4, $5P_3$-activated kinase (Fig. 1).

The role of PKB in insulin action is presently under intensive investigation. Several reports have described the ability of recombinant, constitutively active PKBs to activate glucose uptake, Glut4 translocation and glycogen synthase in 3T3-L1 adipocytes [120, 121] Persuasive evidence for an important contribution by PKB to the acute regulation of metabolism is the demonstration that PKB can directly phosphorylate and inactivate glycogen synthase kinase 3 (GSK-3), a potent inhibitor of glycogen synthase [83].

Glycogen synthase kinase-3

Glycogen synthase kinase-3 (GSK-3) was first identified as a protein kinase capable of inhibiting glycogen synthase (GS) *in vitro* [122], through the phosphorylation of a series of serines near the GS carboxyterminus, so called sites 4, 3a, b, c [123]. These phosphorylations require a prior phosphorylation of GS by casein kinase 2 at site 5, which establishes a binding site for GSK3; phosphorylation by GSK-3 then creates the motif (SXXXS(P)) that enables the next GSK3 phosphorylation, resulting in a processive, so-called 'hierarchical' phosphorylation [124]. This series of phosphorylations strongly inhibits GS activity much more effectively than GS phosphorylation by kinase A, which phosphorylates other sites on GS, specifically sites 1a and 2 [125]. The activation of GS by insulin *in vivo* is accompanied by dephosphorylation of sites 3 [126] and 2, which is not a GSK-3 site [127, 128]. It was therefore attractive to hypothesize that insulin inhibition of GSK-3 might underlie, in part, the insulin activation of glycogen synthase. GSK-3 is a 55 KDa polypeptide expressed as two isoforms, α and β [129]. The enzyme requires a tyrosine phosphorylation in the catalytic subdomain VIII for activity, but like kinase A, this phosphorylation is constitutive rather than post-translationally regulated [130]. In contrast to ERK and MEK, GSK-3 is active in unstimulated cells; following on initial observations by Benjamin and coworkers [131, 132], several groups established that insulin and growth factors can induce a rapid, partial and transient inhibition GSK-3 activity *in situ* [133, 134]. This inhibition is attributable to the phosphorylation of a single serine residue near the GSK-3 aminoterminus, a phosphorylation that can be catalyzed *in vitro* by Rsk-2, p70 S6 kinase [82, 135] and as reported most recently, by PKB [83]. In L6 myoblasts, neither rapamycin (a selective inhibitor of p70 S6 kinase) nor PD098059 (a MEK inhibitor, which inhibits MAPK and Rsk) prevent the insulin inhibition of GSK-3, whereas wortmannin prevents entirely the insulin inhibition of GSK-3. Extracts prepared from L6 myoblasts treated with rapamycin and PD098059 exhibit two peaks of insulin-stimulated GSK-3 (N-terminal) kinase on Mono Q chromatography, which comigrate with immunoreactive PKB and abolished by treatment of the cells with wortmannin prior to extraction [83]. GSK-3 is an enzyme of broad specificity, and many of its candidate substrates in addition to GS undergo dephosphorylation in response to insulin, e.g. (eIF-2B, c-myc, c-myb, c-jun at sites adjacent to DNA binding domain etc.). Consequently, insulin inhibition of GSK-3 is potentially an important aspect of

insulin action. Nevertheless, questions remain as to the contribution of GSK-3 in the insulin regulation of glycogen synthase as compared to the insulin activation of glycogen synthase phosphatase (i.e. glycogen-bound protein phosphatase 1) or inhibition of other glycogen synthase kinases [136].

p70 S6 kinase

This enzyme was the first insulin-stimulated protein (ser/thr) kinase to be detected [137], and among the first to be purified [138, 139] and molecularly cloned [140, 141]. The immunosuppressant drug rapamycin was identified in 1992 as a selective inhibitor of the p70 S6 kinase activity and 40S S6 phosphorylation *in situ* [78, 142, 143], an effect subsequently shown to occur through a functional inhibition of the mTOR kinase [144]. The availability of a relatively selective inhibitor has facilitated the survey of the role of p70 in the program of insulin action. Rapamycin strongly inhibits IL-2 induced cell cycle progression in T cells, but is only somewhat inhibitory to G1 progression in most other cells [145]. Inhibition of p70 *in situ* by microinjection of anti p70 antibodies into REF52 cells arrests cell cycle progression in late G1 [146, 147], establishing the importance of p70 itself. Whether this cell cycle arrest is attributable to the loss of S6 phosphorylation or to interference with the p70-catalyzed phosphorylation of other, as yet unidentified substrates is not known. Although a facilitatory role for S6 phosphorylation in translation has been proposed for some time, direct supportive evidence for an important action on overall mRNA translation is lacking. Nevertheless, rapamycin selectively inhibits translation of mRNAs that encode a polypyrimidine tract at their immediate 5′ end [148]; such mRNAs include those for the elongation factors and many ribosomal proteins [149], and the synthesis of these polypeptides is known to be especially upregulated in response to insulin [150]. Rapamycin has no effect on insulin-stimulated glucose transport, antilipolysis or GSK-3 inhibition [3]. Some, but not all reports [3] observe a partial inhibition by rapamycin of insulin activation of glycogen synthase in 3T3 L1 cells [151] and skeletal muscle [152]. Despite the apparently narrow role of the p70 S6 kinase in insulin action, the kinase is a reliable target of insulin regulation *in situ*, exhibiting a rapid and robust activation by insulin both in cell culture and in human skeletal muscle, during insulin infusion *in vivo* [153]. Moreover, insulin-resistant native American (Pima) subjects exhibit, during insulin infusion, defective activation of both skeletal muscle glycogen synthase and p70 S6 kinase despite normal activation of skeletal muscle insulin receptor kinase [153]. Thus, although the evidence in support of a role for p70 itself in the regulation of glycogen synthase activity is scant, it appears that some of the insulin-directed signals for GS activation are also critical for the insulin activation of p70 S6 kinase.

The p70 S6 kinase is expressed as two polypeptides (α1 and α2) from a single gene, through alternative mRNA splicing and the alternative utilization of two translational start sites [154]. The two p70 polypeptides differ only at their aminoterminus, where the 525 amino acid α1 contains a 23 amino acid extension not found in the 502 amino acid α2 isoform [154]. The N-terminal extension in α1 encodes a polybasic nuclear localization motif that restricts this isoform to the nucleus [155, 156]. Both p70 S6 kinase polypeptides exhibit a centrally placed catalytic domain that is most closely related to the Rsk aminoterminal catalytic domain, PKB and the PKCs, followed by a 104 amino acid, noncatalytic carboxyterminal tail [141]. The activation of p70 occurs through an unusually complex multisite phosphorylation; at least three independently regulated sets of phosphorylations, directed at separate domains are required for p70 activation. One set of phosphorylations is directed to a psuedosubstrate, autoinhibitory domain situated in the carboxyterminal tail [157, 158]. This segment contains a cluster of four (SP/TP) sites which are phosphorylated by an array of insulin-stimulated, proline directed kinases [66]; phosphorylation of this domain is thought to abrogate its inhibitory effect. Nevertheless, deletion of the carboxy-terminal tail does not *per se*, activate the p70 S6 kinase; the 'tailless' enzymes still exhibits a low basal activity, that is activated by insulin in situ, and inhibited by wortmannin and rapamycin [159, 160]. This regulation derives from the phosphorylation of two, perhaps three additional sites; two of these p70 sites, thr252 (229 in α2) and thr412 (389 in α2) [162] correspond in location to PKB ser308 (in catalytic subdomain VIII) and ser473, respectively. The phosphorylation of these p70 sites is stimulated by co-expressed recombinant PI-3 kinase and is inhibited by wortmannin [161–163]; thus they are controlled by Ptd Ins 3P regulated kinases. Whether both sites are phosphorylated by one kinase, or whether, like PKB, a separate kinase acts at each site is not known. Coexpression of p70 with an oncogenic, constitutively active v-Akt results in partial activation of p70 [108], however PKB does not appear to phosphorylate p70 directly *in vitro*. Thus one or both of the Ptd Ins 3P-activated, p70 kinase-kinases may be downstream of PKB. The p70 S6 kinase has been reported to associate with the small GTPases, cdc42 and rac1, both *in vitro* and *in situ* [164], however confirmation of this observation is lacking.

The mechanism by which mTOR, the direct target of the rapamycin-FKBP complex, regulates the p70 kinase remains unclear. Rapamycin inhibition of the mTOR input results in the dephosphorylation of the wortmannin-sensitive sites (thr252 and thr412) on p70 [159, 162, 163], however a mutant p70, rendered resistant to inhibition by rapamycin by deletion of N- and C- terminal sequences, is still activated by insulin and inhibited by wortmannin *in situ* [159, 160, 165]. Thus although rapamycin and wortmannin both result in dephosphorylation of the same p70 sites, this is achieved through the inhibition of two separate pathways. One simple model envisions rapamycin promoting the dephosphorylation of the wortmannin-sensitive sites, whereas PI-3 kinase activates p70 kinase-kinases which phosphorylate these sites. A recently uncovered aspect of p70 regulation is that the basal activity of the enzyme, and its ability to be further activated by insulin are completely inhibited by removal of amino acids from the medium. In contrast, this maneuver does not modify insulin activation of PI-3 kinase or of the MAPK pathway. Thus the regulation of p70 is directed by multiple RTK activated pathways, including PI-3 kinase, and also by signals that depend on amino acid sufficiency. The elucidation of the biochemical mechanisms that underlie regulation of the p70 kinase continues to provide an exceptional challenge.

Based on the effects of PI-3 kinase inhibitors, it seems clear that many important insulin-sensitive regulators of energy metabolism will be found downstream of the PI-3 kinase. Whether all the regulatory apparatus relevant to the rapid control of metabolism by insulin is situated downstream of PI-3 kinase, or whether there exist fundamentally different signal outputs that contribute to acute metabolic regulation cannot yet be stated. It is also not known whether any of the candidate Ptd Ins 3P-activated protein (ser/thr) kinases identified thusfar (such as PKB) are central players in the ability of PI-3 kinase to rapidly regulate nutrient metabolism, in the way that Raf is for the Ras regulation of mitogenesis. Moreover, although it is certain that protein (ser/thr) kinases will play an important role in the signalling apparatus downstream of PI-3 kinase, it is not yet clear whether the major immediate effectors of PI-3 kinase will be Ptd Ins 3P regulated protein (ser/thr) kinases, or other catalysts from among the large number of candidate signalling proteins that contain PH domains. Conceivably, the major mode of protein kinase regulation downstream of PI-3 kinase may be achieved indirectly, e.g. through the Ptd Ins 3P recruitment of guanyl nucleotide exchange factors (GNEFs), such as those of the PH/dbl domain GNEFs that control the rho family of small GTPases [166].

Other ERK-based protein kinase cascades

As first established for *S. cervisiae* [167], mammalian cells contain multiple protein cascades with the general architecture, MEKK-MEK-ERK. Based on the sequence similarity of the ERK components, at least 4 ERK sub-families have been identified, in addition to the classical MAPKs, erk1 and erk2; these include the SAPKs (also known as JNKs), two sets of enzymes most closely related to yeast HOG1, the first containing p38 (also known as RK, CSBP and Mxi2) and p38β, the other contains SAPK3/ERK6 and SAPK4, and a fourth subfamily characterized by ERK5/BMK1. A fifth type of probable ERK, the fos kinase FRK, is not yet cloned. Obviously the nomenclature of these enzymes is chaotic at present. Grouping the ERKs by overall sequence similarity rather than just the *TXY* phosphorylation site motif in the catalytic loop (which is TEY for MAPKs/ERKs and ERK5/BMK1; TPY for SAPKs/JNKs; TGY for both HOG-1 type subfamilies) is preferable, inasmuch as it assembles enzymes that exhibit very similar substrate specificity and common upstream activators (i.e. MEKs). The best characterized of the newer ERK pathways are those converging on the SAPK/JNK and p38/RK/HOG-1 kinases, reviewed in [168, 169].

The regulation of the SAPK and p38 pathways was first elaborated in cell lines of fibroblastic origin; in this cell background, insulin and other ligands that act through receptor tyrosine kinases *do not* cause significant activation of SAPK and p38; rather, cellular stresses, and inflammatory cytokines such as TNFα and/or IL-1β are the potent SAPK/p38 activators [170, 171]. In other cell backgrounds, robust responses to stress and inflammatory cytokines are also consistently observed, however RTK stimulation, concomitant with activation of the MAPKs, often also recruits the SAPK and p38 cascades effectively. Such responses are consistently observed in epithelial (e.g. EGF in hepatocytes [170, 172]) and neuronal (e.g. FGF in neuroblastoma and myoblastic [173]) cells. Only a few reports are available addressing the ability of insulin to activate the SAPKs and p38 in primary target cells, and the role, if any, of these cascades in insulin action. Administration of insulin *in vivo* to mice gave a rapid activation of p38 and SAPK as well as ERK1 and 2 in skeletal muscle homogenates, coincident with activation of glycogen synthase [176]. Anisomycin, a protein synthesis inhibitor that consistently activates SAPK > p38 in cell culture, when given *in vivo*, selectively activated skeletal muscle SAPK, but not p38 or ERK1 and 2, and also activated skeletal muscle glycogen synthase. *In vivo* insulin administration also activated skeletal muscle Rsk3 and mixing experiments using partially purified enzymes, demonstrated the *in vitro* activation of Rsk3 by column fractions containing SAPK. Proceeding from the earlier proposal that Rsk2 might regulate glycogen synthase phosphatase activity in skeletal muscle (see above), Moxham *et al.* [76], proposed that insulin activation of SAPK/Rsk3 in skeletal muscle might underlie, in part, the insulin activation of glycogen synthase. Confirmation and extension of these interesting observations is awaited.

Insulin activates the p38 kinase in cultures of L6 myotubes, however the potential significance of such activation to insulin action is obscure [174]. It should be noted that the p38 kinases control the MAPKAP-2 [175], MAPKAP-3 [176] and Mnk [65, 177] families of protein kinases, much as the MAPKs (and ? SAPK) regulate the Rsks. Thus, just as the proline-directed substrate specificity of the MAPKs is diversified by their control of the Rsks, which possesses a very different specificity, the proline-directed specificity of the p38 family is diversified by their control of the Mnk and MAPKAP-2 isoforms. Little is known at present concerning Mnk substrates, save for eIF4E [65]. The MAPKAP-2 and -3 kinase, like Rsk prefer an arginine at –3 but a bulky hydrophobic (Phe, Trp, Tyr) residue at –5. The best characterized substrates of the MAPKAP-2 and MAPKAP-3 kinases are the p25/27 small heatshock proteins [178]. Like SAPK and p38, the regulation of the MAPKAP2 isoforms has been examined primarily in response to cellular stress, and no data is available bearing on insulin regulation.

The discussion thusfar has considered the limited evidence as to whether the SAPK/p38 kinases are likely to function as positive effector elements in insulin signal transduction; the possibility that these cascades serve to downregulate insulin signalling is worthy of consideration. Certainly the stimuli that reliably activate the SAPKs/p38 i.e. cellular stress and inflammatory cytokines, in general are associated with antagonism of the ability of insulin to promote anabolism. An interesting body of work points to the possibility that TNFα, which is overproduced in adipose tissue as a consequence of obesity [179], is capable of antagonizing insulin signal transduction by converting IRS-1 into an inhibitor of the insulin receptor kinase. This TNFα-induced modification of IRS-1 appear to be the result of a TNFα-directed IRS-1 (ser/thr) phosphorylation (mediated by a sphingomyelinase/ceramide directed pathway [180, 181]. No direct evidence implicates SAPK and/or p38 as the mediator of this TNFα/ceramide effect, but these TNFα/ceramide activated kinases are clearly plausible candidates. The possibility that SAPK/p38 play a negative effector role in insulin signalling is not incompatible with a coexisting positive function. Prolonged insulin stimulation is known to reduce the response to subsequent rechallenge with insulin; certainly sustained MAPK activation is known to downregulate the operation of the Ras-MAPK pathway at several levels, e.g. by phosphorylation and desensitization of mSOS [182], or activation of the expression of genes encoding the dual specificity phosphatases (MKPs) that can deactivate the MAPKs [183].

The more recently elucidated cascades, although they retain the MEKK-MEK-ERK cassette, exhibit several architectural features not seen with the Ras-MAPK cascade. The newer pathways identified thusfar are not regulated directly by Ras or the Ras subfamily (i.e. rap, ral, TC21, etc.) of GTPases, but rather by members of the rho subfamily, specifically cdc42, rac-1 and rhoa [184]. Each of these rho family GTPases has now been shown to directly regulate the activity of one or more protein (ser/thr) kinases. Rac-1 and cdc42, when overexpressed in active forms, can each activate the SAPK and p38 pathways [185, 186]; no report of a rhoa regulated ERK cascade has appeared thusfar. Indirect evidence indicates that the activity of these GTPases is regulated by receptors; e.g. insulin can alter cytoskeletal function in specific cell backgrounds, through pathways that require activation of rac1 and cdc42 [187]. Nevertheless, the biochemical pathways by which receptors control the activity of the rho subfamily are not well defined. Over-expression of constitutively active forms of Ras can recruit the rho GTPases to an active state [188], but it is likely that receptors can recruit the rho GTPases through mechanisms independent of Ras. By analogy with Ras, receptors probably control the rho GTPases through regulation of GTP/GDP exchange. The prototype guanyl nucleotide exchange protein for the rho subfamily is the oncogene dbl. Remarkably, over 40 polypeptides have been identified thusfar that contain guanyl nucleotide exchange domains structurally and/or functionally related to the dbl catalytic domain. Every protein identified thusfar that contains a dbl domain also contains a pleckstrin homology (PH) domain immediately adjacent [166]. The functional significance of this organization is not known, but is especially intriguing in view of the ability of PH domains to bind Ptd Ins 3P lipids, i.e. products of the PI-3 kinase. A variety of data point to important functional interactions between PI-3 kinase and the rho family of GTPases. Examples of PI-3 kinase control of Ras [189] and rho [190] family GTPase function have been reported, as well as the opposite, i.e. control of PI-3 kinase activity by Ras [191] and rho [192] subfamily of GTPases. The specific relationships appear to be very cell-dependent, and the wiring between the relevant components in skeletal muscle, liver and adipocytes is not known.

Another novel feature of the newer erk pathways is the presence of one or more additional layer of kinases upstream of the MEKK; the kinases thusfar identified can be assembled into three subfamilies based on homology with the *S. cerevisiae* protein (ser/thr) kinases known as Ste20, PKC1, and SPS1 [168, 169]. Ste20 participates in the yeast mating pathway, and genetic evidence places it downstream of cdc42 and upstream of the MEKK, Stell [192]. Ste20 is directly activated by cdc42 [193], and is the prototype of the mammalian PAK subfamily [194], 62–68 KDa kinases that contain a carboxyterminal catalytic domain and like Ste20 bind to cdc42 and rac-1 through their aminoterminal noncatalytic segment, and are activated thereby. Some reports [195, 196] indicate that the PAKs, when overexpressed or microinjected, can activate the SAPK and/or p38 kinases. Physiologic substrates of the PAKs have not yet been identified however, so that SAPK/p38 activation may yet prove to be very indirect, i.e. the PAKs may well couple cdc42 and/or rac1 primarily to targets other than the MEKKS that control SAPK/p38.

Insulin rapidly activates a p65 PAK in L6 myotubes, concomitant with the activation of p38 kinase [174]. Interestingly, wortmannin blocks insulin-induced PAK activation but not p38 activation. Thus insulin activates both these novel kinases, but through different pathways; the PAK isoform activated by insulin is clearly not upstream of p38. As to the mechanism by which insulin activates PAK, insulin may control the activity of rho family GTPases as suggested by indirect evidence. In addition, several of the PAKs contain proline-rich SH3 binding motifs (in addition to cdc42/rac1 binding sites) in their aminoterminal noncatalytic segment, bind *in vitro* to Nck through its SH3 domain, and are also activated *in situ* by coexpression with Nck [197, 199]. These observations provide another plausible linkage between receptor tyrosine kinase activation and PAK activity. The requirement for PI-3 kinase activity in PAK activation could be due to Ptd Ins 3P regulation of a cdc42/rac guanyl nucleotide exchange activity (through recruitment of the PH domain). Alternatively, PAK activation requires PAK phosphorylation; whereas PAK autophosphorylation apparently suffices on binding cdc42, recruitment of PAK through Nck may require another (? Ptd Ins 3P sensitive) kinase for PAK phosphorylation and activation.

Yeast PKC-1 binds to RhoA, and is situated upstream of an ERK cascade that controls yeast cell wall synthesis [200]; the closest mammalian homologs to the yeast PKC1 are the kinases known variously as PRK (*p*rotein kinase C *r*elated-*k*inase) or PKN [201, 202]; these mammalian protein kinases have recently been shown to bind rhoA [201–203] and rac-1 [203] and undergo activation. In mammalian systems, rhoa has been shown to specifically regulate the activity of the SRF transcription factor crucial to c-fos expression [204], and some evidence supports a role for PRK2 in this action [205]. However in contrast to the situation in *S. cervisiae* the operation of an ERK cascade between PRK and SRF has not yet been observed. Insulin and all other ligands that act through RTKs regulate c-fos expression in many systems; evidence as to whether insulin regulates the putative rhoA/PRK2/SRF pathway is not yet available.

The mammalian homologs of the yeast enzyme SPS1, which controls the onset of sporulation, include the protein kinases, GCK, MST-1 and others. These kinases are all capable of activating the SAPK on cotransfection [206]. Nevertheless, evidence as regards the modes of regulation or native substrates of the mammalian SPS-1 homologs is not available.

The status of the newer erk pathways in insulin action is uncertain, however new information of these pathways is accumulating very rapidly, as are the reagents required to address their role in insulin signal transduction. As is evident from the foregoing, remarkable progress in the understanding of insulin signal transduction has occurred in the last decade (e.g. see [207, 208] for comparison). Nevertheless, it is still not possible to construct a map from the insulin receptor to the activation of glucose transport, glycogen synthase, acetyl coA carboxylase, cAMP phosphodiesterase mRNA translation, or gene expression. Nevertheless, the likelihood that such a map will emerge in the next decade seems very high. As in many areas of cell regulation, the study of insulin signal transduction is in an exceptionally vigorous phase.

Note added in proof

The protein kinase PDK1 has recently been identified as one of the enzymes responsible for the Ptd Ins 3P-dependent activation of both PKB [209] and the P70 Sb kinase [210] through the phosphorylation of these kinases in their activation loop (PKB ser 308 and P70 thr 252). The mechanism of activation of Rsk through the combined effects of MAPK-catalyzed phosphorylation and intramolecular autophosphorylation was recently described [211].

References

1. Cahill GF, Steiner DF (eds).: Endocrinology 1 Handbook of Physiology. American Physiological Society, 1972
2. White MW, Kahn CR: The insulin signalling system. J Biol Chem 269: 1–4, 1994
3. Avruch J: Small GTPases and (serine/threonine) protein kinase cascades in insulin signal transduction. In: D LeRoith, J Olefsky, S Taylor (eds). Diabetes Mellitus: A Fundamental and Clinical Text. J.P. Lippincott Co., PA, USA, 1996
4. Bliss R: The discovery of insulin. University of Chicago Press, 1982
5. Larner J: Insulin-signalling mechanisms. Lessons from the old testament of glycogen metabolism and the new testament of molecular biology. Diabetes 37: 262–275, 1988
6. Cuatrecasas P: Interaction of insulin with the cell membrane: The primary action of insulin. Proc Natl Acad Sci 63: 450–457, 1969
7. Robison GA, Butcher RW, Sutherland EW: Cyclic AMP, Academic Press, New York, 1971, pp 1–23
8. Krebs EG: Protein kinases. Curr Top Cell Reg 5: 99–133, 1972
9. Butcher RW, Sneyd S, Park CR, Sutherland EW: Effect of insulin on adenosine 3′ 5′ monophosphate in raf epidydimal fat pads. J Biol Chem 242: 1651–1656, 1996
10. Ullrich A, Schlessinger J: Signal transduction by receptors with tyr kinase activity. Cell 61: 203–206, 1990
11. Fantl WJ, Johnson DE, Williams LT: Signalling by receptor tyrosine kinases. Ann Rev Biochem 62: 453–481, 1993
12. Koch CA, Anderson D, Moran MF, Ellis C, Pawson T: SH2 and SH3 domains: Elements that control interactions of cytoplasmic signalling proteins. Science 252: 668–674, 1991
13. Kavanaugh WM, Williams LT: An alternative to SH2 domains for binding tyrosine-phosphorylated proteins. Science 266: 1862?–18655, 1994
14. Bork P, Margolis B: A phosphotyrosine interaction domain Cell. 80: 693–594, 1995
15. Matsuda M, Mayer BJ, Fukui Y, Hanafusa H: Binding of transforming protein, P47gag-crk, to a broad range of phosphotyrosine-containing proteins. Science 248(4962): 1537–1539, 1990
16. Tornqvist HE, Avruch J: Relationship of site- specific β subunit tyrosine autophosphorylation to insulin activation of the insulin receptor (tyrosine) kinase activity. J Biol Chem 263: 4593–4601, 1988
17. Avruch J, Nemenoff RA, Pierce M, Kwok YC, Blackshear PJ: Protein phosphorylations as a mode of insulin action. In: MP Czech (ed). Molecular Basis for Insulin Action. Plenum Press, New York, 1985, pp 263–296
18. Kyriakis JM, Avruch J: S6 kinases and MAP kinases: Sequential intermediates in insulin/mitogen-activated protein kinase cascades. In: JR Woodgett (ed). Protein Kinases: Frontiers in Molecular Biology. Oxford University Press, Oxford, 1994, pp 85–148
19. Avruch J, Zhang XF, Kyriakis JM: Raf meets Ras: Closing a frontier in signal transduction. TIBS 19: 274–283, 1994
20. Lowy DR, Willumsen BM: Function and regulation of Ras. Ann Rev Biochem 62: 851–891, 1993
21. Avruch J, Kyriakis JM, Zhang XF: Raf-1 kinase. In: B Draznin, D Leroith (eds). Molecular Biology of Diabetes. Humana Press Inc., New Jersey, USA, 1994, pp 179–207
22. Ahn NG, Seger R, Krebs EC: The mitogen-activated protein kinase activator. Curr Opin Cell Biol 4: 992–999, 1992
23. Cobb MH, Goldsmith EJ: How MAP kinases are regulated. J Biol Chem 270: 14843–14846, 1995
24. Denton RM, Tavare JM: Does mitogen-activated-protein kinase have a role in insulin action? The cases for and against. Eur J Biochem 227: 597–611, 1995
25. Valencia A, Chardin P, Wittinghofer A, Sander C: The Ras protein family: Evolutionary tree and role of conserved amino acids. Biochem 30: 4637–4648, 1991
26. Zhang FL, Casey PJ: Protein prenylation: Molecular mechanisms and functional consequences. Ann Rev Biochem 65: 241–269, 1996
27. Schlessinger J: How receptor tyrosine kinases activate Ras. TIBS 18: 273–275, 1993
28. Boguski MS, McCormick F: Proteins regulating Ras and its relatives. Nature 366: 643–654, 1993
29. Marshall MS: The effector interactions of p21Ras. TIBS 18: 250–254, 1993
30. Zhang XF, Settleman J, Kyriakis JM, Takeuchi-Suzuki E, Elledge SJ, Marshall MS, Bruder JT, Rapp UR, Avruch J: Normal and oncogenic p21^{Ras} binds to the amino-terminal regulatory domain of c-Raf-1. Nature 364: 308–313, 1993
31. Satoh T, Nakafuku M, Kaziro Y: Function of Ras as a molecular switch in signal transduction: J Biol Chem 267: 24149–24152, 1992
32. Marshall CJ: Ras effectors. Curr Opin Cell Biol 8: 197–204, 1996
33. Rapp UR, Heidecker C, Huleihel M et al.: Raf family serine/threonine protein kinases in mitogen signal transduction. Cold Spring Harbor, Symp Quant Biol 53: 173–184, 1988
34. Chuang E, Barnard D, Hettich L, Zhang XF, Avruch J, Marshall MS: Critical binding and regulatory interactions between Ras and Raf occur through a small, stable N-terminal domain of Raf and specific Ras effector residues. Mol Cell Biol 14: 5318–5325, 1994
35. Barnard D, Diaz B, Hettich L, Chuang E, Zhang XF, Avruch J, Marshall M: Identification of the sites of interaction between c-Raf-1 and Ras-GTP. Oncogene 10: 1283–1290, 1995
36. Herrmann C, Martin GA, Wittinghoefer A: Quantitative analysis of the complex between p21^{Ras} and the Ras-binding domain of the human raf-1 protein kinase. J Cell Biol 272: 2901–2905, 1995
37. Hu CD, Kariya K, Tamada M, Akasaka K, Shirouzu M, Yokoyama S, Kataoka T: Cysteine-rich region of Raf-1 interacts with activator domain of post-translationally modified Ha-Ras. J Biol Chem 270: 30274–30277, 1995
38. Luo Z, Diaz B, Marshall MS, Avruch J: An intact raf zinc finger is required for optimal binding to processed Ras and for Ras-dependent raf activation in situ. Mol Cell Biol 17: 46–53, 1997

39. Leevers SJ, Paterson HF, Marshall CJ: Requirement for Ras in Raf activation is overcome by targeting Raf to the plasma membrane. Nature 369: 411–414, 1994

40. Stokoe D, Macdonald SC, Cadwallader K, Symons M, Hancock JF: Activation of Raf as a result of recruitment to the plasma. Science 264: 1463–1467, 1994

41. Mineo C, Anderson RG, White M: Physical association with Ras enhances activation of membrane-bound Raf (Raf CAAX). J Biol Chem 272: 10345–10348, 1997

42. Luo Z, Zhang X-f, Rapp U, Avruch J: Identification of the 14.3.3 zeta domains important for self association and Raf binding. J Biol Chem 270: 23681–23687, 1995

43. Fantl WJ, Muslin AJ, Kikuchi A, Martin JA, MacNicol AM, Gross RW, Williams LT: Activation of Raf-1 by 14-3-3 proteins. Nature 371: 612–614, 1994

44. Muslin AJ, Tanner JW, Allen PM, Shaw AS: Interaction of 14-3-3 with signalling proteins is mediated by the recognition of phosphoserine. Cell 84: 889–897, 1996

45. Morrison DK, Heidecker C, Rapp UR, Copeland TD: Identification of the major phosphorylation sites of the Raf-1 kinase. J Biol Chem 268: 17309–17316, 1993

46. Luo Z, Tzivion C, Belshaw PJ, Marshall M, Avruch J: Oligomerization activates c-Raf-1 through a Ras-dependent mechanism. Nature 383: 181–185, 1996

47. Farrar MA, Alberol-Ila, Perlmutter RM: Activation of the Raf-1 kinase cascade by coumermycin-induced dimerization. Nature 383: 178–181, 1996

48. Kornfeld K, Hom DB, Horvitz HR: The ksr-1 gene encodes a novel protein kinase involved in Ras-mediated signalling in C. elegans Cell 83: 903–913, 1995

49. Therrien M, Chang HC, Solomon NM, Karim FD, Wassarman DA, Rubin CM: KSR, a novel protein kinase required for RAS signal transduction. Cell 83: 879–888, 1995

50. Zhang Y, Yao B, Delikat S, Bayoumy S, Lin XH, Basu S, McCinley M, Chan-Hui PY, Lichenstein H, Kolesnick R: Kinase suppressor of Ras is ceramide-activated protein kinase. Cell 89: 63–72, 1997

51. Kolch W, Heidecker G, Kochs G: Protein kinase Cα activates RAF-1 by direct phosphorylation. Nature 364: 249–252, 1993

52. Marais R, Light Y, Paterson HF, Marshall CJ: Ras recruits Raf-1 to the plasma membrane for activation by tyrosine phosphorylation. EMBO J 14: 3136–3145, 1995

53. Kyriakis JM, Force TL, Rapp UR, Bonventre JV, Avruch J: Mitogen regulation of c-Raf-1 protein kinase activity toward mitogen-activated protein kinase-kinase. J Biol Chem 268: 16009–16019, 1993

54. Wu J, Dent P, Jelinek T, Wolfman A, Weber MJ, Sturgill TJ: Inhibition of the EGF-activated MAP kinase signalling pathway by adenosine 3′,5′-monophosphate. Science 262: 1065–1069, 1993

55. Altschuler DL, Peterson SN, Ostrowski MC, Lapetina EG: Cyclic AMP-dependent activation of Rap1b. J Cell Biol 270: 10373–10376, 1995

56. Cook SJ, Rubinfeld B, Albert I, McCormick F: RapV12 antagonizes Ras-dependent activation of ERK1 and ERK2 by LPA and EGF in Rat-1 fibroblasts: EMBO J 12: 3475–3485, 1993

57. Vossler NM, Yao H, York RD, Pan MC, Rim CS, Stork PJ: cAMP activates MAP kinase and Elk-1 through a B-Raf- and Rap1-dependent pathway. Cell 89: 73–82, 1997

58. Yamamori B, Kuroda S, Shimizu K, Fukui K, Ohtsuka T, Takai Y: Purification of a Ras-dependent mitogen-activated protein kinase kinase kinase from bovine brain cytosol and its identification as a complex of B-Raf and 14-3-3 proteins. J Biol Chem 270: 11723–11726, 1995

59. Okada T, Masuda T, Shinkai M, Kariya K, Kataoka T: Post-translational modification of H-Ras is required for activation of, but not for association with, B-Raf. J Biol Chem 271: 4671–4678, 1996

60. Kyriakis JM, App H, Zhang X-F, Banerjee P, Brautigan DL, Rapp UR, Avruch J: Raf-1 activates MAP kinase-kinase. Nature 358: 417–421, 1992

61. Galaktino K, Jessus C, Beach D: Raf-1 interaction with Cdc25 phosphatase ties mitogenic signal transduction to cell cycle activation. Genes Devel 9: 1046–1052, 1995

62. Posada J, Yew N, Ahn NG, Vandewoude CF, Cooper JA: Mos stimulates MAP kinase in Xenopus oocytes and activates a MAP kinase kinase in vitro. Mol Cell Biol: 132546–132553, 1993

63. Lange-Carter CA, Pleiman CM, Cardner AM, Blumer KJ, Johnson CL: A divergence in the MAP kinase regulatory network defined by MEK kinase and Raf. Science 260: 315–319, 1993

64. Sturgill TW, Ray LB, Erikson E, Maller JL: Insulin-stimulated MAP-2 kinase phosphorylates and activates ribosomal protein S6 kinase II. Nature 334: 715–718, 1988

65. Waskiewicz AJ, Flynn A, Proud CC, Cooper JA: Mitogen-activated protein kinases activate the serine/threonine kinases Mnk1 and Mnk2. EMBO J 16: 1909–1929, 1997

66. Mukhopadhyay NK, Price DJ, Kyriakis JM, Pelech S, Sanghera J, Avruch J: An array of insulin-activated, proline-directed (Ser/Thr) protein kinases phosphorylate the p70 S6 kinase. J Biol Chem 267: 3325–3335, 1992

67. Hsiao KM, Chou SY, Shih SJ, Ferrell JE Jr: Evidence that inactive p42 mitogen-activated protein kinase and inactive Rsk exist as a heterodimer in vivo. Proc Natl Acad Sci 91: 5480–5484, 1994

68. Dai T, Rubie E, Franklin CC, Kraft A, Gillespie DA, Avruch J, Kyriakis JM, Woodgett JR: Stress-activated protein kinases bind directly to the delta domain of c-jun in resting cells: Implications for repression of c-jun function. Oncogene 10: 849–855, 1995

69. Kallunki T, Deng T, Hibi M, Karin M: c-jun can recruit JNK to phosphorylate dimerization partners via specific docking interactions. Cell 87: 929–939, 1996

70. Erikson E, Maller JL: Purification and characterization of a protein kinase from Xenopus eggs highly specific for ribosomal protein S6. J Biol Chem 261: 350–355, 1986

71. Alcorta DA, Crews CM, Sweet LJ, Bankston L, Jones SW, Erikson RL: Sequence and expression of chicken and mouse rsk: Homologs of Xenopus laevis ribosomal S6 kinase. Mol Cell Biol 9: 3850–3859, 1989

72. Zhao Y, Bjorbaek C, Weremowicz S, Morton CC, Moller DE: RSK3 encodes a novel pp90rsk isoform with a unique N-terminal sequence: Growth factor-stimulated kinase function and nuclear translocation. Mol Cell Biol 15: 4353–4363, 1995

73. Sutherland C, Campbell DG, Cohen P: Identification of insulin-stimulated protein kinase-1 as the rabbit equivalent of rskmo- 2; identification of two threonines phosphorylated during activation by MAP kinases. Eur J Biochem 212: 581–588, 1993

78. Price DJ, Grove, JR, Calvo V, Avruch J, Bierer BE: Rapamycin-induced inhibition of the 70-kilodalton S6 protein kinase. Science 257: 973–977, 1992

79. Erikson E, Maller JL: Substrate specificity of ribosomal protein S6 kinase H from Xenopus eggs. Second Mess and Phosphoprotein 12: 135–143, 1988

80. Blenis J: Signal transduction via the MAP kinases: Proceed at your own risk. Proc Natl Acad Sci USA 90: 5889–5992, 1993

81. Dent P, Lavoinne A, Nakielny S, Caudwell FB, Watt P, Cohen P: The molecular mechanism by which insulin stimulates glycogen synthesis in mammalian skeletal muscle. Nature 348: 302–308, 1990

82. Sutherland C, Leighton I, Cohen P: Inactivation of glycogen synthase kinase-3b by MAP kinase-activated protein kinase-1 (RSK-2) and p70 S6 kinase; new kinase connections in insulin and growth factor signalling. Biochem J 296: 15–19, 1993

83. Cross DA, Alessi DR, Cohen P. Andjelkovich M. Hemmings BA: Inhibition of glycogen synthase kinase-3 by insulin mediated by protein kinase B. Nature 378: 785–789, 1995

84. Eldar-Finkelman H, Seger R, Vandenheede JR, Krebs EG: Inactivation of glycogen synthase kinase-3 by epidermal growth factor is mediated by mitogen-activated protein kinase/p90 ribosomal protein S6 kinase signalling pathway in NIH/3T3 cells. J Biol Chem 270: 987–990, 1995

85. Benito M, Porras A, Nebreda AR, Santos E: Differentiation of 3T3 fibroblasts to adipocytes induced by transfection of Ras oncogenes. Science 253: 565–568, 1991

86. Porras A, Maszynski K, Rapp UR, Santos E: Dissociation between activation of Raf-1 and the 42-kDa mitogen-activated protein kinase/90-kDa S6 kinase (MAPK/RSK) cascade in the insulin/Ras pathway of adipocytic differentiation of 3T3 L1 cells. J Biol Chem 269: 12741–12748, 1994

90. Yano H, Nakanishi S, Kimura K et al.: Inhibition of histamine secretion by wortmannin through the blockade of phosphatidylinositol 3-kinase in RBL-2H3 cells. J Biol Chem 268: 25846–25856, 1993

91. Cheatham B, Vlahos CJ, Cheatham L, Wang L, Blenis J, Kahn CR: Phosphatidylinositol 3-kinase activation is required for insulin stimulation of pp70 S6 kinase, DNA synthesis, and glucose transporter translocation. Mol Cell Biol 14: 4902–4911, 1994

92. Hara K, Yonezawa K, Sakave H, Ando A, Kotani K, Kitamura T, Kitamura Y, Ueda H, Stephens L, Jackson TR et al.: 1-Phosphatidylinositol 3-kinase activity is required for insulin stimulated glucose transport but not for RAS activation in CHO cells. Proc Natl Acad Sci 91: 7415–7419, 1994

93. Robinson LA, Razzack ZF, Lawrence JCJ, James DE: Mitogen-activated protein kinase activation is not sufficient for stimulation of glucose transport or glycogen synthase in 3T3-L1 adipocytes. J Biol Chem 268: 26422–26427, 1993

94. Lin TA, Lawrence JC: Activation of ribosomal protein S6 kinases does not increase glycogen synthesis or glucose transport in rat adipocytes. J Biol Chem 269: 21255–21261, 1994

95. Carpenter CL, Cantley LC: Phosphoinositide kinases. Curr Opin Cell Biol 8: 153–158, 1996

96. Toker A, Meyer M, Reddy K et al.: Activation of protein kinase C family members by the novel polyphosphoinositides PtdIns-3,4-P2 and PtdIns-3,4,5-P3. J Biol Chem 269: 32358–32367, 1994

97. Nakanishi H, Brewer KA, Exton JH: Activation of the z isozyme of protein kinase C by phosphatidylinositol 3,4,5- triphosphate. J Biol Chem 268: 13–16, 1993

98. Franke TF, Kaplan DR, Cantley LC: P13K: Downstream AKTion blocks apoptosis. Cell 88: 435–437, 1997

99. Farese RV: In: D LeRoith, J Olefsky, S Taylor (eds). Diabetes Mellitus: A Fundamental and Clinical Text. J.P. Lippincott Co., PA, USA, 1996

100. Blackshear PJ: In: B Draznin, D Leroith (eds). The Role (or Lack Thereof) of Protein Kinase C in Insulin Action. Humana Press Inc., New Jersey, USA, 1994, pp 229–244

101. Rameh LE, Chen CS, Cantley LC: Phosphatidylinositol (3,4,5)P3 interacts with SH2 domains and modulates PI 3-kinase association with tyrosine-phosphorylated proteins. Cell 83: 821–830, 1995

102. Jones PF, Jakubowicz T, Pitossi FJ, Maurer F, Hemmings BA: Molecular cloning and identification of a serine/threonine protein kinase of the second-messenger subfamily. Proc Natl Acad Sci 88: 4171–4175, 1991

103. Coffer PJ, Woodgett JR: Molecular cloning and characterisation of a novel putative protein-serine kinase related to the cAMP-dependent and protein kinase C families. Euro J Biochem 201: 475–481, 1991

104. Bellacosa A, Testa JR, Staal SP, Tsichlis PN: A retroviral oncogene, akt, encoding a serine-threonine kinase containing an SH2-like region. Science 254: 274–277, 1991

105. Staal SP, Hartley JW, Rowe WP: Isolation of transforming murine leukemia viruses from mice with a high incidence of spontaneous lymphoma. Proc Natl Acad Sci 74: 3065–3067, 1977

106. Franke TF, Yang SI, Chan TO, Datta K, Kazlauskas A, Morrison DK, Kaplan DR, Tsichlis PN: The protein kinase encoded by the Akt protooncogene is a target of the PDGF-activated phosphatidylinositol 3-kinase. Cell 81: 727–736, 1995

107. Kohn AD, Kovacina KS, Roth RA: Insulin stimulates the kinase activity of RAC-PK, a pleckstrin homology domain containing ser/thr kinase. EMBO J 14: 4288–4295, 1995

108. Burgering BM, Coffer PJ: Protein kinase B (c-Akt) in phosphatidyl-inositol-3-OH kinase signal transduction. Nature 376: 599–602, 1995

109. Klippel A, Reinhard C, Kavanaugh WM, Apell G, Escobedo MA, Williams LT: Membrane localization of phosphatidylinositol 3-kinase is sufficient to activate multiple signal-transducing kinase pathways. Mol Cell Biol 16: 4117–4127, 1996

110. Marte BM, Rodriguez-Viciana P, Wennstrom S, Warne PH, Downward J: R-Ras can activate the phosphoinositide 3-kinase but not the MAP kinase arm of the Ras effector pathways. Curr Biol 7: 63–70, 1997

111. Franke TF, Kaplan DR, Cantley LC, Toker A: Direct regulation of the Akt proto-oncogene product by phosphatidylinositol-3,4-bisphosphate Science 275: 665–668, 1997

112. Klippel A, Kavanaugh WM, Pot D, Williams LT: A specific product of phosphatidylinositol 3-kinase directly activates the protein kinase Akt through its pleckstrin homology domain. Mol Cell Biol 17: 338–344, 1997

113. Andjelkovic M, Jakubowicz T, Cron P, Ming XF, Han JW, Hemmings BA: Activation and phosphorylation of a pleckstrin homology domain containing protein kinase (RAC-PK/PKB) promoted by serum and protein phosphatase inhibitors. Proc Natl Acad Sci 93: 5699–5704, 1996

114. Konishi H, Matsuzaki H, Tanaka M, Ono Y, Tokunaga C, Kuroda S, Kikkawa U: Activation of RAC-protein kinase by heat shock and hyperosmolarity stress through a pathway independent of phosphatidylinositol 3-kinase. Proc Natl Acad Sci 93: 7639–7643, 1996

115. Kohn AD, Takeuchi F, Roth RA: Akt, a pleckstrin homology domain containing kinase, is activated primarily by phosphorylation. J Biol Chem 271: 21920–21926, 1996

116. Alessi DR, Andjelkovic M, Caudwell B, Cron P, Morrice N, Cohen P, Hemmings BA: Mechanism of activation of protein kinase B by insulin and IGF-1. EMBO J 15: 6541–6551, 1996

117. Keranen LM, Dutil EM, Newton AC: Protein kinase C is regulated in vivo by three functionally distinct phosphorylations. Curr Biol 5: 1394–1403, 1995

118. Pearson RB, Dennis PB, Han JW, Williamson NA, Kozma SC, Wettenhall RE, Thomas C: The principal target of rapamycin-induced p70s6k inactivation is a novel phosphorylation site within a conserved hydrophobic domain. EMBO J 14: 5279–5287, 1995

119. Alessi DR, james SR, Downes CP, Holmes AB, Gafrey PRJ, Reese CB, Cohen P: Characterization of a 3-phosphoinositide- dependent protein kinase which phosphorylates and activates protein kinase Ba. Curr Biol 7: 261–269, 1997

120. Frevert EU, Kahn BB: Differential effects of constitutively active phosphatidylinositol 3-kinase on glucose transport, glycogen synthase activity, and DNA synthesis in 3T3-L1 adipocytes. Mol Cell Biol 17: 190–198, 1997

121. Kohn AD, Summers SA, Bimbaum MJ, Roth RA: Expression of a constitutively active Akt Ser/Thr kinase in 3T3-L1 adipocytes stimulates glucose uptake and glucose transporter 4 translocation. J Biol Chem 271: 31372–31378, 1996

122. Embi N, Rylatt DB, Cohen P: Glycogen synthase kinase-3 from rabbit skeletal muscle. Separation from cyclic-AMP- dependent protein kinase and phosphorylase kinase. Eur J Biochem 107: 519–527, 1980

123. Parker PJ, Embi N, Caudwell FB, Cohen P: Glycogen synthase from rabbit skeletal muscle. State of phosphorylation of the seven phosphoserine residues in vivo in the presence and absence of adrenaline. Eur J Biochem 124: 47–55, 1982

124. Roach PJ: Multisite and hierarchal protein phosphorylation. J Biol Chem 266: 14139–14142, 1991

125. Cohen P: In: PD Boyer, EG Krebs (eds). The Enzymes. Academic Press, London, 1996, p 462

126. Poulter L, Ang SC, Gibson BW, Williams DH, Holmes CF, Caudwell FB, Pitcher J, Cohen P: Analysis of the *in vivo* phosphorylation state of rabbit skeletal muscle glycogen synthase by fast-atom-bombardment mass spectrometry. Eur J Biochem 175: 497–510, 1988

127. Sheorain VS, Juhl H, Bass M, Soderling TR: Effects of epinephrine, diabetes, and insulin on rabbit skeletal muscle glycogen synthase. Phosphorylation site occupancies. J Biol Chem 259: 7024–7030, 1984

128. Lawrence JC, Hiken JF, DePaoli Roach AA, Roach PJ: Hormonal control of glycogen synthase in rat hemidiaphragms. Effects of two cyanogen bromide fragments. Biol Chem 258: 10710–10719, 1983

129. Woodgett JR: Molecular cloning and expression of glycogen synthase kinase-3/factor A. EMBO J 9: 2431–2438, 1990

130. Hughes K, Nikolakaki E, E PS, Totty NF, Woodgett JR: Modulation of the glycogen synthase kinase-3 family by tyrosine phosphorylation. EMBO J 12: 803–808, 1993

131. Ramakrishna S, Benjamin WB: Insulin action rapidly decreases multifunctional protein kinase activity in rat adipose tissue. J Biol Chem 263: 12677–12681, 1988

132. Hughes K, Ramakrishna SB, Benjamin WB, Woodgett JR: Identification of multifunction ATP-citrate lyase kinase as the alpha-isoform of glycogen synthase kinase-3. Biochem J 288: 309–314, 1992.

133. Welsh GI, Proud CG: Glycogen synthase kinase-3 is rapidly inactivated in response to insulin and phosphorylates eukaryotic initiation factor EIF-2B. Biochem J 294: 625–629, 1993

134. Stambolic V, Woodgett JR: Mitogen inactivation of glycogen synthase kinase-3 beta in intact cells via serine 9 phosphorylation. Biochem J 303: 701–704, 1994

135. Sutherland C, Cohen P: The alpha-isoform of glycogen synthase kinase-3 from rabbit skeletal muscle is inactivated by p70 S6 kinase or MAP kinase-activated protein kinase-1 *in vitro*. FEBS Lett 338: 37–42, 1994

136. Lawrence JC, Roach PJ: New insights into the role and mechanism of glycogen synthase activation by insulin. Diabetes 46: 541–547, 1997

137. Smith CJ/ Rubin CS, Rosen OM: Insulin-treated 3T3-L1 adipocytes and cell-free extracts derived from them incorporated ^{32}P into ribosomal protein S6. Proc Natl Acad Sci USA 77: 2641–2645, 1980

138. Price DJ, Nemenoff RA, Avruch J: Purification of a hepatic S6 kinase from cycloheximide-treated rats. J Biol Chem 264: 13825–13833, 1989

139. Kozma SC, Lane HA, Ferrari S, Luther H, Siegmaun M, Thomas G: A stimulated S6 kinase from rat liver: Identity with the mitogen activated S6 kinase of 3T3 cells. EMBO J 8: 4125–4132, 1989

140. Kozma SC, Ferrari S, Bassand P, Siegmann M, Totty N, Thomas G: Cloning of the mitogen-activated S6 kinase from rat liver reveals an enzyme of the second messenger subfamily. Proc Natl Acad Sci USA 87: 7365–7369, 1990

141. Banerjee P, Ahmad MF, Grove JR, Kozlosky C, Price DJ, Avruch J: Molecular structure of a major insulin/mitogen-activated 70 kDa S6 protein kinase. Proc Natl Acad Sci USA 87: 8550–8554, 1990

142. Calvo V, Crews CM, Vik TA, Bierer BE: Interleukin 2 stimulation of p70 S6 kinase activity is inhibited by the immunosuppressant rapamycin. Proc Natl Acad Sci USA 89: 7571–7575, 1992

143. Chung J, Kuo CJ, Crabtree CR, Blenis J: Rapamycin-FKBP specifically blocks growth-dependent activation of a signalling by the 70 kd S6 protein kinases. Cell 69: 1227–1236, 1992

144. Brown EJ, Beal PA, Keith CT, Chen J, Shin TB, Schreiber SL: Control of p70 s6 kinase bykinase activity of FRAP *in vivo*. Nature 377: 441–446, 1995

145. Dumont FJ, Su Q: Mechanism of action of the immunosuppressant rapamycin. Life Sciences 58(5): 373–395, 1996

146. Lane HA, Fernandez A, Lamb NJC, Thomas G: p70s6k function is essential for G1 progression. Nature 363: 170–172, 1993

147. Reinhard C, Fernandez A, Lamb NJ, Thomas G: Nuclear localization of p85s6k: Functional requirement for entry into S phase. EMBO J 13: 1557–1565, 1994

148. Jefferies HB, Reinhard C, Kozma SC, Thomas G: Rapamycin selectively represses translation of the 'polypyrimidine tract' MRNA family. Proc Natl Acad Sci USA 91: 4441–4445, 1994

149. Meyuhas D, Avri P, Shama S: In: JWB Hershey, MB Mathews, N Sonenberg (eds). Translational Control. Cold Spring Harbor Laboratory Press, 1996, p 363

150. DePhilip RM, Rudert WA, Lieberman I: Preferential stimulation of ribosomal protein synthesis by insulin and in the absence of ribosomal and messenger ribonucleic acid formation. Biochem 19: 1662–1669, 1980

151. Shepherd PR, Nave BT, Siddle K: Insulin stimulation of glycogen synthesis and glycogen synthase activity is blocked by wortmannin and rapamycin in 3T3-L1 adipocytes: Evidence for the involvement of phosphoinositide 3-kinase and p70 ribosomal protein-S6 kinase. Biochem J 305: 25–28, 1995

152. Azpiazu I, Saltiel AR, DePaoli-Roach AA, Lawrence JC: Regulation of both glycogen synthase and PHAS-I by insulin in rat skeletal muscle involves mitogen-activated protein kinase-independent and rapamycin-sensitive pathways. J Biol Chem 271: 5033–5039, 1996

153. Sommercorn J, Fields R, Raz I, Maeda R: Abnormal regulation of ribosomal protein S6 kinase by insulin in skeletal muscle of insulin-resistant humans. J Clin Invest 91: 509–514, 1993

154. Grove JR, Banerjee P, Balasubramanyam A *et al*.: Cloning and expression of two human p70 S6 kinase polypeptides differing only at their amino termini. Mol Cell Biol 11: 5541–5550, 1991

155. Reinhard C, Thomas G, Kozma SC: A single gene encodes two isoforms of the p70 S6 kinase: activation upon mitogenic stimulation. Proc Natl Acad Sci USA 89: 4052–4056, 1992

156. Coffer PJ, Woodgett JR: Differential subcellular localisation of two isoforms of p70 S6 protein kinase. Biochem Biophys Res Comm 198: 7806, 1994

157. Price DJ, Mukhopadhyay NK, Avruch J: Insulin-activated protein kinases phosphorylate a pseudosubstrate synthetic peptide inhibitor of the p70 S6 kinase. J Biol Chem 266: 16281–16284, 1991

158. Ferrari S, Bannwarth W, Morley SJ, Totty NF, Thomas G: Activation of p70^{s6k} is associated with phosphorylation of four clustered sites displaying Ser/Thr-Pro motifs. Proc Natl Acad Sci USA 89: 7282–7286, 1992

159. Weng Q-P, Andrabi K, Kozlowski MT, Grove JR, Avruch J: Multiple independent inputs are required for activation of the p70 S6 kinase. Mol Cell Biol 15: 2333–2340, 1995

160. Cheatham L, Monfar M, Chou MM, Blenis J: Structural and functional analysis of pp70S6k. Proc Natl Acad Sci USA 92: 11696–11700, 1995

161. Weng Q-P, Andrabi K, Klippel A, Kozlowski MT, Williams LT, Avruch J: Phosphatidylinositol-3 kinase signals activation of p70 S6 kinase in situ through site-specific p70 phosphorylation. Proc Natl Acad Sci USA 15: 5744–5748, 1995

162. Han J-W, Pearson RB, Dennis PB, Thomas G: Rapamycin, wortmannin, and the methylxanthine SQ20006 inactivate p70^{S6K} by inducing dephosphorylation of the same subset of sites. J Biol Chem 270: 21396–21403, 1995

163. Chung J, Grammer TC, Lemon KP, Kazlauskas A, Blenis J: PDGF- and insulin-dependent pp70S6k activation mediated by phosphatidyl-inositol-3-OH kinase. Nature 370: 71–75, 1994

164. Chou MM, Blenis J: The 70 kDa S6 kinase complexes with and is activated by the Rho family G proteins Cdc42 and Rac1. Cell 85: 573–583, 1996

165. Dennis PB, Pullen N, Kozma SC, Thomas G: The principal rapamycin-sensitive p70(s6k) phosphorylation sites, T-229 and T-389, are differentially regulated by rapamycin-insensitive kinase kinases. Mol Cell Biol 16: 6242–6251, 1996

166. Whitehead IP, Campbell S, Rossman KL, Der CJ: Dbl family proteins. Biochim Biophys Acta 133: 1–23, 1997

167. Levin DE, Errede B: The proliferation of MAP kinase signalling pathways in yeast. Curr Opin Cell Biol 7: 197–202, 1995

168. Kyriakis JM, Avruch J: Protein kinase cascades activated by stress and inflammatory cytokines. BioEssays 18: 567–577, 1996

169. Kyriakis JM, Avruch J: Sounding the alarm: Protein kinase cascades activated by stress and inflammation. 271: 24313–24316, 1996

170. Kyriakis JM, Banerjee P, Nikolakaki E, Dai T, Rubie EA, Ahmad MF, Avruch J, Woodgett JR: The stress-activated protein kinase subfamily of c-jun kinases. Nature 369: 156–160, 1994

171. Kyriakis JM, Woodgett JR, Avruch J: The stress-activated protein kinases; A novel ERK subfamily responsive to cellular stress inflammatory cytokines. NY Acad Sci 766: 303–319, 1995

172. Westwick JK, Weitzel C, Leffert HL, Brenner DA: Activation of Jun kinase is an early event in hepatic regeneration. J Clin Invest 95: 803–810, 1995

173. 1. Tan Y, Rouse J, Zhang A, Cariati S, Cohen P, Comb MJ: FGF and stress regulate CREB and ATF-1 via a pathway involving p38 MAP kinase and MAPKAP kinase-2. EMBO J 15: 4629–4642, 1996

174. Tsakiridis T, Taha C, Crinstein S, Klip A: Insulin activates a p21-activated kinase in muscle cells via phosphatidylinositol 3-kinase. J Biol Chem 271(33): 19664–19667, 1996

175. Rouse J, Cohen P, Trigon S, Morange M, Alonso-Llamazares A, Zamanillo D, Hunt T, Nebreda AR: A novel kinase cascade triggered by stress and heat shock that stimulates MAPKAP kinase-2 and phosphorylation of the small heat shock proteins. Cell 78: 1027–1037, 1994

176. McLaughlin MM, Kumar S, McDonnell PC, Van Horn S, Lee JC, Livi CP, Young PR: Identification of mitogen-activated protein (MAP) kinase-activated protein kinase-3, a novel substrate of CSBP p38 MAP kinase. J Biol Chem 271: 8488–8492, 1996

177. Fukunaga R, Hunter T: MAK1, a new MAP kinase-activated protein kinase, isolated by a novel expression screening method for identifying protein kinase substrates. EMBO J 16: 1921–1933, 1977

178. Chfton AD, Young PR, Cohen P: A comparison of the substrate specificity of MAPKAP kinase-2 and MAPKAP kinase-3 and their activation by cytokines and cellular stress. FEBS Lett 392: 209–214, 1996

179. Spiegelman BM, Hotamisligil GS: Through thick and thin: Wasting, obesity, and TNF alpha. Cell 73: 625–627, 1993

180. Hotamisligil CS, Peraldi P, Budavari A, Ellis R, White MF, Spiegelman BM: IRS-1-mediated inhibition of insulin receptor tyrosine kinase activity in TNF-alpha- and obesity-induced insulin resistance. Science 271: 665–668, 1996

181. Kanety H, Hemi R, Papa MZ, Karasik A: Sphingomyelinase and ceramide suppress insulin-induced tyrosine phosphorylation of the insulin receptor substrate-1. J Biol Chem 271: 9895–9897, 1996

182. Dong Chen, Waters SB, Holt KH, Pessin JE: SOS phosphorylation and disassociation of the Grb2-SOS complex by the ERK and JNK signalling pathways. J Biol Chem 271: 6328–6332, 1996

183. Chu Y, Solski PA, Khosravi-Far R, Der CJ, Kelly K: The mitogen-activated protein kinase phosphatases PAC1, MKP-1, and MKP-2 have unique substrate specificities and reduced activity *in vivo* toward the ERK2 sevenmaker mutation. J Biol Chem 271: 6497–6591, 1996

184. Vojtek AB, Cooper JA: Rho family members: activators of MAP kinase cascades. Cell 82: 527–529, 1995

185. Coso OA, Chiariello M, Yu JC, Teramoto H, Crespo P, Xu N, Miki T, Gutkind JS: The small GTP-binding proteins Rac1 and Cdc42 regulate the activity of the JNK/SAPK signalling pathway. Cell 81: 1137–1146, 1995

186. Minden A, Lin A, Claret FX, Abo A, Karin M: Selective activation of the JNK signalling cascade and c-jun transcriptional activity by the small GTPases Rac and Cdc42Hs. Cell 81:1147–1157, 1995

187. Kotani K, Yonezawa K, Hara K, Ueda H, Kitamura Y, Sakaue H, Ando A, Chavanieu A, Calas B, Grigorescu F *et al.*: Involvement of phosphoinositide 3-kinase in insulin- or IGF-1-induced membrane ruffling. EMBO J 13: 2313–2321, 1994

188. Ridley AJ, Paterson HF, Johnston CL, Diekmann D, Hall A: The small GTP-binding protein rac regulates growth factor-induced membrane ruffling. Cell 70: 401–410, 1992

189. Hu Q, Klippel A, Muslin AJ, Fantl WJ, Williams LT: Ras-dependent induction of cellular responses by constitutively active phosphatidyl-inositol-3 kinase. Science 268: 100–102, 1995

190. Reif K, Nobes CD, Thomas G, Hall A, Cantrell DA: Phosphatidylinositol 3-kinase signals activate a selective subset of Rac/Rho-dependent effector pathways. Curr Biol 6: 1445–1455, 1996

191. Rodriguez-Viciana P, Warne PH, Dhand R, Vanhaesebroeck B, Gout I, Fry NU, Waterfield MD, Downward J: Phosphatidylinositol-3-OH kinase as a direct target of Ras. Nature 370: 527–532, 1994

192. Leberer E, Dignard D, Harcus D, Thomas DY, Whiteway M: The protein kinase homologue Ste20p is required to link the yeast pheromone response G-protein beta gamma subunits to downstream signalling components. EMBO J 11: 4815–4824, 1992

193. Simon MN, De Virgilio C, Souza B, Pringle JR, Abo A, Reed SI: Role for the Rho-family GTPase Cdc42 in yeast mating-pheromone signal pathway. Nature 376: 702–705, 1995

194. Manser E, Leung T, Salihuddin H, Zhao ZS, Lim L: A brain serine/threonine protein kinase activated by Cdc42 and Rac1. Nature 367: 40–46, 1994

195. Zhang S, Han J, Sells MA, Chernoff J, Knaus UG, Ulevitch RJ, Bokoch GM: Rho family GTPases regulate p38 mitogen-activated protein kinase through the downstream mediator Pak1. J Biol Chem 270: 23934–23936, 1995

196. Polverino A, Frost J, Yang P, Hutchison M, Neiman AM, Cobb MH, Marcus S: Activation of mitogen-activated protein kinase cascades by p21-activated protein kinases in cell-free extracts of *Xenopus* oocytes. J Biol Chem 270: 26067–26070, 1995

197. Galisteo ML, Chernoff J, Su YC, Skolnik EY, Schlessinger J: The adaptor protein Nck links receptor tyrosine kinases with the serine-threonine kinase Pak1. J Biol Chem 271: 20997–21000, 1996

198. Bokoch GM, Wang Y, Bohl BP, Sells MA, Quilliam LA, Knaus UG: Interaction of the Nck adapter protein with p21-activated kinase (PAK1). J Biol Chem 271(42): 25746–25749, 1996

199. Lu W, Katz S, Cupta R, Mayer BJ: Activation of Pak by membrane localization mediated by an SH3 domain from the adaptor protein Nck. Curr Biol 7: 85–94, 1997

200. Nonaka H, Tanaka K, Hirano H, Fujiwara T, Kohno H, Umikawa M, Mino A, Takai Y: A downstream target of RHO1 small GTP-binding protein is PKC1, a homolog of protein kinase C, which leads to activation of the MAP kinase cascade in *Saccharomyces cerevisiae*. EMBO J 14: 5931–5938, 1995

201. Watanabe G, Saito Y, Madaule P, Ishizaki T, Fujisawa K, Morii N, Mukai H, Ono Y, Kakizuka A, Narumiya S: Protein kinase N (PKN) and PKN-related protein rhophilin as targets of small GTPase Rho. Science 271: 645–648, 1996

202. Amano M, Mukai H, Ono Y, Chihara K, Matsui T, Hamajima Y, Okawa K, Iwamatsu A, Kaibuchi K: Identification of a putative target for Rho as the serine-threonine kinase protein kinase N. Science 271: 648–650, 1996

203. Vincent S, Settleman J: The PRK2 kinase is a potential effector target of both Rho and Rac GTPases and regulates actin cytoskeletal organization. Mol Cell Biol 17: 2247–2256, 1997

204. Hill CS, Wynne J, Treisman R: The Rho family GTPases RhoA, Rac1, and CDC42Hs regulate transcriptional activation by SRF. Cell 81: 1159–1170, 1995

205. Quilliam LA, Lambert QT, Mickelson-Young LA, Westwick JK, Sparks AB, Kay BK, Jenkins NA, Gilbert DJ, Copeland NG, Der CJ: Isolation of a NCK-associated kinase, PRK2, an SH3-binding protein and potential effector of Rho protein signalling. J Biol Chem 271: 28772–28776, 1996

48

206. Pombo CM, Kehrl JH, Sanchez I, Katz P, Avruch J, Zon LI, Woodgett JR, Force T, Kyriakis JM: Activation of the SAPK pathway by the human STE20 homologue germinal centre kinase. Nature 377: 750–754, 1995

207. Avruch J, Tornqvist HE, Gunsalus JR, Yurkow EJ, Kyriakis JM, Price DJ: Insulin regulation of protein phosphorylation. In: P. Cuatrecasas, S Jacob (eds). Handbook of Experimental Pharmacology. Vol. 92, Chapter 15, Berlin: Springer-Verlag, 1990, pp 313–366

208. Czech MP, Klarlund JK, Yagaloff KA, Bradford AP, Lewis RE: Insulin receptor signalling. Activation of multiple serine kinases. J Biol Chem 263: 1017–11020, 1988

209. Alessi DR, James SR, Downes CP, Holmes AB, Gaffney PR, Reese CB, Cohen P: Characterization of a 3-phosphoinositide-dependent protein kinase which phosphoylates and activates PKBα. Curr Biol 7: 261-269, 1997

210. Alessi DR, Kozlowski MT, Weng QP, Morrice N, Avruch J: 3-phosphoinositide-dependent protein kinase 1 (PDK1) phosphorylates and activates the P70 Sb kinase *in vivo* and *in vitro*. Curr Biol 8: 69-81, 1997

211. Dalby KN, Morrice N, Caudwell FB, Avruch J, Cohen P: Identification of regulatory phosphorylation sites in mitogen-activated protein kinase (MAPK)-activated protein kinase 2a/P90 rsk that are inducible by MAPK. J Biol Chem 273: 1496-1502, 1998

Molecular and Cellular Biochemistry **182**: 49–58, 1998.
© 1998 *Kluwer Academic Publishers. Printed in the Netherlands.*

Protein phosphatase-1 and insulin action

Louis Ragolia[1] and Najma Begum[1,2]

[1]*The Diabetes Research Laboratory, Winthrop University Hospital, Mineola, NY;* [2]*School of Medicine, State University of New York, Stony Brook, New York, USA*

Abstract

Protein Phosphatase-1 (PP-1) appears to be the key component of the insulin signalling pathway which is responsible for bridging the initial insulin-simulated phosphorylation cascade with the ultimate dephosphorylation of insulin sensitive substrates. Dephosphorylations catalyzed by PP-1 activate glycogen synthase (GS) and simultaneously inactivate phosphorylase a and phosphorylase kinase promoting glycogen synthesis. Our *in vivo* studies using L6 rat skeletal muscle cells and freshly isolated adipocytes indicate that insulin stimulates PP-1 by increasing the phosphorylation status of its regulatory subunit (PP-1$_G$). PP-1 activation is accompanied by an inactivation of Protein Phosphatase-2A (PP-2A) activity. To gain insight into the upstream kinases that mediate insulin-stimulated PP-1$_G$ phosphorylation, we employed inhibitors of the ras/MAPK, PI3-kinase, and PKC signalling pathways. These inhibitor studies suggest that PP-1$_G$ phosphorylation is mediated via a complex, cell type specific mechanism involving PI3-kinase/PKC/PKB and/or the ras/MAP kinase/Rsk kinase cascade. cAMP agonists such as SpcAMP (via PKA) and TNF-α (recently identified as endogenous inhibitor of insulin action via ceramide) block insulin-stimulated PP-1$_G$ phosphorylation with a parallel decrease of PP-1 activity, presumably due to the dissociation of the PP-1 catalytic subunit from the regulatory G-subunit. It appears that any agent or condition which interferes with the insulin-induced phosphorylation and activation of PP-1, will decrease the magnitude of insulin's effect on downstream metabolic processes. Therefore, regulation of the PP-1$_G$ subunit by site-specific phosphorylation plays an important role in insulin signal transduction in target cells. Mechanistic and functional studies with cell lines expressing PP-1$_G$ subunit site-specific mutations will help clarify the exact role and regulation of PP-1$_G$ site-specific phosphorylations on PP-1 catalytic function. (Mol Cell Biochem **182**: 49–58, 1998)

Key words: protein phosphatase-1, insulin, glycogen synthesis

Introduction

Insulin is the dominant hormone responsible for the synthesis and storage of carbohydrates, lipids, and proteins, as well as the inhibition of their degradation. The precise molecular mechanism by which insulin regulates cellular metabolism is however, not fully understood. Insulin action begins with the binding of insulin to the α-subunit of its receptor, located on the surface of the plasma membrane, which stimulates the intrinsic tyrosine kinase activity of the β subunit [1–4] and results in the autophosphorylation of the receptor and its substrates, IRS-1 (insulin receptor substrate-1) and Shc [1–4]. This is followed by the recruitment of SH2/SH3-domain containing molecules [24] to the membrane which thereby initiates a complex series of phosphorylations that involve multiple intracellular sub-

strates [2]. Although tyrosine phosphorylations of the above signalling molecules play a significant role in the proximal events of insulin action, much of the intracellular signalling is mediated by serine/threonine (Ser/Thr) phosphorylations [1–3]. Remarkably, at a certain point along the insulin signalling cascade the phosphorylation reactions are followed by dephosphorylation via an activated phosphatase, and this dephosphorylation controls downstream metabolic reactions by either inhibiting or stimulating several intracellular enzymes and proteins [14]. The best examples are the dephosphorylation and activation of glycogen synthase (GS), pyruvate dehydrogenase, and hormone-sensitive lipoprotein lipase, and the dephosphorylation and inactivation of phosphorylase kinase and glycogen phosphorylase [5–7]. Thus, Ser/Thr protein phosphatases assume a major regulatory role in the transmission

Address for offprints: N. Begum, The Diabetes Research Laboratory, Winthrop University Hospital, 259 First Street, Mineola, NY 11501, USA

of insulin-generated signals. How the cell regulates the dephosphorylation of these proteins and which phosphatases are involved in specific signals are poorly understood. In fact, the difficulty in defining the specific regulatory roles for different phosphatases is their ubiquitous expression and the multitude of pathways controlled by phosphorylation/dephosphorylation.

In skeletal muscle, glycogen metabolism has a central role in the control of blood glucose levels in response to insulin [8]. This hormonal effect involves the uptake of glucose and the activation of glycogen synthase (GS). Although the exact mechanism by which insulin regulates glycogen synthase is not completely understood, it is well established that changes in the phosphorylation state of GS alters its enzymatic activity. Thus, GS activation by insulin may be achieved through the inhibition of protein kinases which phosphorylate GS, such as glycogen synthase kinase-3 (GSK-3) or cAMP dependent protein kinase [9, 10], as well as the stimulation of protein phosphatases that dephosphorylate GS [9, 10]. PP-1 seems to fit nicely in this scheme of GS dephosphorylation since insulin activates PP-1 and this phosphatase is able to dephosphorylate multiple sites of glycogen synthase at three serine residues (positions 3A, 3B, and 3C) which are phosphorylated by the multisubstrate protein kinase, GSK-3 [11]. In addition, an impairment in glycogen synthase and PP-1 *catalytic activity*, despite elevations in the amount of the enzyme, have been reported in the skeletal muscles of insulin resistant subjects [12, 13] and animal models of insulin deficient diabetes [14]. Also, protein phosphatase inhibitors such as okadaic acid can potently block many of the metabolic effects of insulin [15]. *Therefore, in order to have a normal insulin effect, both kinases and phosphatases must be appropriately regulated.*

These observations in conjunction with the recent suggestion that the glycogen-associated regulatory subunit of PP-1 (PP-1$_G$) may be a candidate gene for inherited insulin resistance [16], underscore the importance of examining, in detail, the exact role of PP-1 in insulin action and the importance of the PP-1$_G$ subunit in the catalytic and regulatory functions of PP-1 in response to insulin and specifically its impact on glucose metabolism.

Over the last few years, we have extensively investigated the roles and regulation of PP-1 and PP-2A in response to insulin, counter-regulatory hormones, and inhibitors of insulin signalling (i.e. TNF-α), using cultured L6 rat skeletal muscle cells, and freshly isolated rat adipocytes as model systems. In this review article we will give a brief background of protein phosphatases including their structure, function, and regulation as well as review recent findings from our laboratory relating to the role of PP-1 in insulin action and the mechanism of its regulation by the PP-1$_G$ subunit.

Background

Protein phosphatases

Mammalian Ser/Thr protein phosphatases are divided into four major classes (PP-1, PP-2A, PP-2B, PP-2C) depending upon their substrate specificity, sensitivity to endogenous inhibitor proteins and okadaic acid, as well as other enzymatic properties [17–19]. The cloning of these biochemically distinct catalytic subunits in the late 1980's indicated that PP-1, PP-2A, and PP-2B belonged to one gene family, while PP-2C formed another. Both PP-1 and PP-2A show strong evolutionary conservation, with a greater than 80% amino acid identity between plants, yeasts, and mammals [17–19]. Biochemical and genetic evidence suggest that these enzymes exert a multitude of effects and therefore must undergo tight regulation [20]. This review, however, will focus primarily on the regulation of PP-1.

Protein phosphatase-1

PP-1 was discovered more than 50 years ago as an enzyme that inactivated phosphorylase *a* [21], although at the time it was not realized that PP-1 was operating as a phosphatase. The PP-1 catalytic subunit (PP-1C) is a protein with a molecular mass of 37 kDa and cDNA cloning of this subunit has uncovered three mammalian isoforms (α, β or δ, and γ) which may arise from alternate splicing of the same gene [22]. PP-1C exists *in vivo* in a variety of multimeric forms depending on the regulatory proteins it complexes with [22, 23]. The endogenous low molecular weight thermostable proteins, inhibitor-1 and inhibitor-2 can associate with PP-1C and specifically block PP-1 catalytic activity [22]. Inhibitor-1 association with PP-1C is dependent upon prior phosphorylation of inhibitor-1 by cAMP dependent protein kinase A.

In rabbit skeletal muscle, PP-1 is found associated with glycogen particles (PP-1$_G$), the sarcoplasmic reticulum (PP-1s$_R$), myofibrils (PP-1$_M$), the cytosolic inhibitor-2 protein (PP-1$_I$), and the nucleus (PP-1$_N$) through a set of nuclear inhibitor proteins [17–20, 22]. Cohen and Cohen [23] have introduced the concept of targeting subunits as a mechanism for directing protein phosphatases to particular locations within a cell and thus selectively enhancing phosphatase activity toward specific substrates. The paradigm for this model is the glycogen/SR-associated form of PP-1 (PP-1$_G$) found in rabbit skeletal muscle. PP-1$_G$ holoenzyme is a heterodimer composed of the 37 kDa PP-1Cδ and a 160 kDa glycogen/SR binding regulatory subunit (PP-1$_G$), which is responsible for the association with glycogen and the sarcoplasmic reticulum [24]. PP-1$_G$ holoenzyme is 5–10 times more active than the C-subunit alone in terms of GS dephosphorylation but *only* when both PP-1$_G$ and GS are bound to glycogen. In contrast, PP-1$_G$ holoenzyme does not show increased activity on substrates not involved in glycogen metabolism when compared to free C-subunit [25]. PP-1$_{SR}$ in skeletal and cardiac muscle is similar to PP-1$_G$, suggesting this subunit may have a role in targeting the C-subunit to membranes [26]. The GACl gene of *Saccharomyces cerevisiae* [27] is homologous to the PP-1$_G$ subunit cloned by Tang *et al.* [28]. Elevated levels of GACl protein cause hyper accumulation of glycogen, while disruption of the gene causes reduced glycogen levels. Expression of PP-1$_G$ in yeast strains defective in GACl rescues the glycogen phenotype, indicating that PP-1$_G$ subunit and its yeast homologue play direct roles in glycogen metabolism [22, 27]. The predicted amino acid sequence of PP-1$_G$ derived from the cDNA sequence reveals a hydrophobic region near the carboxyl terminus which could serve as a membrane anchoring domain [28]. The binding sites for PP-1C and glycogen, as well as the phosphorylation sites are located towards the amino terminus of the G-subunit. There are five serine residues within a 28 amino acid stretch that are phosphorylated by cAMP-dependent protein kinase A (PKA), GSK-3, casein kinase 2, and insulin-stimulated protein kinase (ISPK; a mammalian homologue of S6 kinase II) [29].

It is widely believed that regulation of glycogen metabolism by insulin is mediated by PP-1. Indeed, PP-1 is activated upon insulin treatment of normal or diabetic rats [30] or in tissue culture cells [31, 32]. This effect is not due to changes in the activity and/or phosphorylation status of inhibitors-1 and 2 but rather due to an increased association of PP-1C with PP-1$_G$, and specifically, an increase in site-1 phosphorylation. In an elegant series of in vitro experiments, Cohen's laboratory demonstrated that PP-1$_G$ phosphorylation at site-1 by ISPK increased PP-1 activity 2–3 fold towards GS and phosphorylase kinase under physiological assay conditions [29]. The increase in site-1 phosphorylation was also seen after insulin administration to rabbits [29]. In contrast to insulin, PKA increased the phosphorylation of PP-1$_G$ at sites 1 and 2, and this dual phosphorylation resulted in the dissociation of catalytic subunit from the G-subunit [33–35]. It appears that serine phosphorylation at site-2 dominates the release of the catalytic subunit from PP-1$_G$ by lowering the binding affinity by a factor of 10^4 [33–36]. Following this phosphorylation event, the G-subunit remains bound to glycogen, whereas the catalytic subunit is released into the cytosol where it is immediately bound by inhibitors 1 and 2. Thus phosphorylation of PP-1$_G$ by PKA removes PP-1 from the spontaneously 'active' pool and transfers it to a 'latent' pool. Adrenergic stimuli (acting via PKA) also increase phosphorylation at site-2 of PP-1$_G$ [37] and may release PP-1C into the cytosol and diminish overall PP-1 activity. Acting in concert is the phosphorylation of inhibitor-1 by PKA which activates this inhibitor and promotes its binding, and inactivation, of PP-1C. In contrast to inhibitor-1, phosphorylation of inhibitor-2 on Thr 72 initiates reactivation of PP-1 (PP-1$_I$) [22, 38].

Inhibitor-2 phosphorylation is mediated by GSK-3 which induces a conformational change of PP-1C resulting in the exposure of a metal binding site. Subsequent Mg^{2+} binding to this site converts the inactive PP-1C into its active conformation [39, 40]. A recent *in vitro* study by Wang *et al.* [41] suggests that inhibitor-2 is phosphorylated by MAPK and this leads to the activation of PP-1$_I$. The physiologic role of PP-1$_I$ and its *in vivo* regulation by hormones remains unknown.

The G-subunit is also phosphorylated by GSK-3 on sites 3A and 3B, however, prior site-1 phosphorylation is required. Casein kinase-2 phosphorylates site-4 on the G-subunit [42]. The occurrence and functional significance of site-specific phosphorylation in vivo and its relevance to insulin action remains to be established.

It has recently been demonstrated that 90 kDa S6 kinase II (Rsk-2) *in vitro*, and protein kinase B (a.k.a. cAkt/Rac, a downstream target of PI3-kinase) *in vivo*, inhibit GSK-3 by increasing the phosphorylation of a specific amino terminal serine residue [43–46]. The combined inhibition of GSK-3 and activation of PP-1$_G$ may contribute to insulin-induced dephosphorylation and activation of GS, the rate limiting enzyme in glycogen synthesis. These observations provide a potential *link* between tyrosine kinase-mediated MAP Kinase and/or PI3-kinase activation and stimulation of glycogen synthesis. Several recent studies suggest that ras/MAPK activation is *not* necessary for insulin-mediated GSK-3 inhibition and/or PP-1 activation and glycogen synthesis [47–49], indicating participation of alternative pathways in insulin-mediated glycogen synthesis. A recent report suggests the involvement of jun-N-terminal kinase, a recently cloned new member of the MAP kinase family, as the upstream activator of PP-1 [50]. Regardless of the mechanism of PP-1 activation it is clear that PP-1 is a key component of the insulin signalling pathway and responsible for connecting the initial phosphorylation cascade with the ultimate dephosphorylation of insulin sensitive substrates and enzymes.

Other than the *in vitro* studies of Dent *et al.* [25] on purified rabbit skeletal muscle PP-1$_G$, no data on the roles of PP-1 and PP-2A in insulin action in other cell types, tissues or animal models of insulin resistance exists. The nature of the regulation of these enzymes by insulin and other counter-regulatory hormones have not been investigated. An impairment in PP-1 *catalytic activity*, despite elevations in the amount of enzyme protein (C-subunit) have been reported in skeletal muscles of insulin resistant subjects [12]. Whether the observed reductions in PP-1 activity in insulin resistant subjects are due to a reduction in the abundance of the PP-1$_G$ subunit or defects in the regulation of site-specific phosphorylation have not been determined. A widespread amino acid polymorphism (aspartate → tyrosine) at codon 905 of human PP-1$_G$ is accompanied by

52

insulin resistance and hypersecretion of insulin in a general Caucasian population [15]. Whether or not this mutation results in a loss of ability of PP-1$_G$ to associate with PP-1C and/or glycogen is not known. Further mechanistic studies of insulin action on PP-1 in tissue culture models will help understand PP-1 regulation.

Regulation of PP-1 by insulin in cultured L6 rat skeletal muscle cells

Differentiated L6 cells exhibit insulin effects that closely parallel the findings in isolated skeletal muscle preparations [51–54]. In our laboratory, we have examined the ability of insulin to stimulate PP-1 activity in cultured L6 rat skeletal muscle cells. Serum starved L6 cell extracts were assayed for PP-1 activity using [^{32}P]-labelled phosphorylase *a* as a substrate in the presence and absence of insulin. Assays were performed in the presence of 1 nM okadaic acid to differentiate between PP-1 and PP-2A activities. At this concentration, okadaic acid inhibits only PP-2A and the remaining activity is attributed to PP-1 [55]. As reported in skeletal muscle, PP-1 activity was present predominantly (70–80%) in the particulate fraction of a 100,000 × g centrifugation. In contrast, absolutely all of the PP-2A activity was found in the cytosolic fraction. Insulin treatment increased particulate PP-1 activity (35–40% over the basal values) without altering the cytosolic activity [56]. The insulin effect on PP-1 activation was differentiation dependent and was absent in proliferating myoblasts [56]. Additionally, the insulin responsiveness of PP-1 in differentiated L6 cells coincided with the appearance of PP-1$_G$, which is believed to regulate PP-1 activation [56]. Earlier studies [51–54] have shown that differentiation of L6 rat skeletal muscle cells is associated with the development of insulin receptors, Glut-4 (the insulin-sensitive glucose transporter), and the capacity to respond to physiological concentrations of the hormone in terms of glucose transport. Therefore, the unresponsiveness of PP-1 to insulin in the proliferating myoblasts could be due to the absence of insulin receptors, PP-1$_G$ subunit, or both. It is not known presently whether the appearance of insulin receptors precedes the appearance of PP-1$_G$ in this cell system.

Insulin stimulation of PP-1 in L6 myotubes was accompanied by a concomitant decrease in PP-2A activity [56–58]. The effect of insulin on these two phosphatases was time and dose dependent with a maximal effect seen at 10 nM insulin for 10 min, followed by a gradual decline in PP-1 activation to basal levels within 60 min. Unlike PP-1, the effect of insulin on PP-2A was sustained when the incubation time and insulin dose was increased to 60 min and 100 nM respectively (Figs 1A and 1B).

Analysis of the PP-1 activity in the immunoprecipitates of rabbit skeletal muscle PP-1$_G$ with an antibody directed against site-1, revealed that 40–45% of the activity present in particulate fractions of L6 cells was due to the glycogen/SR associated form of PP-1, PP-1$_G$. Insulin treatment resulted in a 2 fold increase in the activity of the bound form of PP-1C (Table 1). The activity of the other forms of PP-1 present in the immunodepleted supernatants were not affected by insulin. These results clearly indicate that in whole cells, insulin increases PP-1 activity by promoting its association with the regulatory G-subunit [56, 59].

A significantly larger insulin effect on PP-1 was observed when [^{32}P]-labelled phosphorylase kinase and glycogen synthase were used as substrates, yielding a 78 and 51% stimulation respectively [56]. However, these substrates were highly unstable and therefore not suitable for extensive analyses. In addition, unlike glycogen phosphorylase *a*, which is phosphorylated on a single serine residue, the above substrates are phosphorylated at multiple sites, thereby complicating the analysis of the dephosphorylation kinetics. [60]. The increased activation of PP-1 in L6 cells was accompanied by a greater than 2 fold increase in the phosphorylation of PP-1$_G$ immunoprecipitated from [^{32}P]-labelled L6 cells (Fig. 2).

Further studies with PP-1$_G$ site-1 and site-2 mutants will clearly define the role of site-specific phosphorylation in the regulation of PP-1 catalysis *in vivo*.

Table 1. Insulin stimulates the bound form of PP-1C in the immunoprecipitates

Treatment	Sample	PP-1 activity (nmol/min/mg)	% insulin effect
None	Extract	1.26 ± 0.18	
	Immunoprecipitate	0.51 ± 0.04	
	Supernatant	0.74 ± 0.10	
Insulin	Extract	1.70 ± 0.20	
	Immunoprecipitate	1.06 ± 0.21	120
	Supernatant	0.64 ± 0.21	
SpcAMP + Insulin	Extract	1.20 ± 0.20	
	Immunoprecipitate	0.37 ± 0.06	–27
	Supernatant	0.71 ± 0.12	
Insulin + SpcAMP	Extract	1.20 ± 0.20	
	Immunoprecipitate	0.42 ± 0.05	–29
	Supernatant	0.78 ± 0.16	

Equal amounts of extract proteins (50 µg) prepared from control and insulin treated L6 myotubes were immunoprecipitated with PP-1$_G$ site-1 antibody (µg/ml) for 1 h at 4°C followed by addition of 50 µl of protein A-Sepharose (50% v/v). After 1 h, the supernatant was removed and assayed for unbound PP-1 activity. The immunoprecipitates were washed four times with wash buffer, brought back to the original volume, the bound enzyme was released by incubating the immunoprecipitates with excess site-1 peptide (20 µg) for 1 h, and the activity released in the supernatants was assayed for PP-1 activity. Results are the mean ± S.E.M. of 3 independent experiments. (Data reproduced with permission from M Srinivasan, N Begum: J Biol Chem 269: 12514–12520, 1994).

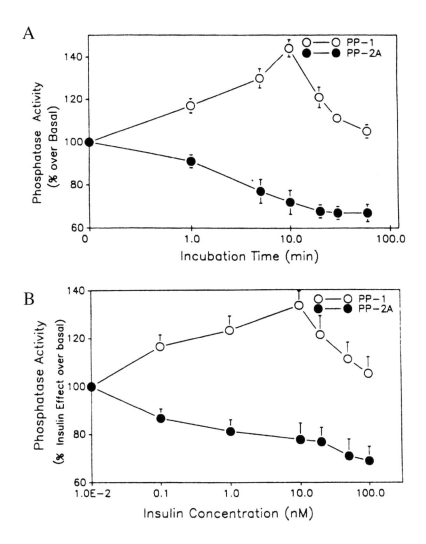

Fig. 1. (A) Effect of insulin on PP-1 and PP-2A activities over time. Serum starved 14 day old L6 myotubes were pulsed with insulin (10 nM) for the indicated times. Enzyme activities were assayed in the particulate fractions (PP-1) or cytosol (PP-2A). Results are the mean ± S.E.M. of 4–5 separate experiments performed in triplicate. Insulin effect was significant at all time points except 30 and 60 min (p < 0.01 vs. time 0); (B) Effect of insulin concentration on PP-1 and PP-2A activities. Serum starved cells were incubated with insulin (0.1–100 nM) for 10 min at 37°C followed by the assay of PP-1 and PP-2A. Results are the mean ± S.E.M. of 4 independent experiments performed in duplicate. (Data reproduced with permission from reference [55]).

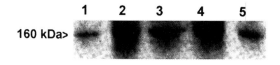

160 kDa>

Fig. 2. Effect of insulin and SpcAMP on PP-1C phosphorylation. Serum starved L6 cells were incubated with [^{32}P]-orthophosphate (0.5 μCi/rnl) for 4 h, treated with or without SpcAMP for 20 min followed by insulin (10 nM) for 10 min. The extracts were precleared by the incubation with 5 μg/ml of pre-immune Ig-G for 1 h at 4°C followed by incubation with 50 μl of protein A-Sepharose (50% v/v). Non-specific proteins were pelleted and the supernatants used for immunoprecipitation with PP-1$_G$ antibody. The immunoprecipitates were subjected to 7.5% SDS/PAGE followed by autoradiography. An autoradiogram from a representative experiment is shown. Lanes 1 and 5 – control cells; lane 2 – insulin; lane 3 – SpcAMP; lane 4 – SpcAMP followed by insulin. (Data reproduced with permission from reference [55]).

Counter-regulatory hormones and inhibitors of insulin action block insulin-stimulation of PP-1 activity by inhibiting PP-1$_G$ phosphorylation

To further investigate whether PP-1 activity could be differentially regulated *in vivo* by insulin and other agents, the effects of counter-regulatory hormones and TNF-α (a recently identified inhibitor of insulin action) were studied [60, 61]. L6 cells were exposed to SpcAMP (a cAMP agonist) or TNF-α for 30–60 min prior to insulin treatment. While neither SpcAMP or TNF-α significantly altered basal PP-1 activity, both completely prevented insulin stimulated PP-1 activation [56, 61]. Similar impairment in PP-1 activation was observed in adipocytes isolated from insulin resistant Goto-Kakizaki diabetic rats, a model for NIDDM in humans. All the three conditions also resulted in elevated basal PP-2A activities. Insulin treatment failed to decrease PP-2A activities.

Immunoprecipitation of PP-1$_G$ from [^{32}P]-labelled cells followed by SDS/PAGE and autoradiography revealed that both SpcAMP and TNF-α blocked insulin mediated phosphorylation of the PP-1$_G$ subunit. SpcAMP alone caused a small increase in basal PP-1$_G$ phosphorylation when compared to control cells (Fig. 2). The increase might be due to site-2 phosphorylation mediated by the activation of protein kinase A (Fig. 2) and was not due to variations in the amounts of PP-1$_G$ immunoprecipitated. These results indicate that in cultured rat L6 skeletal muscle cells PP-1C is complexed to a G-subunit of molecular weight 160 kDa and insulin increases the activity of PP-1C by promoting the phosphorylation of the G-subunit and its association with the PP-1C-subunit (Table 1 and Fig. 2). Additionally, insulin neither altered total (trypsin-released) enzyme activity, nor the amount of PP-1C, suggesting that insulin stimulates PP-1 activity by converting the inactive form of the enzyme into an active form. cAMP agonist and TNF-α blocked insulin-stimulated PP-1$_G$ phosphorylation and PP-1 activation presumably by blocking the upstream kinases that phosphorylate PP-1$_G$ at site-1 (see Fig. 3).

The present observations coincide with earlier *in vitro* findings of Dent *et al.* [25] on purified preparations of rabbit skeletal muscle PP-1$_G$ who demonstrated that insulin, via ISPK, stimulates phosphorylation of PP-1 at a specific site (site-1) on the G-subunit and this effect was accompanied by a greater than 2 fold increase in the rate of dephosphorylation of phosphorylase kinase and GS when assayed under physiological conditions. Contrary to insulin, cAMP agonists, via PKA, mediate phosphorylation of sites 1 and 2 on PP-1$_G$ which leads to the dissociation of the catalytic subunit from the G-subunit. PP-1$_G$ is then released into the cytosol where it binds activated inhibitor 1 or 2, contributing to the inactive (latent) pool of PP-1. This study emphasized the importance of site-specific phosphorylation on the activity state of the enzyme, even though most of the work was performed on a purified enzyme *in vitro*.

Our results on cell extracts prepared from control, insulin, TNF-α and SpcAMP treated cells suggest that a similar mechanism of PP-1 regulation may be involved in rat L6 skeletal muscle cells. The time course of 160 kDa protein phosphorylation by insulin correlated well (r = 0.866) with the time course of PP-1 stimulation by insulin. In contrast to insulin, cAMP agonist and TNF-α, added prior to insulin, inhibited phosphorylation of the 160 kDa protein and this resulted in decreased immunoprecipitable enzyme activity. The identity of the 160 kDa band as the glycogen associated regulatory subunit of PP-1 was established by immunoprecipitation with an antibody raised against a synthetic peptide corresponding to the site-1 sequence of rabbit skeletal muscle PP-1$_G$. Preincubation of the antibody with this peptide blocked immunoprecipitation and the detection of a 160 kDa protein by Western blot analysis.

Upstream activators of PP-1
It is clear from the above *in vivo* studies of L6 cells, that insulin regulates the catalytic activity of PP-1 by promoting phosphorylation on its regulatory subunit (PP-1$_G$). A number of earlier studies have implicated a role for PKC in insulin mediated activation of glycogen synthesis and glucose transport [62–64]. To gain an insight into the potential upstream activators of PP-1 that might mediate G-subunit phosphorylation, we employed synthetic inhibitors of protein kinase C and the MAP kinase/Rsk pathways.

TPA, a potent PKC activator, mimicked insulin's effect on PP-1 activation with kinetics comparable to insulin ($t_{1/2}$ = 1 min and EC_{50} = 5 nM). The effect of TPA on PP-1 activation was accompanied by a 2 fold increase in PP-1$_G$ phosphorylation (Fig. 4). Both insulin and TPA increased cellular MAPK activity towards myelin basic protein with a time course slightly preceding the activation of PP-1. Inhibition of PKC by chronic treatment with TPA blocked subsequent effects of insulin and TPA on MAPK and PP-1 activation. Furthermore, two selective inhibitors of PKC, calphostin and chelerythrine blocked insulin's effect on MAPK, PP-1 activation and G-subunit phosphorylation. The presence of tyrosine kinase inhibitors, herbimycin or erbstatin A, blocked insulin's effect on PP-1 activation without affecting the TPA response [62].

These observations together with our findings that insulin caused a rapid translocation of PKCα and PKCβ from the cytosol to the plasma membrane suggests that insulin stimulation of PP-1 may be mediated through a PKC dependent mechanism in a normal cell [62].

Insulin regulation of PP-1 in isolated rat adipocytes
Isolated rat adipocytes have provided a useful model system for investigating the hormonal regulation of glycogen metabolism. As in skeletal muscle, insulin stimulates glycogen synthesis in adipocytes by activating both glucose transport and glycogen synthase [65]. In fact, adipocytes and skeletal muscle express the same glycogen synthase isoforms and glucose transporter [65]. We therefore examined insulin regulation of PP-1 activity in freshly isolated rat adipocytes.

Consistent with L6 rat skeletal muscle cells, insulin also caused a rapid activation of PP-1 in the particulate fractions of rat adipocytes in a time and dose dependent manner, with maximum stimulation at 5 min and an insulin concentration of 4 nM. The insulin effect was accompanied by the phosphorylation of the 160 kDa regulatory subunit of PP-1 (PP-1$_G$). Treatment of α-toxin permeabilized rat adipocytes with GTPγS (a GTP analogue previously reported to stimulate MAP kinase/S6 kinase pathway) mimicked insulin's effect on

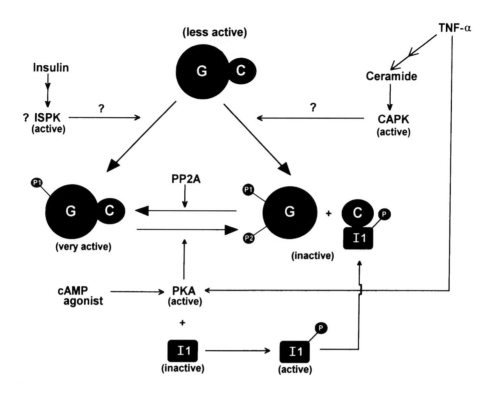

Fig. 3. The counter-regulatory effects of cAMP agonists and TNF-α on PP-1 activation and glycogen synthesis are mediated through the phosphorylation of PP-1$_G$. Phosphorylation of the G-subunit at site P1 increases in response to insulin *in vivo* and *in vitro* by an insulin stimulated protein kinase (a mammalian homologue of 90 kDa rsk kinase). This activates the G-C complex, increasing the rate at which it dephosphorylates GS, PHK, and phosphorylase. cAMP agonist (SpcAMP) activates PKA which then phosphorylates the G-subunit at site P2. Site-2 phosphorylation triggers the dissociation of the C-subunit from the G-subunit which releases it from glycogen. TNF-α via sphingomyelin hydrolysis, increases cellular ceramide levels which may activate a ceramide kinase (CAPK). CAPK can act directly on the G-subunit or via PKA to increase the dissociation of the G-subunit from the C-subunit by phosphorylation at P2. Both cAMP and TNF-α can also activate PP-2A. This may dephosphorylate site P2 of the G-subunit to reactivate PP-1 and reinitiate glycogen synthesis when the stimulation of TNF-α and SpcAMP ceases.

PP-1 activity and MAPK. Inclusion of insulin along with the GTP analogue did not further increase the stimulation of PP-1. In contrast, the presence of a GTP antagonist, GDPβS, during insulin exposure completely blocked the effect of insulin on PP-1 activation and MAPK. The inhibitors of MAPK (viz. cAMP agonists, SpcAMP and ML-9, a myosin light chain kinase inhibitor) also blocked PP-1 activation by insulin. Wortmannin, a PI3-kinase inhibitor, did not abolish insulin's effect on PP-1 activation. Our results are in agreement with Kahn's data regarding LY 294008 (another

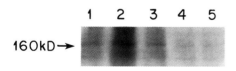

Fig. 4. PKC inhibitors block insulin and TPA-induced phosphorylation of PP-1$_G$. [^{32}P]-labeled cells were treated with calphostin or chelerythrine followed by exposure to insulin and TPA. PP-1$_G$ was immunoprecipitated as detailed in Fig. 2. Lane 1, control; lane 2, TPA; Lane 3, TPA + chelerythrine; lane 4, insulin + chelerythrine; lane 5, insulin + calphostin.

potent PI3-kinase inhibitor) inability to stimulate MAPK in 3T3 L1 adipocytes [66], but differ from the results of Cross *et al.* [45] who demonstrate a blockade of insulin's effect on MAPK activation and GSK3 by wortmannin in rat L6 cells. Presently it is not known whether the effect of wortmannin is cell type specific.

Role of phosphatases in regulation of ion channels
It is well accepted that insulin acutely regulates the activity of the Na$^+$/K$^+$ ATPase pump, however, the exact mechanism is unknown. Insulin resistance associated with diabetes, fasting, and hypertension are accompanied by a reduction in the αII isoform of the enzyme [67, 68]. Insulin stimulates this αII isoform in skeletal muscle as well as adipocytes [69, 70]. A number of recent studies using purified preparations of the *a* subunit of Na$^+$/K$^+$ ATPase have shown that cAMP dependent PKA and PKC inactivate the enzyme by increasing phosphorylation of the α subunit, suggesting that the enzyme may be regulated by phosphorylation/dephosphorylation [71, 72]. We examined the mechanism of Na$^+$/K$^+$ ATPase activation using cultured rat L6 skeletal muscle cells. Our preliminary results

56

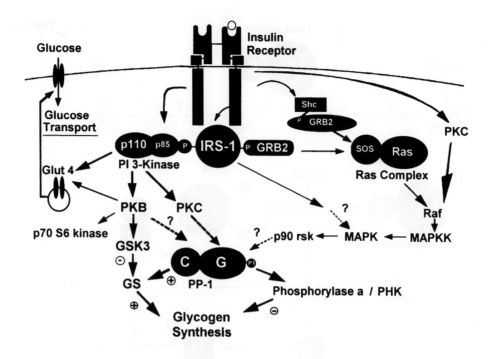

Fig. 5. Schematic representation of the proposed pathway for insulin activation of PP-1. Insulin-stimulated receptor tyrosine kinase mediates the activation of IRS-1/PI3-kinase/PKB, PKC and the ras/MAP kinase cascade leading to activation of rsk kinases. The activated PKB and/or rsk kinase may phosphorylate PP-1$_G$ leading to stimulation of PP-1. Adapted from Reference 3.

indicate that insulin causes a 2 fold increase in the ouabain sensitive αII isoform of the enzyme. Pretreatment of L6 cells with calyculin (1 μM) or okadaic (0.1 μM) for 30 min block insulin's activation of Na$^+$/K$^+$ ATPase. Immunoprecipitation of the enzyme from [^{32}P]-labelled cells with αII subunit antibody detect a 112 kDa protein. Phosphorylation of this protein is decreased by 60% in the plasma membranes of insulin treated cells. Okadaic acid and calyculin prevent the dephosphorylation and block enzyme activation by insulin (data not shown). Low concentrations of okadaic acid which inhibit PP-2A were ineffective. These observations were further confirmed in transfected L6 skeletal muscle cells over-expressing PP-1$_G$ by 4 fold. In these cells PP-1$_G$ over-expression was accompanied by a 3 fold increase in Na$^+$/K$^+$ ATPase activity in response to insulin. In contrast, depletion of PP-1$_G$ caused a complete lack of insulin stimulated Na$^+$/K$^+$ ATPase activity. A comparison of the insulin stimulated Na$^+$/K$^+$ ATPase activation kinetics with those of PP-1 suggest that insulin may stimulate Na$^+$/K$^+$ ATPase activity by dephosphorylation via PP-1. Further mechanistic studies using cell lines containing reduced or elevated PP-1 activity will be helpful in understanding the exact role of PP-1 in insulin regulation of Na$^+$/K$^+$ ATPase activity.

Future perspectives
Over the past few years, considerable progress has been made in understanding the role and regulation of protein serine/

threonine phosphatases, especially PP-1 and PP-2A. However, little is known about the physiological role of these two phosphatases in the control of dephosphory-lation of insulin signalling molecules. We have shown that both rat skeletal muscle and adipose tissue express a form of PP-1 which is similar to rabbit skeletal muscle PP-1$_G$. Insulin, counter-regulatory hormones, and TNF-α regulate the activity of particulate PP-1 in both cell types by altering the phosphorylation status of PP-1$_G$. Regulation is through a complex cell specific mechanism involving the PKC/PI3-kinase/PKB cascade and/or the ras/MAP kinase cascade, with a resultant activation of GS and glycogen synthesis via an activated PP-1 and an inhibited GSK3 (Fig. 5). Functional studies with cell lines expressing wild type and mutant PP-1$_G$ will further our understanding of the exact function of PP-1$_G$, and the importance of its site-specific phosphorylation in *in vivo* insulin action and its impact on downstream metabolic effects.

Acknowledgements

This work was supported by the Winthrop medical education research fund. Part of the experiments were performed by Malathi Srinivasan.

References

1. Saltiel AR: Diverse signalling pathways in the cellular actions of insulin. Am J Physiol 270: E375–E385, 1996

2. Cheatham B, Kahn RC: Insulin action and the insulin signalling network. Endo Rev 16: 117–142, 1995

3. White MF, Kahn CR: The insulin signalling system. J Biol Chem 269: 1–4, 1994

4. Keller SR, Lienhard GE: Insulin signalling. Trends Cell Biol 4: 115–119, 1994

5. Zhang J-n, Hiken J, Davis AK, Lawrence JC Jr: Insulin stimulates dephosphorylation of phosphorylase in rat epitrochlearis muscle. J Biol Chem 264: 17513–17523, 1989

6. Mandarino LJ, Wright KS, Vesity LS, Nicholas J, Bell JM, Kolterman OG, Beck-Nielsen H: Effects of insulin infusion on human skeletal muscle pyruvate dehydrogenase, phosphofructokinase and glycogen synthase. J Clin Invest 80: 655–663, 1987

7. Bogardus C, Lillioja S, Stone K, Mott D: Correlation between muscle glycogen synthase activity and *in vivo* insulin action in man. J Clin Invest 73: 1185–1190, 1984

8. Baron AD, Brechtel G, Wallace P, Edelman SV: Rates and tissue sites of non-insulin and insulin-mediated glucose uptake in humans. Am J Physiol 255: E769–E774, 1992

9. Cohen P. In: P Boyer, EC Krebs (eds). The Enzymes. Academic Press, Orlando, FL, 1986, pp 461–497

10. Lawrence JC Jr: Signal transduction and protein phosphorylation in the regulation of cellular metabolism. Ann Rev Physiol 54: 177–193, 1992

11. Cohen P: Adv. Second. Messenger Phosphoprotein Res 24: 230–235, 1990

12. Nyomba BL, Brautigan DL, Schlender KK, Wang W, Bogardus C, Mott DM: Deficiency in phosphorylase phosphatase activity despite elevated protein phosphatase Type 1-catalytic subunit in skeletal muscle from insulin-resistant subjects. J Clin Invest 88: 1540–1545, 1991

13. Lofman M, Hannele Yki-Jarvinen, Parkkonen M, Lindstrom J, Koranyi L, Camilla Schalin-Janti, Groop L: Increased concentrations of glycogen synthase protein in skeletal muscle of patients with NIDDM. Am J Physiol 269: E27–E32, 1995

14. Begum N, Sussman KE, Draznin B: Differential effects of diabetes on adipocyte and liver phosphotyrosine and phosphoserine phosphatase activities. Diabetes 40: 1620–1629, 1991

15. Hess SL, Suchin CR, Saltiel AR: The specific protein phosphatase inhibitor okadaic acid differentially modulates insulin action. J Cell Biochem 45: 374–380, 1991

16. Hansen L, Hansen T, Vestergaard H, Bjorbaek C, Echwald SM, Clausen JO, Chen YH, Chen MX, Cohen PTW, Pedersen O: A widespread amino acid polymorphism at codon 905 of the glycogen associated regulatory subunit of protein phosphatase-1 is associated with insulin resistance and hypersecretion of insulin. Human Mol Genet 4: 1313–1320, 1995

17. Mumby MC, Walter GI: Protein serine/threonine phosphatases: Structure, regulation and functions in cell growth. Physiol Rev 73: 673–699, 1993

18. Cohen P: The structure and regulation of protein phosphatases. Ann Rev Biochem 58: 453–508, 1989

19. Cohen P: Structure, function and regulation of protein phosphatases: In: Nishizuka (ed). The Biology and Medicine of Signal Transduction. Raven Press, NY, 1990, pp 230–235

20. Bollen M, Stalmans W: The structure, role and regulation of type-1 protein phosphatases. Crit Rev Biochem Mol Biol 27: 227–281, 1992

21. Cori GT, Green AA: J Biol Chem 151: 31–38, 1943

22. Depaoli-Roach AA, In-kyung Park, Cerovsky V, Csortos C, Durbin SD, Kuntz MJ, Sitikov A, Tang PM, Verin A, Zolinierowicz S: Serine/threonine protein phosphatases in the control of cell regulation. Adv Enzyme Reg 34: 199–224, 1994

23. Cohen P, Cohen PTW: Protein phosphatases come of age. J Biol Chem 264: 21435–21438, 1989

24. Stralfors P, Hiraga A, Cohen P: The protein phosphatases involved in cellular regulation: Purification and characterization of the glycogen associated form of PP-1 from rabbit skeletal muscle. Eur J Biochem 149: 295–303, 1985

25. Dent P, Lavoinne A, Nakielny S, Caudwell FB, Watt P, Cohe P: The molecular mechanism by which insulin stimulates glycogen synthesis in mammalian skeletal muscle. Nature 348: 302–308, 1990

26. Hubbard MJ, Dent P, Smythe C, Cohen P. Targeting of protein phosphatase 1 to the sarcoplasmic reticulum of rabbit skeletal muscle by a protein that is very similar or identical to the G subunit that directs the enzyme to glycogen. Eur J Biochem 189: 243–249, 1990

27. Francois JM, Jaeger ST, Skroch J, Zellenka U, Spevak W, Tatchell K: GAC1 may encode a regulatory subunit for protein phosphatase type 1 in *Saccharomyces cerevisiae*. EMBO J 11: 87–96, 1992

28. Tang PM, Bondor JA, Swiderek KM, DePaoli-Roach AA: Molecular cloning and expression of the regulatory (R_{GI}) subunit of the glycogen-associated protein phosphatase. J Biol Chem 266: 15782–15789, 1991

29. Lavoinne A, Erikson E, Maller JL, Price DJ, Avruch J, Cohen P: Purification and characterization of the insulin-stimulated protein kinase from rabbit skeletal muscle; close similarity to S6 kinase II. Eur J Biochem 199: 723–728, 1991

30. Toth B, Bollen M, Stalmans W: Acute regulation of hepatic protein phosphatases by glucagon, insulin and glucose. J Biol Chem 263: 14061–14066, 1988

31. Olivier AR, Ballou LM, Thomas G: Differential regulation of S6 phosphorylation by insulin and EGF in Swiss mouse 3T3 cells: Insulin activation of type 1 phosphatase. Proc Natl Acad Sci USA 85: 4720–4724, 1988

32. Olivier AR, Thomas G: Three forms of phosphatase type-1 in Swiss 3T3 fibroblasts. J Biol Chem 265: 22460–22466, 1990

33. Hubbard MJ, Cohen P: Regulation of protein phosphatase-1G from rabbit skeletal muscle. 1. Phosphorylation by cAMP-dependent protein kinase at site-2 releases catalytic subunit from glycogen-bound holoenzyme. Eur J Biochem 186: 701–709, 1989

34. Dent P, Campbell DG, Hubbard MJ, Cohen P: Multisite phosphorylation of the glycogen-binding subunit of protein phosphatase-1_G by cAMP-dependent protein kinase and glycogen synthase kinase-3. FEBS Lett 248: 76–72, 1989

35. Hubbard MJ, Cohen P: Targeting subunits for protein phosphatases. Meth Enzymol 201: 414–427, 1991

36. Hubbard MJ, Cohen P: Regulation of protein phosphatase 1_G from rabbit skeletal muscle. 2. Catalytic subunit translocation is a mechanism for reversible inhibition of activity toward glycogen-bound substrates. Eur J Biochem 186: 711–716, 1989

37. Hubbard MJ, Cohen P: Regulation of protein phosphatase-1_G from rabbit skeletal muscle. Phosphorylation of cAMP-dependent protein kinase at site 2 releases catalytic subunit from the glycogen bound holoenzyme. Eur J Biochem 186: 701–709, 1989

38. Ballow LM, Brautigan DL, Fischer EH: Subunit structure and activation of inactive phosphorylase phosphatase. Biochemistry 22: 3393–3399, 1983

39. Hemmings BA, Resinek TJ, Cohen P: Reconstitution of a Mg-ATP-dependent protein phosphatase and its activation through a phosphorylation mechanism. FEBS Lett 150: 319–324, 1982

40. In-kyung Park, Depaoli-Roach, AA: Domains of phosphatase inhibitor-2 involved in the control of the ATP-Mg-dependent protein phosphatase. J Biol Chem 46: 28919–28928, 1994

41. Wang MQ, Guan KL, Roach PJ, Depaoli-Roach AA: Phosphorylation and activation of the ATP-Mg-dependent protein phosphatase by the mitogen-activated protein kinase. J Biol Chem 270: 18352–18358, 1995

58

42. Fiol CJ, Haesman JH, Wang Y, Roach PJ, Roeske RW, Kowalczuk M: Phosphoserine as a recognition determinant for glycogen synthase kinase-3: Phosphorylation of a synthetic peptide based on the G-component of protein phosphatase-1. Arch Biochem Biophys 267: 797–802, 1988

43. Hagit Eldar-Finkelman, Seger R, Vandenheede JR, Krebs EG: Inactivation oaf glycogen synthase kinase-3 by epidermal growth factor is mediated by mitogen-activated protein kinase/p90 ribosomal protein S6 kinase signalling pathway in NIH/3T3 cells. J Biol Chem 270: 987–990, 1995

44. Sutherland C, Leighton IA, Cohen P: Inactivation of glycogen synthase kinase-3 beta by phosphorylation: New kinase connections in insulin and growth factor signalling. J Biochem 296: 15–19, 1993

45. Cross AK, Alessi DR, Cohen P, Andjelkovich M, Hemmings BA: Inhibition of glycogen synthase-3 by insulin is mediated by protein kinase B. Nature 378: 785–789, 1995

46. Nina van Den Berghe, Ouwens MD, Maassen AJ, Michelle GH van Mackelenbergh, Sips HTC, Krans MH: Activation of the ras/MAP kinase signalling pathway alone is not sufficient to induce glucose uptake in 3T3-L1 adipocytes. Mol Cell Biol 14: 2372–2377, 1994

47. Robinson LJ, Razzack ZF, Lawrence JC, James DE: Mitogen-activated protein kinase is not sufficient for stimulation of glucose transport or glycogen synthase in 3T3-L1 adipocytes. J Biol Chem 268: 26422–26427, 1993

48. Lazar DF, Wiese, RJ, Brady MJ, Mastik CC, Waters SB, Yamuchi K, Pessin JE, Cuatrecasas P, Saltiel AR: MAP kinase Kinase inhibition does not block the stimulation of glucose utilization by insulin. J Biol Chem 270: 20801–20807, 1995

49. Azpiazu I, Saltiel AR, DePaoli-Roach AA, Lawrence JC Jr: Regulation of both glycogen synthase and PHAS-1 by insulin in rat skeletal muscle involves mitogen-activated protein kinase-independent and rapamycin-sensitive pathways. J Biol Chem 271: 5033–5039, 1996

50. Moxham CM, Tabrizchi A, Davis RJ, Malbon CC: jun N-terminal kinase mediates activation of skeletal muscle glycogen synthase by insulin *in vivo*. J Biol Chem 271: 30765–30773, 1996

51. Ramlal T, Sarabia V, Bilan PJ, Klip A: Insulin mediated translocation of glucose transporters from intracellular membranes to plasma membranes: Sole mechanism of stimulation of glucose transport in L6 muscle cells. Biochem Biophys Res Comm 157: 1329–1335, 1988

52. Walker P, Ramlal T, Donovan JA, Doering TP, Sandra A, Klip A, Pessin JE: Insulin and glucose dependent regulation of the glucose transport in the rat L6 skeletal muscle cell line. J Biol Chem 264: 6587–6595, 1989

53. Beguino F, Kahn RC, Moses AL, Smith RJ: Distinct biologically active receptors for insulin, insulin like growth factor 1, and insulin like growth factor 2 in cultured skeletal muscle cells. J Biol Chem 260: 15892–15898, 1985

54. Beguino F, Khan RC, Moses AC, Smith RJ: The development of insulin receptors and responsiveness is an early marker of differentiation in the muscle cell line L6. Endocrinology 18: 446–455, 1986

55. Cohen P, Holmes CFB, Tsukitani Y: Okadaic acid: A new probe for the study of regulation. TIBS 15: 98–102, 1990

56. Srinivasan M, Begum N: Regulation of protein phosphatase-1 and 2A activities by insulin during myogenesis in rat skeletal muscle cells in culture. J Biol Chem 269: 12514–12520, 1994

57. Begum N, Ragolia L: CAMP counter-regulates insulin-mediated inactivation of PP-2A. J Biol Chem 271: 31166–31171, 1996

58. Begum N, Ragolia L, Srinivasan M: Effect of tumor necrosis Factor-α on insulin-stimulated mitogen-activated protein kinase cascade in rat skeletal muscle cells. Eur J Biochem 238: 214–220, 1996

59. Begum N: Stimulation of PP-1 activity by insulin: Evaluation of the role of MAP kinase cascade. J Biol Chem 270: 709–714, 1995

60. Chan CP, McNall SJ, Krebs EG, Fisher EH: Stimulation of protein phosphatase activity by insulin and growth factors in 3T3 cells. Proc Natl Acad Sci USA 85: 6257–6261, 1988

61. Begum N, Ragolia L: Effect of tumor necrosis factor-α on insulin action in cultured rat skeletal muscle cells. Endocrinology 137: 2441–2445, 1996

62. Srinivasan M, Begum N: Stimulation of protein phosphatase-1 activity by phorbol esters: Evaluation of the regulatory role of protein kinase C in insulin action. J Biol Chem 269: 16662–16667, 1994

63. Farese RV: Insulin-sensitive phospholipid signalling systems and glucose transport: An update. Proc Soc Exp Biol Med 213: 1–12, 1996

64. Standaert ML, Bandyopadhay G, Zhou X, Galloway L, Farese RV: Insulin stimulates phospholipase D-dependent phosphatidylcholine hydrolysis, *de novo* phospholipid synthesis, and diacylglycerol/protein kinase C signalling in 16 myotubes. Endocrinology 137: 3014–3020, 1996

65. Lawrence JC Jr, Hiken JF, James DE: Stimulation of glucose transport and glucose transporter phosphorylation by okadaic acid in rat adipocytes. J Biol Chem 265: 19768–19776, 1990

66. Cheatham B, Vlahos CJ, Cheatham L, Wang L, Benis J, Kahn CR: Phosphatidyl inositol 3 kinase activation is required for insulin stimulation of pp70 S6 kinase, DNA synthesis and glucose transporter translocation. Diabetes 43(suppl)(abstr): 259, 1994

67. Moore RD, Munford JW, Pillsworth TJ Jr: Effect of streptozotocin diabetes and fasting on intracellular sodium and adenosine triphosphate in rat soleus muscle. J Physiol (Lond) 338: 277–294, 1983

68. Draznin B, Reusch J, Begum N, Sussman K, Byyny R, Ohara T: Calcium, insulin action, and insulin resistance. In: U Smith, NE Bruun, T Hedner, B Hokfelt (eds). Hypertension as an Insulin Resistant Disorder. Novo Symp, no 5, Excerpta Medica, 1991, pp 225–245

69. Brodsky JL: Characterization of the Na⁺/K⁺ ATPase from 3T3-F442A fibroblasts and adipocytes. Isozymes and insulin sensitivity. J Biol Chem 265: 10458–10468, 1990

70. Rosic NK, Standaert ML, Pollet RJ: The mechanism of insulin stimulation of Na⁺/K⁺ ATPase transport activity in muscle. J Biol Chem 260: 6206–6212, 1985

71. Beguin P, Beggah AT, Chibalin AV, Burgener-Kairuz P, Jaiser F, Mathews PM, Rossier BC, Cotecchia S, Geering K: Phosphorylation of the Na⁺/K⁺ ATPase-α subunit by protein kinase A and C *in vitro* and in intact cells. Identification of a novel motif for PKC-mediated phosphorylation. J Biol Chem 269: 24437–24445, 1994

72. Feschenko MS, Sweadner KJ: Conformation-dependent phosphorylation of Na⁺/K⁺ ATPase by protein kinase A and Protein kinase C. J Biol Chem 269: 30436–30444, 1994

Molecular and Cellular Biochemistry **182**: 59–63, 1998.
© 1998 *Kluwer Academic Publishers. Printed in the Netherlands.*

Insulin receptor internalization and signalling

Gianni M. Di Guglielmo,[1] Paul G. Drake,[3] Patricia C. Baass,[2] François Authier,[4] Barry I. Posner[3] and John J.M. Bergeron[2]
Departments of [1]Biochemistry, [2]Anatomy and Cell Biology, [3]Polypeptide Hormone Laboratory, McGill University, Montreal, Quebec, Canada; [4]INSERM U30, Hôpital des Enfants Malades, Paris, France

Abstract

The insulin receptor kinase (IRK) is a tyrosine kinase whose activation, subsequent to insulin binding, is essential for insulin-signalling in target tissues. Insulin binding to its cell surface receptor is rapidly followed by internalization of insulin-IRK complexes into the endosomal apparatus (EN) of the cell. Internalization of insulin into target organs, especially liver, is implicated in effecting insulin clearance from the circulation. Internalization mediates IRK downregulation and hence attenuation of insulin sensitivity although most internalized IRKs readily recycle to the plasma membrane at physiological levels of insulin. A role for internalization in insulin signalling is indicated by the accumulation of activated IRKs in ENs. Furthermore, the maximal level of IRK activation has been shown to exceed that attained at the cell surface. Using an *in vivo* rat liver model in which endosomal IRKs are exclusively activated has revealed that IRKs at this intracellular locus are able by themselves to promote IRS-1 tyrosine phosphorylation and induce hypoglycemia. Furthermore, studies with isolated rat adipocytes reveal the EN to be the principle site of insulin-stimulated IRS-1 tyrosine phosphorylation and associated PI3K activation. Key steps in the termination of the insulin signal are also operative in ENs. Thus, an endosomal acidic insulinase has been identified which limits the extent of IRK activation. Furthermore, IRK dephosphorylation is effected in ENs by an intimately associated phosphotyrosine phosphatase(s) which, in rat liver, appears to regulate IRK activity in both a positive and negative fashion. Thus, insulin-mediated internalization of IRKs into ENs plays a crucial role in effecting and regulating signal transduction in addition to modulating the levels of circulating insulin and the cellular concentration of IRK in target tissues. (Mol Cell Biochem **182**: 59–63, 1998)

Key words: insulin receptor, internalization, endosomal apparatus, insulin degradation, insulin receptor dephosphorylation

Introduction

Many polypeptide hormones and growth factors, including insulin, mediate their biological effects by binding to and activating the intrinsic tyrosine kinase activity of their respective cell surface receptors. Ligand binding leads to receptor activation and the initiation of signal transduction pathways which control diverse physiological processes, including cell metabolism, differentiation and proliferation. Because these processes are critical for normal development and the maintenance of homeostasis, as well as the role they play in developmental disorders and neoplasia, it is important to determine the specific signal transduction pathways leading from these receptors and the mechanisms involved in their regulation in a physiological context.

The liver is composed primarily of quiescent and highly differentiated parenchymal cells which carry out a range of specialized functions including production of the majority of plasma proteins and the regulation of carbohydrate, urea, fatty acid, and cholesterol metabolism. The liver expresses the highest concentration of insulin receptors (> 10^5 per hepatocyte) [1] of any organ of the body, and is exposed via the portal circulation to major increases in insulin concentration in response to food intake during which it removes over 45% of the circulating insulin in one pass [2].

The insulin receptor is a type 1 transmembrane glycoprotein derived from a precursor that is proteolitically cleaved to yield α (135 kDa) and β (94 kDa) subunits which are linked covalently by disulfide bonds to form a heterotetramer. The

Address for offprints: J.J.M. Bergeron, Department of Anatomy and Cell Biology, Strathcona Anatomy and Dentristry Building, 3640 University Street, Montreal, Quebec, H3A 2B2, Canada

extracellular α-subunits contain the insulin binding site whereas the transmembrane β-subunits contain tyrosine kinase activity in their cytosolic domains. Insulin action in the liver, as in other sensitive target tissues (viz. adipose tissue and skeletal muscle), requires activation of the IRK and the tyrosine phosphorylation of key substrate molecules (viz. IRS-1, -2, etc.). These substrate molecules act as docking proteins which, via their phosphotyrosine motifs bind the SH2 and/or PTB domains of various signalling molecules (viz. PI3 Kinase, Grb2, Syp) to effect the insulin response. The elucidation of the processes by which the IRK promotes signalling is under intense scrutiny. One aspect of this process appears to involve the rapid internalization of activated IRK molecules into ENs.

IRK compartmentalization

In rat hepatocytes, IRKs are preferentially associated with surface microvilli [3]. Upon ligand binding, insulin-IRK complexes are sequestered from the plasma membrane (PM) and concentrate into a heterogeneous non-lysosomal population of tubulovesicular structures referred to as the endosomal apparatus [4, 5]. The endosomal apparatus is positioned temporally and physically between the PM and lysosomes where it carries out a number of functions including; (1) receptor sorting, to the PM (receptor recycling) or to lysosomes (receptor degradation), (2) ligand processing and targeting, and (3) signal transduction and termination [4, 5]. To date it is not known whether the endosomal apparatus exists as a set of discrete entities (viz. small vesicles, early ENs, late ENs), with the transference of material between components achieved by carrier vesicles, or whether there is a maturation of vesicles during the course of their intracellular itinerary.

Initially, ligand-receptor complexes are delivered from small vesicles into early endosomes (within 2–5 min of ligand binding) which are located at the cell periphery and have weakly acidic tubular elements (pH 6–6.5) [6, 7]. As ligand-receptor complexes appear sequentially in small vesicles, early ENs and then late ENs, an endosomal ATP-dependent proton pump generates an increasingly acidic intraluminal environment within the EN. For many ligand-receptor complexes, increasing acidification results in the dissociation of the ligand from its receptor. It appears that it is early in the endosomal pathway that a mechanism exists which sorts the receptors to be recycled to the PM from those targeted to the lysosome for degradation [8]. Late ENs (pH 5–6), accumulate internalized receptor complexes between 10–20 min following ligand binding to its receptor and consist of tubulovesicular structures of varying sizes located in the Golgi-lysosome area of the cell [6, 9].

Regulation of IRK activity by ENs

Insulin degradation

Internalization of the insulin-IRK complex constitutes the major mechanism for insulin degradation and the down-regulation of cell surface receptors. The acidic pH of the endocytic compartment promotes the dissociation of insulin from its receptor. Several studies have demonstrated that the hepatic endosome is a major site of degradation of insulin [10–12]. Degradation is initiated in early ENs as rapidly as 1 min following the intraportal injection of insulin into rats and is carried out by a recently identified endosomal acidic insulinase (EAI) located in the lumen of hepatic endosomes [12]. EAI demonstrates optimal activity between pH 5.0–5.5 and is distinct from insulin degrading enzyme (IDE), now recognized to be a peroxisomal protease, and hence unlikely to metabolize insulin in vivo [13, 14].

Although acidification augments the release of insulin from the receptor, it is not known whether EAI acts on receptor-bound insulin or whether the enzyme requires free insulin. However, the data of Doherty et al. [10] indicate that the inhibition of insulin release from the receptor reduces degradation of insulin. While insulin is degraded in the endosome, the insulin receptor may recycle back to the plasma membrane or translocate into lysosomes for degradation. Prolonged insulin stimulation, or receptor saturating doses of insulin, appears to cause the degradation of the internalized receptor in rat liver [8, 10, 11] leading to receptor down-regulation.

IRK dephosphorylation

Since the IRK maintains its phosphorylation state and tyrosine kinase activity following the dissociation of insulin from the α-subunit [15], dephosphorylation of specific β-subunit tyrosine residues is necessary to deactivate the intrinsic kinase activity of the receptor and attenuate insulin-signalling [16]. To date the identity of the specific protein tyrosine phosphatase(s) (PTPs) that mediate IRK dephosphorylation in vivo remain unclear. However, studies indicate that PTP activity against the IR is predominantly (~70%) located in isolated membrane fractions [17–19] with substantial activity observed in rat hepatic ENs. Treatment of endosomal fractions with Triton X-100 completely abolished IR dephosphorylation suggesting that the observed PTP(s) activity resulted from the action of an intrinsic endosomal enzyme closely associated with the IRK [20]. More recent studies utilizing peroxovanadium insulin-mimetics give further support to the importance of the EN for IRK dephosphorylation in vivo. Peroxovanadium (pV) compounds are potent PTP inhibitors that activate the IRK in a ligand-independent manner through the inhibition of IR-associated PTP activity [21]. When administered in vivo these agents activate the hepatic IRK, promote hypoglycemia [22, 23] and, depending on the compound, stimulate skeletal muscle glycogen synthesis [23]. Interestingly, pV compounds effect virtual complete inhibition of hepatic endosomal IRK dephosphorylation [24] whereas the activity of cytosolic PTP(s) [25] and the dephosphorylation of plasma membrane IRKs [24] are unaffected by this treatment. Although the mechanism by which pV compounds preferentially target endosomal PTP(s) is unclear, the potent insulin-mimetic effects of these compounds suggest that endosomally-located PTP(s) play a key, if not principal, role in mediating IRK dephosphorylation *in vivo*.

Internalization of IRKs into rat hepatic ENs is associated with a transient increase in both IRK autophosphorylating and exogenous tyrosine kinase activity [26, 27]. Similar observations have been made in isolated adipocytes [28, 29]. In rat liver, but not adipocytes, the increase in endosomal IRK activity corresponded with reduced phosphotyrosine content of the IRK β-subunit, compared to IRKs located at the plasma membrane. These observations suggest that IRK internalization in liver is associated with a partial dephosphorylation of phosphotyrosine residues. Thus full activation of the hepatic IRK may require limited dephosphorylation of the internalized IRK β-subunit [30]. Recent work, utilizing bpV(phen), a pV PTP inhibitor, to specifically block rat hepatic endosomal IRK dephosphorylation *in vivo* supports this hypothesis. Thus by inhibiting dephosphorylation of the internalized IRK, augmentation of IRK activity within this intracellular compartment was prevented [24]. The mechanism by which IRK activity is enhanced in ENs is unclear. However, since a number of studies suggest that the carboxy-terminal tyrosines of the IRK β-subunit may play a role in restraining IRK activity [31–33], endosomal PTP(s) may initially act to dephosphorylate these 'inhibitory' residues and hence increase IRK activity before the continuation of IRK β-subunit dephosphorylation leads to eventual inactivation [30]. Thus dephosphorylation of the rat liver IRK appears to be a complex process whereby PTPs may modulate IRK activity in both a positive and negative fashion.

Role of ENs in signalling

Mounting evidence suggests that for a number of growth factors and hormones, including insulin, EGF, TNF, PDGF and NGF, endocytosis plays a critical role in extending and/or initiating signalling events at sites removed from the PM and within the larger cytosolic volume of the cell [34, 35]. By increasing the surface area of activated receptors that come into contact with cytosolic substrates, ENs are thought to amplify and extend the temporal window for signal transduction. This is exemplified by EGF receptor (EGFR) signalling where activation leads to greater and prolonged EGFR tyrosine phosphorylation in rat liver endosomal membranes (compared to EGFRs located at the PM) that is mirrored by elevated SHC phosphorylation and SHC/GRB2 association at this intracellular locus [36]. More recent studies with endocytosis-defective cells reveal that EGFR internalization is also necessary for full activation of mitogen-activated protein kinase (MAPK) [37]. Internalization has the potential to allow access of activated receptors to substrates that may not reside at the PM/cytosol interface and consequently are inaccessible to receptors located at the cell surface. This is illustrated by the TNF receptor (TNFR) where ligand-induced internalization of TNFR in human T-cells (Jurkat cells) stimulates an exclusively endosomal acidic sphingomyelinase resulting in the activation of the transcription factor NF-kB [38].

Studies in rat liver [26, 39], hepatoma cells [27] and isolated adipocytes [28, 29, 40] have established that insulin treatment results in the accumulation and concentration within ENs of IRKs that are both tyrosine phosphorylated and active towards exogenous substrates. Indeed, as discussed above, there exists a transient period in rat liver where endosomal IRK activity is elevated relative to that observed at the plasma membrane. Use of an *in vivo* rat liver model in which endosomally-located IRKs are exclusively activated has revealed that IRKs in this intracellular compartment are able to promote tyrosine phosphorylation of insulin receptor substrate-1 (IRS-1), and induce hypoglycemia [23]. Moreover, studies in adipocytes have implicated ENs as the principle, if not exclusive, site of insulin-stimulated IRS-1 tyrosine phosphorylation and associated PI3 kinase activation *in vivo* [29]. Thus following insulin treatment of isolated rat adipocytes it was observed that; (1) ~75% of phosphotyrosine immunoprecipitable PI3K activity was detected in low density microsomes (viz. ENs) [41], (2) levels of tyrosine phosphorylated IRS-1 and PI3K activity were 10 fold greater in microsomes than at the plasma membrane [41] and, (3) the time course of accumulation of internalized IRKs closely paralleled the time course of IRS-1 tyrosine phosphorylation [29]. Taken together the observations in rat liver and isolated adipocytes provide compelling evidence supporting a critical role for receptor endocytosis in normal insulin signal transduction *in vivo* (Fig. 1) and suggest that ENs play an important part in mediating a number of insulin's biological effects.

Conclusions

These observations suggest that the internalization of insulin-IRK complexes to ENs is key in effecting signal transduction. Taken together our studies and those of others indicate that the endosomal apparatus constitutes a central site at which insulin signal transduction is regulated in that insulin degradation, IRK dephosphorylation, and the targeting of IRK

62

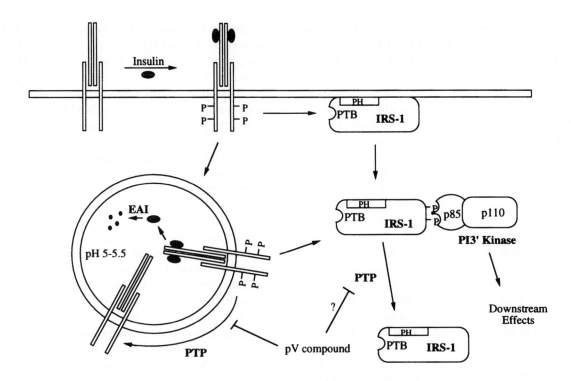

Fig. 1. Compartmentalized insulin receptor signal transduction in rat liver. Insulin binding to its cell surface receptor results in the rapid phosphorylation and activation of the insulin receptor tyrosine kinase (IRK) and internalization of the insulin-IRK complexes into endosomes. IRS-1 may be phosphorylated by activated IRKs at both the PM and ENs. PI3 kinase associates with phosphorylated IRS-1 and initiates further downstream effects. Partial dephosphorylation of the IRK at the EN locus promotes a transient augmentation of IRK activation before IRK activity is attenuated by further dephosphorylation. This process is coupled to insulin degradation by an acidic endosomal insulinase. *Abbreviations*: EAI – endosomal acidic insulinase; PTP – protein tyrosine phosphatase; -P – phosphotyrosine residue; IRK – insulin receptor kinase; IRS-1 – insulin receptor substrate-1; PH – pleckstrin homology domain; PTB – phosphotryrosine binding domain.

to lysosomes all operate at the endosomal level. Thus the internalized activated IRK operates within a temporal window to effect signalling following which mechanisms come into play to attenuate signalling.

Acknowledgements

G.M.D.G. is a research student of the National Cancer Institute of Canada supported with funds provided by the Canadian Cancer Society. P.G.D. was previously supported by a Fellowship from the Royal Victoria Hospital Research Institute (Montreal, Quebec).

References

1. Kahn CR, Freychet P, Roth J, Neville D Jr: Quantitative aspects of the insulin-receptor interaction in liver plasma membranes. J Biol Chem 249: 2249–2257, 1974
2. Jaspan JB, Polonsky KS, Lewis M, Pensler J, Pugh W, Moossa AR, Rubenstein AH: Hepatic metabolism of glucagon in the dog: Contribution of the liver to overall metabolic disposal of glucagon. Am J Physiol 240: E233–E244, 1981
3. Bergeron JJM, Sikstrom R, Hand AR, Posner BI: Binding and uptake of ^{125}I-insulin into rat liver hepatocytes and endothelium. An *in vivo* radioautographic study. J Cell Biol 80: 427–443, 1979
4. Burgess JW, Bevan AP, Bergeron JJM, Posner BI: Intracellular trafficking and processing of ligand receptor complexes in the endosomal system. Exp Clin Endocrinol 11: 67–78, 1992
5. Bergeron JJM, Cruz J, Khan MN, Posner BI: Uptake of insulin and other ligands into receptor-rich endocytic components of target cells: the endosomal apparatus. Ann Rev Physiol 47: 383–403, 1985
6. Wall DA, Hubbard AL: Receptor-mediated endocytosis of asialo-glycoproteins by rat liver hepatocytes: Biochemical characterization of the endosomal compartments. J Cell Biol 101: 2104–2112, 1985
7. Mellman 1, Fuchs R, Helenius A: Acidification of the endocytic and exocytic pathways. Ann Rev Biochem 55: 663–700, 1986
8. Lai WH, Cameron PH, Wada I, Doherty JJ II, Kay DG, Posner BI, Bergeron JJM: Ligand-mediated internalization, recycling, and downregulation of the epidermal growth factor receptor *in vivo*. J Cell Biol 109: 2741–2749, 1989
9. Dunn WA, Hubbard AL: Receptor-mediated endocytosis of epidermal growth factor by hepatocytes in the perfused rat liver: Ligand and receptor dynamics. J Cell Biol 98: 2148–2159, 1984
10. Doherty JJ II, Kay DG, Lai WH, Posner BI, Bergeron JJM: Selective degradation of insulin within rat liver endosomes. J Cell Biol 110: 35–42, 1990
11. Backer JM, Kahn CR, White MF: The dissociation and degradation of internalized insulin occur in the endosomes of rat hepatoma cells. J Biol Chem 265: 14828–14835, 1990

12. Authier F, Rachubinski RA, Posner BI, Bergeron JJM: Endosomal proteolysis of insulin by an acidic thiol metalloprotease unrelated to insulin degrading enzyme. J Biol Chem 269: 3010–3016, 1994

13. Authier F, Bergeron JJM, Ou WJ, Rachubinski RA, Posner BI, Walton PA: Degradation of the cleaved leader peptide of thiolase by a peroxisomal proteinase. Proc Natl Acad Sci USA 92: 3859–3863, 1995

14. Authier F, Posner BI, Bergeron JJM: Insulin-degrading enzyme. Clin Invest Med 19: 149–160, 1996

15. Rosen OM, Herrera R, Olowe Y, Petruzzelli LM, Cobb MH: Phosphorylation activates the IR tyrosine protein kinase. Proc Natl Acad Sci USA 80: 3237–3240, 1983

16. Sale G: Serine/threonine kinases and tyrosine phosphatases that act on the insulin receptor. Biochem Soc Trans 20: 664–670, 1992

17. King MJ, Sale GJ: Insulin-receptor phosphotyrosyl-protein phosphatases. Biochem J 256: 893–902, 1988

18. Ahmad F, Goldstein BJ: Purification, identification and subcellular distribution of three predominant protein-tyrosine phosphatase enzymes in skeletal muscle tissue. Biochim Biophys Acta 1248: 57–69, 1995

19. Meyerovitch J, Backer JM, Csermely P, Shoelson SE, Kahn CR: Insulin differentially regulates protein phosphotyrosine phosphatase activity in rat hepatoma cells. Biochem 31: 10338–10344, 1992

20. Faure R, Baquiran G, Bergeron JJM, Posner BI: The dephosphorylation of insulin and epidermal growth factor receptors. The role of endosome-associated phosphotyrosine phosphatase(s). J Biol Chem 267: 11215–11221, 1992

21. Bevan AP, Drake PG, Yale J-F, Shaver A, Posner BI: Peroxovanadium compounds: Biological actions and mechanism of insulin-mimesis. Mol Cell Biochem 153: 49–58, 1995

22. Posner BI, Faure R, Burgess JW, Bevan AP, Lachance D, Zhang-Sun G, Fantus IG, Ng JN, Hall DA, Soo Lum B, Shaver A: Peroxovanadium compounds. A new class of potent phosphotyrosine phosphatase inhibitors which are insulin mimetics. J Biol Chem 269: 4596–4604, 1994

23. Bevan AP, Burgess JW, Yale J-F, Drake PG, Lachance D, Baquiran G, Shaver A, Posner BI: In vivo insulin mimetic effects of pV compounds: Role for tissue targeting in determining potency. Am J Physiol 268: E60–E66, 1995

24. Drake PG, Bevan AP, Burgess JW, Bergeron JJM, Posner BI: A role for tyrosine phosphorylation in both activation and inhibition of the insulin receptor tyrosine kinase in vivo. Endocrinology 137: 4960–4968, 1996

25. Bevan AP, Burgess JW, Drake PG, Shaver A, Bergeron JJM, Posner BI: Selective activation of the rat hepatic endosomal insulin receptor kinase. Role for the endosome in insulin signalling. J Biol Chem 270: 10784–10791, 1995

26. Khan MN, Baquiran G, Brule C, Burgess J, Foster B, Bergeron JJM, Posner BI: Internalization and activation of the rat liver insulin receptor kinase in vivo. J Biol Chem 264: 12931–12940, 1989

27. Backer JM, Kahn CR, White MF: Tyrosine phosphorylation of the insulin receptor during insulin-stimulated internalization in rat hepatoma cells. J Biol Chem 264: 1694–1701, 1989

28. Klein HH, Freidenberg GR, Matthaei S, Olefsky JM: Insulin receptor kinase following internalization in isolated rat adipocytes. J Biol Chem 262: 10557–10564, 1987

29. Kublaoui B, Lee J, Pilch PF: Dynamics of signalling during insulin-stimulated endocytosis of its receptor in adipocytes. J Biol Chem 270: 59–65, 1995

30. Burgess JW, Wada I, Ling N, Khan MN, Bergeron JJM, Posner BI: Decrease in β-subunit phosphotyrosine correlates with internalization and activation of the endosomal insulin receptor kinase. J Biol Chem 267: 10077–10086, 1992

31. Bemier M, Llotta AS, Kole HK, Shock DD, Roth J: Dynamic regulation of intact and C-terminal truncated insulin receptor phosphorylation in permeablized cells. Biochem 33: 4343–4351, 1994

32. Tavare JM, Ramos P, Ellis L: An assessment of human insulin receptor phosphorylation and exogenous kinase activity following deletion of 69 residues from the carboxy -terminus of the receptor b-subunit. Biochem Biophys Res Comm 188: 86–93, 1992

33. Kaliman P, Baron V, Alengrin F, Takata Y, Webster NJG, Olefsky JM, Van Obberghen E: The insulin receptor C-terminus is involved in the regulation of the receptor kinase activity. Biochemistry 32: 9539–9544, 1993

34. Baass PC, Di Guglielmo GM, Authier F, Posner BI, Bergeron JJM: Compartmentalized signal transduction by receptor tyrosine kinases. Trends Cell Biol 5: 465–470, 1995

35. Bevan AP, Drake PG, Bergeron JJM, Posner BI: Intracellular signal transduction-The role of endosomes. Trends Endocrinol Metab 7: 13–21, 1996

36. Di Guglielmo GM, Baass PC, Ou WJ, Posner BI, Bergeron JJM: Compartmentalization of SHC, GRB2 and mSOS, and hyperphosphorylation of Raf-1 by EGF but not insulin in liver parenchyma. EMBO J 13: 4269–4277, 1994

37. Vieira AV, Lamaze C, Schmid SL: Control of EGF receptor signalling by clathrin-mediated endocytosis. Science 274: 2086–2089, 1996

38. Wiegmann K, Schutze S, Machleidt T, Witte D, Kronke M: Functional dichotomy of neutral and acidic sphingomyelinases in tumor necrosis factor signalling. Cell 78: 1005–1015, 1994

39. Khan MN, Savoie S, Bergeron JJM, Posner BI: Characterization of rat liver endosomal structures. In vivo activation of insulin-stimulable receptor kinase in these structures. J Biol Chem 261: 8462–8472, 1986

40. Wang B, Balba Y, Knutson VP: Insulin-induced in situ phosphorylation of the insulin receptor located in the plasma membrane versus endosomes. Biochem Biophys Res Comm 227: 27–34, 1996

41. Kelly KL, Ruderman NB: Insulin-stimulated phosphatidylinositol 3-kinase: Association with a 185 kDa tyrosine phosphorylated protein (IRS-1) and localization in a low density membrane vesicle. J Biol Chem 268: 4391–4398, 1993

Molecular and Cellular Biochemistry **182**: 65–71, 1998.

Spatial determinants of specificity in insulin action

Cynthia Corley Mastick,[1] Matthew J. Brady,[1] John A. Printen,[1, 2] Vered
Ribon[1, 2] and Alan R. Saltiel[1, 2]
[1]*Department of Cell Biology, Parke-Davis Pharmaceutical Research Division, Warner-Lambert Company, Ann Arbor,
MI;* [2]*Department of Physiology, University of Michigan School of Medicine, Ann Arbor, MI, USA*

Abstract

Insulin is a potent stimulator of intermediary metabolism, however the basis for the remarkable specificity of insulin's stimulation of these pathways remains largely unknown. This review focuses on the role compartmentalization plays in insulin action, both in signal initiation and in signal reception. Two examples are discussed: (1) a novel signalling pathway leading to the phosphorylation of the caveolar coat protein caveolin, and (2) a recently identified scaffolding protein, PTG, involved directly in the regulation of enzymes controlling glycogen metabolism. (Mol Cell Biochem **182**: 65–71, 1998)

Key words: caveolae, caveolin, DARPP32, glycogen, protein phosphorylation, *P*rotein *T*argeting to *G*lycogen (PTG), type I protein serine/threonine phosphatase (PP1)

Introduction

Insulin is the most potent physiological anabolic agent known, promoting the synthesis and storage of carbohydrates, lipids, and proteins, and preventing their degradation and release back into the circulation. Despite years of intense investigation, there remains considerable uncertainty regarding the precise intracellular events that mediate the actions of this hormone. One confounding factor has been the variety of actions of insulin, which depend upon cell type, time of exposure, and the presence of other hormones [1]. Another is the fact that insulin can act as a growth factor for cells in culture, and shares many of the mitogenic signalling pathways elicited by other growth factors. However, the metabolic effects of insulin are unique, and cannot be reproduced with other cellular stimuli [2–5]. Taken together, these findings indicate that signalling mechanisms exist that respond uniquely to insulin, which allow for the specialized effects of insulin on metabolism.

Like receptors for many growth factors, the insulin receptor is a tyrosine kinase that undergoes activation upon insulin binding, leading to the tyrosine phosphorylation of a specific collection of intracellular proteins [6, 7]. Receptor activation leads to a variety of signalling pathways that diverge at or near the insulin receptor itself. While receptor autophosphorylation and the subsequent substrate phosphorylations are clearly required for the full expression of insulin's actions, many of these substrates and subsequent downstream pathways are shared by other receptors that do not mimic insulin's effects on metabolism. A number of hypotheses have been proposed to account for the signalling specificity underlying insulin action, including differences in the strength or duration of signals, combinatorial diversity of signals, or other unique features of these pathways in insulin responsive cells. Another potential mechanism to explain insulin's unique actions may lie in the spatial compartmentalization of signal transduction [8, 9], both regarding signal initiation at the plasma membrane, as well as signal reception inside the cell. In this review we provide two examples of how the subcellular targeting of signalling molecules may contribute a key ingredient in determining the specificity of insulin action.

Specificity in signal initiation: the role of caveolae

One clue to understanding how insulin can elicit its unique actions may lie in the compartmentalization of the signalling molecules themselves. Recent studies have suggested that signal initiation may he functionally segregated into distinct domains of the plasma membrane. Caveolae may represent one type of specialized region of the plasma membrane that

Address for offprints: A.R. Saltiel, Parke-Davis Pharmaceutical Research Division, Warner-Lambert Company, Ann Arbor, MI 48105, USA

is crucial for specificity in signal transduction. Caveolae are small invaginations of the plasma membrane with unique protein and lipid compositions [10–13]. These structures are clearly distinguished from clathrin-coated pits by their characteristic striated coat, which is made up largely of the caveolins [14, 15]. Caveolins are members of a multigene family with three known members [16–20]. In many cell types, caveolin copurifies with or binds to signalling molecules [21–35], suggesting that one potential function of caveolae lies in signal transduction.

Although caveolae have generated much recent interest, their exact function and protein composition remain controversial [36–39]. Other membrane domains with biophysical properties similar to caveolae, such as clathrin coated pits, can be contaminants in purified caveolar fractions [40, 41], so that the precise molecular composition of these structures is uncertain. Moreover, cellular membranes contain sub-domains termed lipid ordered domains or glycolipid rafts [42, 43]. These sub-domains, characterized by their Triton-insolubility, are highly enriched in cholesterol, glycolipids and sphingolipids, and de-enriched in phospholipids. Many proteins are localized to the lipid ordered domains by virtue of post-translational modifications. GPI anchors have been shown to be necessary and sufficient for the localization of proteins to these domains, as has tandem acylation of the Src-family kinases and G-proteins [44–47]. Caveolin itself is also acylated [48]. Although the extent of overlap between caveolae and glycolipid rafts is unknown, it is clear that caveolae form within these lipid ordered domains, perhaps due to the fact that caveolin is a cholesterol binding protein [49]. While glycolipid rafts form in the absence of caveolin [50, 51], expression of caveolins is both necessary and sufficient to induce the formation of caveolae [52–54]. Independently of whether all of the proteins which co-purify with caveolins are actually found within the caveolar structures, the membrane-bound proteins in the lipid ordered domains are found in close proximity to the caveolae, which form within these domains [38].

Caveolae are abundant in adipocytes and muscle, covering a significant fraction of the inner surface of the plasma membrane of adipocytes [55]. Of the four known forms of caveolin, caveolins-1 (α and β) and -2 are most highly expressed in adipocytes, followed by lung and muscle [17, 19, 56], while caveolin-3 appears to be a muscle specific isoform [20, 57]. In addition, caveolin-1 expression increases upon adipocyte differentiation [56, 59]. The abundance of caveolae and caveolins in adipocytes and muscle, together with their potential role in signalling, suggested a possible role in insulin signal transduction. Consistent with an important role for caveolae in insulin action, insulin specifically stimulates the tyrosine phosphorylation of the two forms of caveolin (22 and 24 kD) and an additional unidentified protein of 29 kD that coimmunoprecipitates with caveolin after disruption of the caveolar complexes with octylglucoside [37]. This phosphorylation does not occur in response to other growth factors in adipocytes. In addition, caveolin phosphorylation occurs only in cells in which metabolism is highly responsive to insulin. For example, it occurs only in the fully differentiated 3T3-L1 adipocytes, not in the preadipocytes, despite the expression of both caveolin and the insulin receptor in both cell types [59].

Although insulin specifically increases the tyrosine phosphorylation of caveolin, several lines of evidence indicate that caveolin is not a direct substrate of the insulin receptor. Analysis of caveolin-enriched fractions from 3T3-L1 adipocytes revealed that the caveolin kinase is the Src-family kinase Fyn [37, 59]. Fyn colocalizes with caveolin in low density, Triton-insoluble complexes in both preadipocytes and adipocytes, and it is the only detectable tyrosine kinase in these fractions. Moreover, overexpression of wild type Fyn leads to an increase in the basal level of tyrosine phosphorylation of caveolin, and hyperphosphorylation of caveolin in response to insulin [59]. A dominant negative Fyn construct blocked caveolin phosphorylation, although this might reflect an indirect effect on differentiation. Interestingly, the caveolin kinase in primary adipocytes appears to be the Src-family kinase Lyn, rather than Fyn. Lyn is expressed at high levels in primary adipocytes, and is colocalized with caveolin in low density, Triton-insoluble complexes prepared from these cells (unpublished observations).

How is the caveolar Fyn activated in response to insulin? Fyn and other Src-family kinases can be regulated either through the dephosphorylation of a regulatory tyrosine in the carboxy terminus of the protein [60], or through the occupancy of their SH2 domains by tyrosine phosphorylated proteins [61]. The autophosphorylation of an additional tyrosine is required for full activation of these kinases. No net change in the phosphorylation of Fyn has been observed in response to insulin in adipocytes, although there is a significant increase in constitutive phosphorylation of Fyn after adipocyte differentiation [59]. In addition, tyrosine phosphorylated peptides which bind specifically to the SH2 domains of Fyn stimulate the caveolin kinase activity in caveolar fractions, suggesting that the insulin receptor phosphorylates a specific substrate protein, which in turn activates the caveolar Fyn.

The unique conditions under which caveolin phosphorylation is observed suggest that the insulin receptor substrate responsible for the activation of caveolar Fyn would have several unique properties: (1) phosphorylation showing specificity for insulin in adipocytes; (2) association with Fyn (through an SH2 domain interaction) in response to tyrosine phosphorylation; (3) translocation of the phosphorylated protein into caveolae in response to insulin; and (4) phosphorylation in 3T3-L1 adipocytes, but not in the preadipocytes. While the well characterized insulin receptor substrates IRS-1/-2 and Shc emerged as potential candidates [62, 63], none fulfilled all of these requirements. In contrast, the proto-oncogene product c-Cbl shares many properties with this presumed substrate protein. Insulin specifically stimulates the phosphorylation of c-Cbl in adipocytes [64]. As has been observed in lymphocytes, Cbl constitutively binds to Fyn in unstimulated adipocytes through an SH3 domain-mediated interaction, and insulin-stimulated tyrosine phosphorylation of Cbl increases this association. The Cbl which is phosphorylated in response to insulin binds specifically to fusion proteins containing the SH2 domain of Fyn, but surprisingly not SHP-2 or PI3 kinase. Cbl is translocated into caveolin-enriched Triton-insoluble complexes after insulin stimulation, and the Cbl in these complexes is tyrosine phosphorylated [59]. Most interestingly, unlike the other known substrates of the insulin receptor such as IRS-1/-2 and Shc, Cbl phosphorylation is specific for the differentiated adipocyte phenotype. In addition, tyrosine phosphorylation of Cbl is not observed in tissue culture cell lines which have been engineered to express high levels of the insulin receptor, while the other substrate proteins are readily phosphorylated in these cells [64].

The basis for the cell-type specificity of Cbl phosphorylation is currently unknown. Cbl is expressed at comparable levels in both the preadipocytes and adipocytes. Unlike IRS-1/-2 and Shc, Cbl does not contain a PTB domain, which has been identified in substrates that interact directly with the insulin receptor [65–67]. In addition, in contrast to IRS-1 and Shc, Cbl does not interact with the insulin receptor in the yeast two hybrid system. We hypothesize that tyrosine phosphorylation of Cbl may require a specific adaptor protein which allows for the interaction of Cbl with the insulin receptor, and that it is the regulation of the expression of this protein that accounts for the coordinate regulation of the phosphorylations of Cbl and caveolin in response to insulin in adipocytes. Although the model linking phosphorylation of Cbl to the phosphorylation of caveolin through the activation of Fyn is compelling (Fig. 1), the exact relationship between these three proteins is likely to be complex [68]. For example, Cbl may undergo processive phosphorylation after the activation of Fyn [69, 70]. However, the insulin-dependent association of Cbl and Fyn in caveolae leads to the intriguing possibility that the Fyn/Cbl pathway may have a unique function in adipocytes (phosphorylation of caveolin) due to the localization of these proteins to specialized domains of the plasma membrane.

The precise role of caveolin phosphorylation in insulin action remains uncertain. Preliminary attempts to identify proteins which interact with caveolin in a phosphorylation-dependent manner have thus far proven unsuccessful. However, the correlation between caveolin phosphorylation and the metabolic activities of insulin [37, 59], the lack of

phosphorylation with other tyrosine kinase receptors [37, 59], and a number of reports describing numerous signalling molecules in caveolae or related membrane domains [21–35] suggest that this phosphorylation event may represent an important mechanism for the segregation of early signalling events in insulin action, perhaps as a way to ensure signalling specificity.

Spatial compartmentalization in signal reception: the regulation of glycogen metabolism

It has long been recognized that protein dephosphorylation plays an essential role in regulating the metabolic effects of insulin [71]. Indeed, many of the rate-limiting enzymes involved in glycogen metabolism, such as glycogen synthase, glycogen phosphorylase and phosphorylase kinase are regulated by their phosphorylation state. Insulin causes the dephosphorylation of these enzymes, which results in the activation of glycogen synthase and the inactivation of phosphorylase and phosphorylase kinase. This leads to a significant increase in the rate of glycogen synthesis. The dephosphorylation of these three enzymes is mediated through the activation of the type I serine/threonine phosphatase PP1 in response to insulin. However, insulin promotes the net dephosphorylation of only a specific subset of proteins, while PP1 and additional substrates for this enzyme are found ubiquitously in nearly all cellular compartments. Therefore, a mechanism must exist for the stimulation of discrete PP1 activities by insulin.

The specificity with which insulin stimulates the dephosphorylation of only selected substrates of PP1 suggests that there are distinct functional pools of PP1 in cells. Tissue specific targeting subunits have been identified that localize the catalytic subunit of PP1 (PP1C) to specific sites, conferring both insulin sensitivity to the enzyme, as well as substrate specificity. Several glycogen targeting proteins have been described, including G_m which is expressed in both heart and skeletal muscle [72], and G_L, the hepatic subunit [73]. While the precise functions of these proteins remains uncertain, recent experiments on a related glycogen localizing subunit of PP1C may help to explain how insulin can promote the rapid dephosphorylation of the specific group of proteins involved in glycogen metabolism. This protein, called PTG for *Protein Targeting to Glycogen*, was identified in a two-hybrid screen of a 3T3-L1 adipocyte library using PP1C as the bait (74; also called R5, 75). PTG encodes a 33 kD protein which is found in all insulin responsive tissues. In addition to localizing PP1C to the glycogen pellet, PTG forms specific complexes with the PP1-regulated enzymes that control glycogen metabolism, including glycogen synthase, phosphorylase, and phosphorylase kinase. PTG thus increases PP1 specific activity

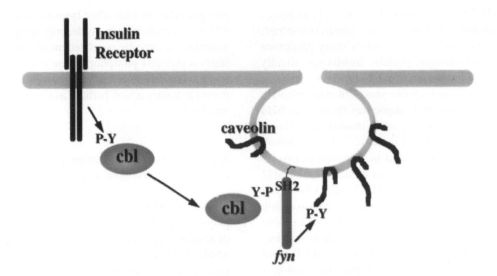

Fig. 1. Stimulation of the tyrosine phosphorylation of caveolin by insulin. Activation of the insulin receptor tyrosine kinase leads to the tyrosine phosphorylation of Cbl. Phosphorylated Cbl binds to the SH2 phosphotyrosine binding domain of Fyn, which is resident in the caveolae. This activates the kinase, resulting in the tyrosine phosphorylation of substrate proteins in these complexes such as caveolin.

against glycogen-bound substrates not only by targeting the PP1 to glycogen, but also by directly interacting with PP1 substrate proteins. This protein, therefore, may assemble glycogen synthase, phosphorylase, phosphorylase kinase, and PP1C onto the glycogen particle, generating a metabolic module for the localized reception of the appropriate intracellular signals [74] (Fig. 2). It is still unclear, however, whether all of the glycogen metabolizing enzymes are bound to a single PTG molecule simultaneously, or if the binding sites are shared between one or more proteins. It is expected that the PP1 at this compartmentalized site is sensitive to activation by insulin, while the non-responsive PP1 in other subcellular compartments is not.

The mechanisms by which insulin specifically activates glycogen-targeted PP1 are still unclear. At one time speculation centered around the phosphorylation of two closely spaced serine residues in the amino terminus of G_m, by the MAP kinase cascade [76 , 77]. However, it is now clear that this model has shortcomings. MAP kinase-activation is neither necessary nor sufficient for the activation of glycogen synthase or PP1 by insulin [2–5, 78]. In addition, the phosphorylation sites in G_m are not conserved in the other glycogen targeting subunits [74]. While the precise mechanisms involved in PP1 activation by insulin remain uncertain, one attractive hypothesis involves dis-inhibition of PP1 by the regulated dissociation of an inhibitory subunit. Studies suggest that regulation of PP1 basal activity contributes to insulin responsiveness, and that the low activity in the basal state is maintained by phosphorylated inhibitory peptides such as inhibitor-1 and DARPP-32 [79, 80]. For example, we have observed that

adipocyte differentiation leads to a decrease in basal PP1 activity, and a significant increase in the stimulation of PP1 activity by insulin [81]. PP1 levels do not change during differentiation, while expression of the PP1 inhibitor peptide DARPP-32 is dramatically increased during adipo-genesis [81], as is the PTG targeting subunit [82]. The DARPP-32 in 3T3-L1 adipocytes is exclusively localized to the particulate fraction [81], which includes the glycogen pellet. We are currently testing the hypothesis that insulin treatment leads to the dissociation of DARPP-32 from glycogen targeted, PTG-bound PP1 (Fig. 3). In addition,

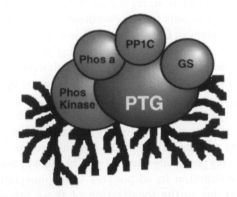

Fig. 2. PTG may act as a molecular scaffold for the hormonal control of glycogen synthesis. The catalytic subunit of PP1 (PP1C) is targeted to glycogen by PTG in 3T3-L1 adipocytes. PTG also serves as a glycogen scaffolding protein, co-localizing PP1 with its substrate enzymes which control glycogen metabolism, including glycogen synthase (GS), phosphorylase a (Phos a), and phosphorylase kinase (Phos Kinase). This may serve as a metabolic module for the localized reception of intracellular signals which regulate glycogen metabolism.

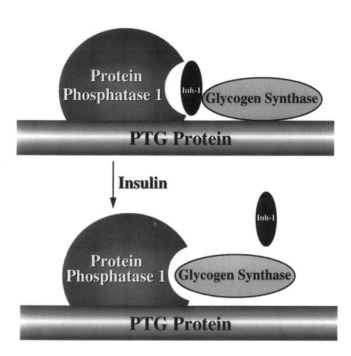

Fig. 3. Activation of glycogen-targeted PP1 by insulin. Glycogen-targeted PP1 activity may be regulated by the binding of inhibitory peptides, such as inhibitor 1 (inh-1) or DARPP-32. Insulin stimulation may cause the dephosphorylation and/or disassociation of bound inhibitor. Dis-inhibition of PP1 would result in the dephosphorylation and activation of glycogen synthase, mediating the insulin-stimulated increase in glucose storage as glycogen.

attempts are underway to determine whether dissociation of DARPP-32 is sufficient to account for the specific insulin-induced dephosphorylation of enzymes involved in glycogen metabolism. It is expected that lipid metabolism is similarly regulated through additional, lipid-specific PP1 scaffolding proteins.

Conclusions

We have focused much of our recent work on trying to understand the mechanisms that account for the remarkable specificity of insulin's regulation of intermediary metabolism. This specificity may be the result of the compartment-alization of both signal initiation at the cell surface as well as signal reception at the glycogen pellet or lipid droplet. We have identified a novel, insulin-stimulated signalling pathway which results in the tyrosine phosphorylation of caveolin, a structural component of unique plasma membrane domains termed caveolae. In addition, we have identified a novel targeting protein, PTG, which forms a distinct, insulin-sensitive pool of PP1 complexed with the enzymes regulating glycogen metabolism. It is still unknown whether these two pathways are linked. However, exploration of the molecular details underlying the specificity of signal initiation and signal reception in insulin action may elucidate the complex biochemical pathways leading from the binding

of insulin at the cell surface to the regulation of intermediary metabolism.

References

1. Saltiel AR: Diverse signaling pathways in the cellular actions of insulin. Am J Physiol 270: E375–E385, 1996
2. Robinson LJ, Razzack ZF, Lawrence JC, James DE: Mitogen-activated protein kinase activation is not sufficient for stimulation of glucose transport or glycogen synthase in 3T3-L1 adipocytes. J Biol Chem 268: 26422–26427, 1993
3. Lin T-A, Lawrence JC: Activation of ribosomal protein S6 kinase does not increase glycogen synthesis or glucose transport in rat adipocytes. J Biol Chem 269: 21255–21261, 1994
4. Wiese RJ, Mastick CC, Lazar DF, Saltiel AR: Activation of mitogen-activated protein kinase and phosphatidylinositol 3'-kinase is not sufficient for the hormonal stimulation of glucose uptake, lipogenesis, or glycogen synthesis in 3T3-L1 adipocytes. J Biol Chem 270: 3442–3446, 1995
5. Azpiazu I, Saltiel AR, DePaoli-Roach AA, Lawrence JC: Regulation of both glycogen synthase and PHAS-1 by insulin in rat skeletal muscle involves mitogen-activated protein kinase-independent and rapamycin-sensitive pathways. J Biol Chem 271: 5033–5039, 1996
6. Kahn CR, White MF: The insulin receptor and the molecular mechanism of insulin action. J Clin Invest 82: 1151–1156, 1988
7. White MF, Kahn CR: The insulin signaling system. J Biol Chem 269: 1–4, 1994
8. Mochly-Rosen D: Localization of protein kinases by anchoring proteins: A theme in signal transduction. Science 268: 247–251, 1995
9. Pawson T, Scott JD: Signaling through scaffold, anchoring, and adaptor proteins. Science 278: 2075–2080, 1997

10. Anderson RGW, Kamen BA, Rothberg KG, Lacey SW: Potocytosis: Sequestration and transport of small molecules by caveolae. Science 255: 410–411, 1992

11. Travis J: Cell biologists explore 'tiny caves'. Science 262: 1208–1209, 1993

12. Anderson RGW: Caveolae: Where incoming and outgoing messengers meet. Proc Natl Acad Sci USA 90: 10909–10913, 1993

13. Parton RG, Simons K: Digging into caveolae. Science 269: 1398–1399, 1995

14. Rothberg KG, Heuser JE, Donzell WC, Ying Y-S, Glenney JR, Anderson RGW: Caveolin, a protein component of caveolae membrane coats. Cell 68: 673–682, 1992

15. Dupree P, Parton RG, Raposo G, Kurzchalia TV, Simons K: Caveolae and sorting in the *trans*-Golgi network of epithelial cells. EMBO J 12: 1597–1605, 1993

16. Glenney JR, Soppet D: Sequence and expression of caveolin, a protein component of caveolae plasma membrane domains phosphorylated on tyrosine in Rous sarcoma virus-transformed fibroblasts. Proc Natl Acad Sci USA 89: 10517–10521, 1992

17. Glenney JR: The sequence of human caveolin reveals identity with VIP21, a component of transport vesicles. FEBS Lett 314: 45–48, 1992

18. Scherer PE, Tang Z, Chun M, Sargiacomo M, Lodish HF, Lisanti MP: Caveolin isoforms differ in their N-terminal protein sequence and subcellular distribution. J Biol Chem 270: 16395–16401, 1995

19. Scherer PE, Okamoto T, Chun M, Nishimoto I, Lodish HF, Lisanti MP: Identification, sequence, and expression of caveolin-2 defines a caveolin gene family. Proc Natl Acad Sci USA 93: 131–135, 1996

20. Tang Z, Scherer PE, Okamoto T, Song K, Chu C, Kohtz DS, Nishimoto I, Lodish HF, Lisanti MP: Molecular cloning of caveolin-3, a novel member of the caveolin gene family expressed predominantly in muscle. J Biol Chem 271: 2255–2261, 1996

21. Sargiacomo M, Sudol M, Tang ZL, Lisanti MP: Signal transducing molecules and glycosyl-phosphatidylinositol-linked proteins form a caveolin-rich insoluble complex in MDCK cells. J Cell Biol 122: 789–807, 1993

22. Lisanti MP, Tang ZL, Sargiacomo M: Caveolin forms a hetero-oligomeric protein complex that interacts with an apical GPI-linked protein: Implications for the biogenesis of caveolae. J Cell Biol 123: 595–604, 1993

23. Lisanti MP, Scherer PE, Vidugiriene J, Tang ZL, Hermanowski-Vosatka A, Tu Y-H, Cook RF, Sargiacomo M: Characterization of caveolin-rich membrane domains isolated from an endothelial-rich source: Implications for human disease. J Cell Biol 126: 111–126, 1994

24. Chang W-J, Ying Y-S, Rothberg KG, Hooper NM, Turner AJ, Gambliel HA, De Gunzburg J, Mumby SM, Gilman AG, Anderson RGW: Purification and characterization of smooth muscle cell caveolae. J Cell Biol 126: 127–138, 1994

25. Chun M, Liyanage UK, Lisanti MP, Lodish HF: Signal transduction of a G protein-coupled receptor in caveolae: Colocalization of endothelin and its receptor with caveolin. Proc Natl Acad Sci USA 91: 11728–11732, 1994

26. Schnitzer JE, Oli P, Jacobson BS, Dvorak AM: Caveolae from luminal plasmalemma of rat lung endothelium: Microdomains enriched in caveolin, Ca^{2+}-ATPase, and inositol triphosphate receptor. Proc Natl Acad Sci USA 92: 1759–1763, 1995

27. Stahl A, Mueller. BM: The urokinase-type plasminogen activator receptor, a GPI-linked protein, is localized in caveolae. J Cell Biol 129: 335–344, 1995

28. Parpal S, Gustavsson J, Stralfors P: Isolation of phosphooligo-saccharide/phosphoinositol glycan from caveolae and cytosol of insulin stimulated cells. J Cell Biol 131: 125–135, 1995

29. Liu P, Anderson RGW: Compartmentalized production of ceramide at the cell surface. J Biol Chem 270: 27179–27815, 1995

30. Liu P, Ying Y, Ko Y-G, Anderson RGW: Localization of platelet-derived growth factors-stimulated phosphorylation cascade to caveolae. J Biol Chem 271: 10299–10303, 1996

31. Mineo C, James GL, Smart EJ, Anderson RGW: Localization of epidermal growth factor-stimulated Ras/Raf-1 interaction to caveolae membrane. J Biol Chem 271: 11930–11935, 1996

32. Hope HR, Pike LJ: Phosphoinositides and phosphoinositide-utilizing enzymes in detergent-insoluble lipid domains. Mol Cell Biol 7: 843–851, 1996

33. Pike LJ, Casey L: Localization and turnover of phosphatidylinositol 4,5-bisphosphate in caveolin-enriched membrane domains. J Biol Chem 271: 26453–26456, 1996

34. Feron O, Belhassen L, Kobzik L, Smith TW, Kelly RA, Michel T: Endothelial nitric oxide synthase targeting to caveolae. J Biol Chem 271: 22810–22814, 1996

35. Li S, Couet J, Lisanti MP: Src tyrosine kinases, G_α subunits, and H-ras share a common membrane-anchored scaffolding protein, caveolin. J Biol Chem 271: 29182–29190, 1996

36. Mayor S, Rothberg KG, Maxfield FR: Sequestration of GPI-anchored proteins in caveolae triggered by cross-linking. Science 264: 1948–1951, 1994

37. Mastick CC, Brady MJ, Saltiel AR: Insulin stimulates the tyrosine phosphorylation of caveolin. J Cell Biol 129: 1523–1531, 1995

38. Schnitzer JE, McIntosh DP, Dvorak AM, Liu J, Oh P: Separation of caveolae from associated microdomains of GPI-anchored proteins. Science 269: 1435–1439, 1995

39. Schnitzer JE, Oh P, McIntosh DP: Role of GTP hydrolysis in fission of caveolae directly from plasma membranes. Science 274: 239–242, 1996

40. Corvera S, Folander K, Clairmont KB, Czech MP: A highly phos-phorylated subpopulation of insulin-like growth factor II/mannose 6-phosphate receptors is concentrated in a clathrin-enriched plasma membrane fraction. Proc Natl Acad Sci USA 85: 7567–7571, 1988

41. Corvera S, Capocasale RJ: Enhanced phosphorylation of a coated vesicle polypeptide in response to insulin stimulation of rat adipocytes. J Biol Chem 265: 15963–15969, 1990

42. Brown DA, Rose JK: Sorting of GPI-anchored proteins to glycolipid-enriched membrane subdomains during transport to the apical cell surface. Cell 68: 533–544, 1992

43. Schroeder R, London E, Brown D: Interactions between saturated acyl chains confer detergent resistance on lipids and glycosylphos-phatidylinositol (GPI)-anchored proteins: GPI-anchored proteins in liposomes and cells show similar behavior. Proc Natl Acad Sci USA 91: 12130–12134, 1994

44. Rodgers W, Crise B, Rose JK: Signals determining protein tyrosine kinase and glycosylphosphatidylinositol-anchored protein targeting to a glycolipid-enriched membrane fraction. Mol Cell Biol 1994: 5384–5391, 1994

45. Shenoy-Scaria AM, Dietzen DJ, Kwong J, Link DC, Lublin DM: Cysteine[3] of Src family protein tyrosine kinases determines palmitoylation and localization in caveolae. J Cell Biol 126: 353–363, 1994

46. Resh MD: Regulation of cellular signalling by fatty acid acylation and prenylation of signal transduction proteins. Cell Sig 8: 403–412, 1996

47. Garcia-Cardena G, Oh P, Liu J, Schnitzer JE, Sessa WC: Targeting of nitric oxide synthase to endothelial cell caveolae via palmitoylation: Implications for nitric oxide signaling. Proc Natl Acad Sci USA 93: 6448–6453, 1996

48. Dietzen DJ, Hastings WR, Lublin DM: Caveolin in palmitoylated on multiple cysteine residues. J Biol Chem 270: 6838–6842, 1995

49. Murata M, Peranen J, Schreiner R, Wieland F, Kurzchalia TV, Simons K: VIP21/caveolin is a cholesterol-binding protein. Proc Natl Acad Sci USA 92: 10339–10343, 1995

50. Fra AM, Williamson E, Simons K, Parton RG: Detergent-insoluble glycolipid microdomains in lymphocytes in the absence of caveolae. J Biol Chem 269: 30745–30748, 1994

51. Gorodinsky A, Harris DA: Glycolipid-anchored proteins in neuro-blastoma cells form detergent-resistant complexes without caveolin. J Cell Biol 129: 619–627, 1995

52. Fra AM, Williamson E, Simons K, Parton RG: *De novo* formation of caveolae in lymphocytes by expression of VIP21-caveolin. Proc Natl Acad Sci USA 92: 8655–8659, 1995

53. Sargiacomo M, Scherer PE, Tang Z, Kubler E, Song KS, Sanders MC, Lisanti MP: Oligomeric structure of caveolin: Implications for caveolae membrane organization. Proc Natl Acad Sci USA 92: 9407–9411, 1995

54. Chung KN, Roth R, Morisaki JH, Levy MA, Elwood PC, Heuser JE: Transfection with the protein caveolin creates plasmalemmal caveolae. 1997 (submitted)

55. Robinson LJ, Pang S, Harris DS, Heuser J, James DE: Translocation of glucose transporter (GLUT4) to the cell surface in permeabilized 3T3-L1 adipocytes: Effects of ATP, insulin, and GTPγS and localization of GLUT4 to clathrin lattices. J Biol Chem 117: 1181–1196, 1992

56. Scherer PE, Lisanti MP, Baldini G, Sargiacomo M, Mastick CC, Lodish HF: Induction of caveolin during adipogenesis and association of GLUT4 with caveolin-rich vesicles. J Cell Biol 127: 1233–1243, 1994

57. Song KS, Scherer PE, Tang Z, Okamoto T, Li S, Chafel M, Cliu C, Kohtz DS, Lisanti MP: Expression of caveolin-3 in skeletal, cardiac, and smooth muscle cells. J Biol Chem 271: 15160–15165, 1996

58. Kandror KV, Stephens JM, Pilch PF: Expression and compartment-alization of caveolin in adipose cells: Coordinate regulation with and structural segregation from GLUT4. J Cell Biol 129: 999–1006, 1995

59. Mastick CC, Saltiel AR: Insulin-stimulated tyrosine phosphorylation of caveolin is specific for the differentiated adipocyte phenotype in 3T3-L1 cells. J Biol Chem 272: 20706–20714, 1997

60. Cantley L, Auger KR, Carpenter C, Duckworth B, Graziani A, Kapeller R, Soltoff S: Oncogenes and signal transduction. Cell 64: 281–302, 1991

61. Alonso G, Koegl M, Mazurenko N, Courtneidge S: Sequence require-ments for binding of Src family tyrosine kinases to activated growth factor receptors. J Biol Chem 270: 9840–9848, 1995

62. Ptasznik A, Traynor-Kaplan A, Bokoch GM: G protein-coupled chemoattractant receptors regulate Lyn tyrosine kinase-Shc adaptor protein signaling complexes. J Biol Chem 270: 19969–19973, 1995

63. Sun XJ, Pons S, Asano T, Myers MG, Glasheen E, White MF: The fyn tyrosine kinase binds Irs-1 and forms a distinct signaling complex during insulin stimulation. J Biol Chem 271: 10583–10587, 1996

64. Ribon V, Saltiel AR: Insulin stimulates the tyrosine phosphorylation of the proto-oncogene product of c-Cbl in 3T3-L1 adipocytes. Biochem J 324: 839–846, 1997

65. O'Neill TJ, Craparo A, Gustafson TA: Characterization of an interaction between insulin receptor substrate 1 and the insulin receptor using the two-hybrid system. Mol Cell Biol 14: 6433–6442, 1994

66. Gustafson TA, He W, Craparo A, Schuab CD, O'Neill TJ: Phospho-tyrosine-dependent interaction of SHC and insulin receptor substrate 1 with the NPXY motif of the insulin receptor via a novel non-SH2 domain. Mol Cell Biol 15: 2500–2508, 1995

67. Wolf G, Trub T, Ottinger L, Lynch A, White MF, Miyazaki M, Lee J, Schoelson SE: PTB domains of IRS-1 and SHC have distinct but overlapping binding specificities. J Biol Chem 271: 27407–27410, 1995

68. Langdon WY: The *cbl* oncogene: A novel substrate of protein tyrosine kinases. Aust NZ J Mod 25: 859–864, 1995

69. Mayer BJ, Hirai H, Sakai R: Evidence that SH2 domains promote processive phosphorylation by protein-tyrosine kinases. Curr Biol 5: 296–305, 1995

70. Ruzzene M, Brunati AM, Marin O, Donella-Deana A, Pinna LA: SH2 domains mediate the sequential phosphorylation of HS1 protein by p72syk and Src-related protein tyrosine kinases. Biochemistry 35: 5327–5332, 1996

71. Lawrence JC: Signal transduction and protein phosphorylation in the regulation of cellular metabolism by insulin. Annu Rev Physiol 54: 177–193, 1992

72. Tang PM, Bondot JA, Swiderek KM, DePaoli-Roach AA: Molecular cloning and expression of the regulatory (R_{G1}) subunit of the glycogen-associated protein phosphatase. J Biol Chem 266: 15782–15789, 1991

73. Doherty MJ, Moorhead G, Morrice N, Cohen P, Cohen PTW: Amino acid sequence and expression of the hepatic glycogen- binding (G_L)-subunit of protein phosphatase-1. FEBS Lett 375: 294–298, 1995

74. Printen JA, Brady MJ, Saltiel AR: PTG, a protein phosphatase 1 binding protein with a role in glycogen metabolism. Science 275: 1475–1478, 1997

75. Doherty MJ, Young PR, Cohen PTW: Amino acid sequence of a novel protein phosphatase 1 binding protein (R5) which is related to the liver- and muscle-specific glycogen binding subunits of protein phosphatase 1. FEBS Lett 399: 339–343, 1996

76. Dent P, Lavoinne A, Nakielny S, Caudwell FB, Watt P, Cohen P: The molecular mechanism by which insulin stimulates glycogen synthesis in mammalian skeletal muscle. Nature 348: 302–308, 1990

77. Lavoinne A, Erikson E, Maller JL, Price DJ, Avruch J, Colien P: Purification and characterization of the insulin-stimulated protein kinase from rabbit skeletal muscle; close similarity to S6 kinase II. Eur J Biochem 199: 723–728, 1991

78. Lazar DF, Weise RJ, Brady MJ, Mastick CC, Waters SB, Yamauchi K, Pessin JE, Cuatrecasas P, Saltiel AR: MEK inhibition does not block the stimulation of glucose utilization by insulin. J Biol Chem 270: 20801–20807, 1995

79. Shenolikar S, Nairn AC: Protein phosphatases: Recent progress. In: P Greengard, GA Robinson (eds). Advances in Second Messenger and Phosphoprotein Research. Raven Press Ltd., New York, 1991

80. Shenolikar S: Protein phosphatase regulation by endogenous inhibitors. Sem Cancer Biol 6: 219–227, 1995

81. Brady MJ, Nairn AC, Saltiel AR: The regulation of glycogen synthase by protein phosphatase 1 in 3T3-L1 adipocytes: Evidence for a role for DARPP-32 in insulin action. J Biol Chem 272: 29698–29703, 1997

82. Brady MJ, Printen JA, Mastick, CC, Saltiel, AR: The role of PTG in the regulation of protein phosphatase-1 activity. J Biol Chem 272: 20198–20204, 1997

Molecular and Cellular Biochemistry **182**: 73–78, 1998.

Binding of SH2 containing proteins to the insulin receptor: A new way for modulating insulin signalling

Feng Liu[1] and Richard A. Roth[2]

[1]*Department of Pharmacology, University of Texas Health Science Center at San Antonio, San Antonio, Texas;*
[2]*Department of Molecular Pharmacology, Stanford University School of Medicine, Stanford, CA, USA*

Abstract

Prior studies have established a role in insulin action for the tyrosine phosphorylation of substrates and their subsequent complexing with SH2 containing proteins. More recently, SH2 proteins have been identified which can tightly bind to the tyrosine phosphorylated insulin receptor. The major protein identified so far (called Grb-IR or Grb10) of this type appears to be present in at least 3 isoforms, varying in the presence of a pleckstrin homology domain and in the sequence of its amino terminus. The binding of this protein to the insulin receptor appears to inhibit signalling by the receptor. The present review will discuss the current knowledge of the structure and function of this protein. (Mol Cell Biochem **182**: 73–78, 1998)

Key words: SH2 domains, insulin receptor, tyrosine phosphorylation

Abbreviations: CHO – Chinese hamster ovary; EGF – epidermal growth factor; GST – glutathione S-transferase; IGF-1 – insulin-like growth factor-1; IR – insulin receptor; IRS-1 – insulin receptor substrate 1; MAP – mitogen-activated protein; PAGE – polyacrylamide gel electrophoresis; PDGF – platelet-derived growth factor receptor; PH – pleckstrin homology; PI 3-kinase – phosphatidylinositol 3-kinase; PMSF – phenylmethanesulphonyl fluoride; SH2 – Src homology 2

Introduction

Insulin receptor (IR) and insulin-like growth factor receptor (IGF-IR) are members of the receptor tyrosine kinase family that regulate cell metabolism, development, and growth. The binding of insulin or IGF-I to their receptors results in receptor autophosphorylation which leads to two intracellular events: first, it activates receptor tyrosine kinase and results in the phosphorylation of various cellular substrates. Second, it generates docking sites for downstream Src-homology 2 (SH2) domain-containing proteins to bind. Unlike other members of the receptor tyrosine kinase family such as the platelet-derived growth factor receptor (PDGFR) and epidermal growth factor receptor (EGFR), which interact directly with SH2 domain-containing proteins, the transmission of the signal from the IR and IGF-IR to the downstream SH2 domain-containing proteins

has been shown to be mainly through an intermediate protein named the insulin receptor substrate-1 or IRS-1 [1] (Fig. 1). Tyrosine phosphorylation of IRS-1 serves as docking sites to recruit multiple downstream proteins in the cascade of insulin action, including adapter proteins Grb2, a tyrosine phosphatase Syp, and the p85 subunit of phosphatidylinositol (PI) 3-kinase [2]. The association of the p85 subunit of PI 3-kinase with IRS-1 results in activation of the enzyme which has been implicated in the regulation of many cellular processes, including insulin or IGF-I-induced membrane ruffling [3], serine kinase Akt activation [4–6], muscle cell differentiation [7], and GLUT4 translocation and glucose uptake [8–10]. The association of IRS-1 with Grb2 results in the recruitment of a guanine nucleotide exchange factor Sos to the membrane and transduces the signal to the MAP kinase [11–13]. Another signalling molecule which may play a role in the link of insulin and IGF-1 signalling to the MAP

Address for offprints: R. Roth, Department of Molecular Pharmacology, Stanford University School of Medicine, Stanford, CA 94305, USA

74

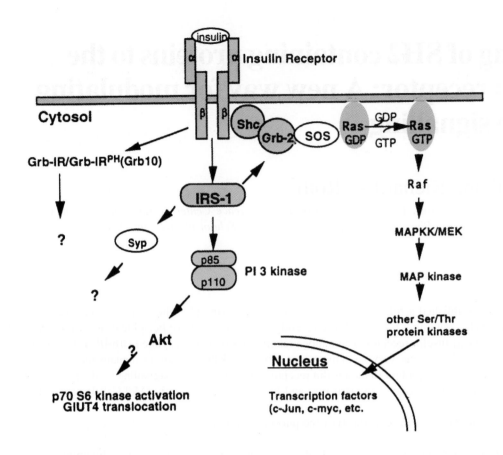

Fig. 1. Insulin receptor signal transduction pathway.

kinase pathway is the SH2/ct-collagen related protein Shc [12, 14]. Insulin activation of MAP kinase has been implicated in cell differentiation, gene regulation, and protein synthesis [15–18].

Although IRS (including IRS-1 and the recently identified IRS-1-like protein IRS-2 [19–21]) and Shc are important in insulin and IGF signalling, these proteins are also activated by a variety of receptors that are not regulated by insulin and IGF [22] and therefore it is unlikely that they are sufficient to mediate all the insulin or IGF regulated biological events such as glucose uptake, glycogen synthesis, and lipid metabolism. It is possible that other specific signalling molecules may exist to transduce and regulate the signals from the insulin receptor to downstream targets. In an attempt to identify such signalling proteins, we (and others) have used the yeast two-hybrid technique with the IR cytoplasmic domain as bait to find potential interacting protein(s). We identified an SH2 domain-containing protein (Grb-IR, renamed as hGrb10α) that binds specifically to the tyrosine phosphorylated insulin receptor [23]. Reverse-transcription (RT)-PCR experiments showed that there are at least two isoforms of Grb-IR/hGrb10a which differ in their PH domains [23]. Grb-IR shows a high homology in

sequence with several recently identified SH2 and PH domain-containing proteins including Grb7 [24], Grb10 [25], and Grb14 [26]. We will focus the present discussion on recent findings on hGrb10 isoforms and their possible roles in receptor tyrosine kinase signal transduction initiated by insulin and other growth factors.

Grb-IR/hGrb10α and its isoforms: structure, tissue expression, and interaction with receptor tyrosine kinases

Grb-IR/hGrb10α was originally identified by screening a yeast two-hybrid library derived from HeLa cells using the cytoplasmic domain of the IR as bait [23]. Highest expression of the protein was observed in insulin target tissues such as skeletal muscle and fat cells as well as in the pancreas. Full-length Grb-IR/hGrb10α CDNA isolated from human skeletal muscle CDNA library encodes a protein with a calculated molecular weight of 62 kDa, with a SH2 domain at its extreme carboxyl terminus (Fig. 2A). Grb-IRI/hGrb10α binds with high affinity to autophosphorylated insulin and IGF-I receptors but only weakly to the PDGF and EGF

A.

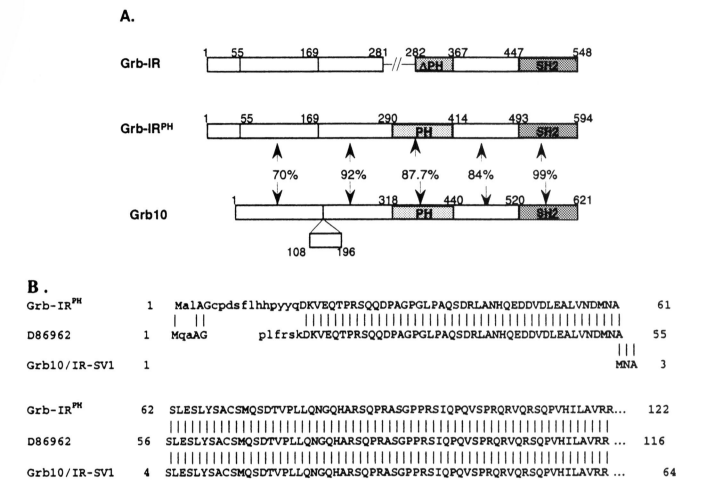

B.

```
Grb-IR^PH    1   MalAGcpdsflhhpyyqDKVEQTPRSQQDPAGPGLPAQSDRLANHQEDDVDLEALVNDMNA   61
                 |  ||          ||||||||||||||||||||||||||||||||||||||||||||||||
D86962       1   MqaAG      plfrskDKVEQTPRSQQDPAGPGLPAQSDRLANHQEDDVDLEALVNDMNA   55
                                                                             |||
Grb10/IR-SV1 1                                                               MNA    3


Grb-IR^PH   62   SLESLYSACSMQSDTVPLLQNGQHARSQPRASGPPRSIQPQVSPRQRVQRSQPVHILAVRR...  122
                 ||||||||||||||||||||||||||||||||||||||||||||||||||||||||||||||
D86962      56   SLESLYSACSMQSDTVPLLQNGQHARSQPRASGPPRSIQPQVSPRQRVQRSQPVHILAVRR ...  116
                 ||||||||||||||||||||||||||||||||||||||||||||||||||||||||||||
Grb10/IR-SV1 4   SLESLYSACSMQSDTVPLLQNGQHARSQPRASGPPRSIQPQVSPRQRVQRSQPVHILAVRR...   64
```

Fig. 2. (A) Schematic diagram comparing the domain structure of human and mouse Grb10; (B) The alignment of the N-terminal amino acid sequences of hGrb10 isoforms.

receptors *in vitro* ([23] and data not shown). Recent yeast two-hybrid studies showed that hGrb10 also binds with a high affinity to Ret, a receptor tyrosine kinase involved in renal and enteric neuron development, thyroid papillary carcinomas and multiple endocrine neoplasia ([27, 28], Dong and Liu, unpublished studies) Grb-IR/hGrb10α is highly homologous in sequence to the mouse SH2 domain-containing protein mGrb10 [25], with a 99% identity in their SH2 domains and 84% identity in their central regions (Fig. 2A). Our recent site-directed mutagenesis and yeast two-hybrid studies showed that, unlike other SH2 domain-containing proteins, hGrb10 binds to the autophosphorylated IR and IGF-1R at the kinase domain, most likely to the autophosphorylated tyrosine residues at activation loop of the IR (Dong *et al.*, manuscript submitted). Three isoforms of hGrb10, including Grb-IR/hGrb10α [23], hGrb10β [29, 29a], and hGrb10γ (Dong *et al.*, manuscript submitted) which differ in their PH domain and N-terminal region, have been found in insulin target tissues such as the skeletal muscle and fat cells (Fig. 2A). A

search of GeneBank databases revealed a newly deposited CDNA sequence identified from human myeloblast (Sequence accession number: D86962, hGrb10δ) which is identical to that of hGrb10γ, except for its N-terminal first 11 amino acids (Fig. 2B). Overexpression of Grb-IR/hGrb10α or microinjection of the SH2 domain of the protein into cells have been shown to inhibit insulin-stimulated PI 3-kinase activity and mitogenesis [23, 29].

Grb7/10/14 gene family

Sequence alignment of several recently identified SH2 and PH domain-containing proteins including Grb7 [24], Grb10 [25] and Grb14 [26] suggest that they belong to a specific family with unique sequence and structural characteristics. All of these proteins contain an SH2 domain at their carboxyl terminals and a PH domain in the central regions. The human Grb-IR/hGrb10α SH2 domain is 99% identical to that of the mouse Grb10, except that the serine residue at position 560 of Grb-IR is replaced by a threonine residue in Grb10. The central regions of these proteins (include the PH domains) are also highly homologous to each other and are similar to the *C. elegans* gene mig-10 that has been shown to be crucial for embryonic neural migration [30]. The third hallmark of the Grb-10/IR family is that all the members of the family contain a highly conserved proline-rich sequence (P(S/A)IPNPFPEL) at their N-terminals, although the sequences surrounding this motif are not conserved (Fig. 3). The presence of several functional domains including the SH2 domain, the PH domain and the proline-rich sequence among Grb7, Grb10, and Grb14 isoforms suggests that these proteins are capable of interacting with different proteins in signalling pathways. It has recently been shown that signal transduction pathways in cells are partly regulated by a mechanism called 'compartmentalization.' This regulation can be achieved by coordinating the localization of multi-enzyme signalling complex through multivalent anchoring or scaffold proteins such as the STE5 [31, 32] and AKAP79 [33]. The presence of multiple functional domains in the Grb7/10/14 family members suggests that these proteins are capable of playing such a role. For example, the proline-rich motif may be involved in the binding of SH3 or SH3-like domain containing proteins. The requirement of an intact PH-domain for insulin-stimulated hGrb10γ serine phosphorylation provides additional evidence that the PH-domain is important for protein-protein interaction. Like other scaffolding or anchoring proteins (reviewed by Faux and Scott [34]), hGrb10γ may bring the downstream signalling molecules close to the autophosphorylated insulin receptor for a reaction to take place.

It should be pointed out that although these proteins are highly related, there are still significant differences in both the structure and sequence of these proteins. The differences in the functional domains of the Grb7/10/14 family members may define the specific roles for these adapter proteins in signalling processes. For example, Grb-IR/

hGrb10α has been shown to bind with high affinity to the IR and IGF-IR but poorly to the EGFR and PDGFR ([23] and data not shown). Grb7, on the other hand, binds with high affinity to the PDGFR [35] and the EGFR-related receptor HER2 [36], The specific binding of these adapter proteins to receptors suggests that they may play roles in the specificity of signal transduction pathways. The structural difference between these proteins may also be important in the determination of the binding specificity to the same receptor. Our laboratory (Dong, Farris and Liu, manuscript submitted) and others [29] have found that Grb-IR binds to autophosphorylated tyrosine residues in the kinase domain of the IR and IGF-IR. On the other hand, Hansen *et al.* [37] showed that Grb10 binds to the phosphorylated tyrosine residue at the C-terminal of the IR. It would be very interesting to see whether the single amino acid difference between Grb-IR and Grb10 provides the specificity for the binding and whether binding to different sites of the same receptor plays different roles in signalling.

Grb7/10/14 family members and their possible roles in receptor signal transduction

Grb7

Grb7 is the first identified member of the Grb7/10/14 family which was cloned by the CORT (for Cloning Of Receptor Targets) method through a screening of bacterial expression libraries using the *in vitro* phosphorylated EGF receptor cytoplasmic domain as a probe [24]. The Grb7 gene was mapped to a region on mouse chromosome 11 which also contains the tyrosine kinase receptor HER2. Although Grb7 has been shown to be coamplified with and binds with a high affinity to the EGFR-related HER2, it binds poorly to the EGFR in cells. Grb7 has also been shown to associate with Shc [36], Ret [38], and the PDGFR [35] through its SH2 domain, probably at a site containing a Y(V/I)N motif. Binding of Grb7 to the PDGF β-receptor was inhibited when tyrosine residues 716 (YSN) or 775 (YDN) of the receptor

```
Grb-IR^PH   55  LVNDMNASLESLYSACSMQS--DTVPLLQNGQHARSQPRASGPPRSIQPQVSPRQRVQRSQPV-   115
Grb7         1        MELDLSPTHLSSSPE---DVCPTPATPPETPP-PPDNPPPG----DVKKR-----SQPL-    46
Grb10        1  MNNDINSSVESLNSACNMQSDTDTAPLLEDGQHASNQGAASSSS-RGQPQASPRQKMQRSQPV-    62
Grb14        1        MTTS------LQDGQS-----AASRAAARDSPLAAQVCGAAQGRGDA    36


Grb-IR     116  ------------------HILAVRRLQEEDQQFRTSSLPAIPNPFPELCG--PGSPPVLTPGS   158
Grb7        47  ------------------PIPSSRKLREEEFQ--ATSLPSIPNPFPELCS--PPSQKPILGGS    87
Grb10       63  ------------------HIL--RRLQEEDQQLRTASLPAIPNPFPELTGAAPGSPPSVAPSP   105
GRB14       37  HDLAPAPWLHARALLPLPDGTRGCAADRRKKKDLDVPEMPSIPNPFPELPITSVLSADLFPKAN   100
```

Fig. 3. The alignment of the N-terminal amino acid sequences of Grb7/10/14 family members.

were replaced by phenylalanine or in the presence of synthetic phosphopeptides containing these residues, suggesting these residues are the binding sites for Grb7 [35]. However, the binding sites for Grb7 on the HER2, Shc, and Ret have not been identified. Because Grb7 is expressed only in kidney, liver and gonad [36], the role of the protein in different signal transduction pathways remains to be clarified.

Grb10

Grb10 CDNA isolated from NIH3T3 library encodes a protein of 621 amino acids with a molecular weight of 70 kDa [25]. Grb10 MRNA is highly expressed in heart, kidney, brain, lung, and NIH3T3 cells. Polyclonal antibodies against Grb10 detected several bands with molecular weights ranging from 65–80 kDa in NIH3T3 cells but not in HeLa cells. Grb10 binds to a variety of receptor tyrosine kinases including the IR [37], the IGF-1R [39, 40], ELK, an Eph-related tyrosine kinase family member [41], and Ret [42]. Phosphopeptide binding studies suggest that Grb10 binds to tyrosine residue 1334 at the C-terminal of the IR [37] This finding, however, was recently challenged by Morrione *et al.* [40] who showed that tyrosine 1316 of the IGF-1R, which is the equivalent of tyrosine 1334 of the human IR, is not the Grb10 binding site. Deletion of the C-terminal sequence of ELK receptor or mutation tyrosine 929, the only tyrosine in this region, disrupted the interaction between the receptor and Grb10 in the yeast two-hybrid system, suggesting that this tyrosine residue may be involved in the binding of Grb10 [41]. The binding site for Grb10 on Ret is currently unknown but competition studies showed Grb10 binds to Ret at the same site at which Grb7 binds [38].

Grb14

Grb14, the newest member of the family, was identified by screening a human breast epithelial cell CDNA library with the tyrosine phosphorylated C-terminus of the EGFR [26]. Grb14 MRNA is highly expressed in the liver, kidney, pancreas, testis, ovary, heart, and skeletal muscles. Grb14 CDNA isolated from human liver encodes a protein of 540 amino acids (60 kDa) which shows a high homology to Grb7 and Grb10 isoforms. The SH2 domain of Grb14 displays 67, 74 and 72% amino acid identity, respectively, with the corresponding domain in Grb7, mGrb10, and hGrb10. A GST fusion protein containing the SH2 domain of Grb14 binds strongly to tyrosine phosphorylated PDGFR *in vitro* and Grb14 undergo PDGF-stimulated serine phosphorylation in cells, suggesting that the protein may play a role PDGF signalling. However, the direct binding of the protein to a receptor could not be detected in cells and further study will be needed to identify the functional role of the protein in receptor tyrosine signal transduction.

Conclusions

Until recently, the paradigm was that insulin stimulated the tyrosine phosphorylation of various endogenous substrates of the insulin receptor kinase (such as IRS-1 and 2 and Shc) which were then bound by various SH2 containing proteins (such as PI 3-kinase and Grb-2). The use of the yeast 2-hybrid system has identified the Grb-IR/10 family of proteins which can bind to the insulin receptor with high affinity such that treatment of mammalian cells with insulin results in the binding of this protein to the IR. This is the first SH2 containing protein with this property. The high affinity of this family of proteins with the IR is also substantiated by the ability of several different groups using different libraries to identify the same protein as a protein able to bind to the insulin receptor.

A number of questions remain on this protein. The first concerns its role in normal insulin signalling. It is possible that the binding of Grb-IR/10 to the insulin serves to bring some partner to the membrane and/or to the IR. This could serve to enhance insulin signalling. The binding of Grb-IR/ 10 to the receptor may alternatively inhibit the ability of the receptor to phosphorylate various endogenous substrates [23]. As such, Grb-IR/10 may then function as a negative regulator of the receptor. It should also be noted that at this time no one has shown that native Grb-IR/10 binds to the native IR in a normal, cell, unlike the studies of the ELK receptor in which it was shown that activation of this receptor results in its association with the endogenous Grb10 in vascular endothelial cells [41]. Thus it is possible that the interaction of Grb-IR/10 with the IR does not interact in a normal system.

Another question concerns the different splice variants of Grb-IR/10 that have been identified. One such splicing event seems to result in the removal of the PH domain, a region involved in membrane localization of various signalling molecules. One study has indicated that the removal of the PH domain of hGrb10 results in a decrease in the ability of the protein to bind to the IR [43]. It is not clear how this, or the other splice variants, will affect the ability of Grb-IR to function in the intact cell.

References

1. White MF, Kahn CR: The insulin signalling system. J Biol Chem 269(1): 1–4, 1994
2. Cheatham B, Kahn CR: Insulin action and the insulin signalling network. Endocrine Rev 16: 117–142, 1995
3. Kotani K, Yonezawa K, Hara K, Ueda H, Kitamura Y, Sakaue H, Ando A, Chavanieu B, Calas B, Grigorescu F, Nishiyama M, Waterfield M, Kasuga M: Involvement of phosphoinositide 3-kinase in insulin- or IGF-I-induced membrane ruffling. EMBO J 13: 2313–2321, 1994

78

4. Burgering BMT, Coffer PJ: Protein kinase B (c-Akt) in phosphatidyl-inositol-3OH kinase signal transduction. Nature 376: 599–602, 1995

5. Franke TF, Yang SI, Chan TO, Kazlauskas A, Morrison DK, Kaplan DR, Tsichlis PN: The protein kinase encoded by the Akt proto-oncogene is a target of the PDGF-activated phosphatidylinositol 3-kinase. Cell 81: 727–736, 1995

6. Kohn AD, Kovacina KS, Roth RA: Insulin stimulates the kinase activity of RACPK, a pleckstrin homology domain containing ser/thr kinase. EMBO J 14: 4288–4295, 1995

7. Kaliman P, Vinals F, Testar X, Palacin M, Zorzano A: Phosphatidyl-inositol 3-kinase inhibitors block differentiation of skeletal muscle cells. J Biol Chem 271: 19146–19151, 1996

8. Evans J, Honer C, Womelsdorf B, Kaplan E, Bell P: The effects of wortmannin, a potent inhibitor of phosphatidylinositol 3-kinase, on insulin-stimulated glucose transport, GLUT4 translocation, anti-lipolysis, and DNA synthesis. Cell Sig 7: 365–376, 1995

9. Hara K, Yonezawa K, Sakaue H, Ando A, Kotani K, Kitamura Y, Ueda H, Stephens L, Jackson T, Hawkins P, Dhand R, Clark A, Holman G, Waterfield M, Kasuga M: 1-Phosphatidylinositol 3-kinase activity is required for insulin-stimulated glucose transport but not for RAS activation in CHO cells. Proc Natl Acad Sci USA (91): 7415–7419, 1994

10. Martin S, Haruta T, Morris A, Klippel A, Williams L, Olefsky J: Activated phosphatidylinositol 3-kinase is sufficient to mediate acting rearrangement and GLUT4 translocation in 3T3-L1 adipocytes. J Biol Chem 271: 17605–17608, 1996

11. Skolnik EY, Lee C-H, Batzer A, Vicentini LM, Zhou M, Daly R, Myers MJ Jr, Backer JM, Ullrich A, White MF, Schlessinger J: The SH2/SH3 domain-containing protein GRB2 interacts with tyrosine-phosphorylated IRS 1 and Shc: Implications for insulin control of ras signalling. EMBO J 12(5): 1929–1936, 1993

12. Skolnik E, Batzer A, Li N, Lee C, Lowenstein E, Mohammadi M, Margolis B, Schlessinger J: The function of GRB2 in linking the insulin receptor to Ras signalling pathways. Science 260: 1953–1955, 1993

13. Yonezawa K, Ando A, Kaburagi Y, R. Y-H, Kitamura T, Hara K, Nakafuku M, Okabayashi Y, Kadowaki T, Kaziro Y, Kasuga M: Signal transduction pathways from insulin receptors to Ras. Analysis by mutant insulin receptors. J Biol Chem 269(6): 4634–4640, 1994

14. Kovacina KS, Roth RA: Identification of SHC as a substrate of the insulin receptor kinase distinct from the GAP-associated 62 kDa tyrosine phosphoprotein. Biochem Biophys Res Comm 192: 1303–1311, 1993

15. Chang P, Le Marchand-Brustel Y, Cheathan L, Moller D: Insulin stimulation of mitogen-activated protein kinase, p90rsk, and p70 S6 kinase in skeletal muscle of normal and insulin-resistant n-iice. Implications for the regulation of glycogen synthase. J Biol Chem 270: 29928–29935, 1995

16. Denton R, Tavare J: Does mitogen-activated-protein kinase have a role in insulin action? The cases for and against. Eur J Biochem 227: 597–611, 1995

17. Sale E, Atkinson P, Sale G: Requirement of MAP kinase for differentiation of fibroblasts to adipocytes, for insulin activation of p90 S6 kinase and for insulin or serum stimulation of DNA synthesis. EMBO J 14: 674–684, 1995

18. Zhang B, Berger J, Zhou G, Elbrecht A, Biswas S, White-Carrington S, Szalkowski D, Moller D: Insulin- and mitogen-activated protein kinase-mediated phosphorylation and activation of peroxisome proliferator-activated receptor γ. J Biol Chem 271: 31771–31774, 1996

19. Patti M, Sun X, Bruening J, Araki E, Lipes M, White M, Kahn C: 4PS/insulin receptor substrate (IRS)-2 is the alternative substrate of the insulin receptor in IRS-1-deficient mice. J Biol Chem 270: 24670–24673, 1995

20. Sun X-J, Wang L-M, Zhang Y, Yenush L, Myers M, Glasheen E, Lane W, Pierce J, White M: Role of IRS-2 in insulin and cytokine signalling. Nature 377: 173–177, 1995

21. Tobe K, Tamemoto H, Yamauchi T, Aizawa S, Yazaki Y, Kadowaki T: Identification of a 190 kDa protein as a novel substrate for the insulin receptor kinase functionally similar to insulin receptor substrate-1. J Biol Chem 270: 5698–5701, 1995

22. Myers MJ, Sun XJ, White MF: The IRS-1 signalling system. Trends Biochem Sci 19(7): 289–293, 1994

23. Liu F, Roth RA: Grb-IR: A SH2-domain-containing protein that binds to and inhibits insulin receptor function. Proc Natl Acad Sci USA 92: 1995

24. Margolis B, Silvennoinen O, Comoglio F, Roonprapunt C, Skolnik E, Ullrich A, Schlessinger J: High-efficiency expression/cloning of epidermal growth factor-receptor-binding proteins with Src homology 2 domains. Proc Natl Acad Sci USA 89: 8894–8898, 1992

25. Ooi J, Yajnik V, Immanuel D, Gordon M, Moskow JJ, Buchberg AM, Margolis B: The cloning of Grb10 reveals a new family of SH2 domain proteins. Oncogene 10: 1610–1630, 1995

26. Daly R, Sanderson G, Janes P, Sutherland R: Cloning and character-ization of Grb14, a novel member of the Grb7 gene family. J Biol Chem 271: 12502–12510, 1996

27. Gagel R: Putting the bits and pieces of the RET proto-oncogene puzzle together. Bone 17: 13S–16S, 1995

28. Takahashi S, Uchida K, Nakagawa A, Miyake Y, Kainosho M, Matsuzaki K, Omura S: Biosynthesis of lactacystin. Jour Antibio 48: 1015–1020, 1995

29. O'Neill T, Rose T, Pillay T, Hotta K, Olefsky J, Gustafson T: Interaction of a GRB-IR splice variant (a human GRB10 homolog) with the insulin and insulin-like growth factor 1 receptors. Evidence for a role in mitogenic signalling. J Biol Chem 271: 22506–22513, 1996

29a. Frantz J, Giorgetti-Peraldi S, Ottinger E, Shoelson S: Human GRB-IRbeta/GRB10. Splice variants of an insulin and growth factor receptor-binding protein with PH and SH2 domains. J Biol Chem 272: 2659–2667, 1997

30. Margolis B: The GRB family of SH2 domain proteins. Prog Biophys Mol Biol 62: 223–244, 1994

31. Choi KY, Satterberg B, Lyons DM, Elion EA: Ste5 tethers multiple protein kinases in the MAP kinase cascade required for mating in S. cerevisiae. Cell 78: 499–512, 1994

32. Marcus S, Polverino A, Barr M, Wigler M: Complexes between STE5 and components of the pheromone-responsive mitogen-activated protein kinase module. Proc Natl Acad Sci USA 91: 7762–7766, 1994

33. Klauck T, Faux M, Labudda K, Langeberg L, Jaken S, Scott J: Coordination of three signalling enzymes by AKAP79, a mammalian scaffold protein. Science 271: 1589–1592, 1996

34. Faux M, Scott J: Molecular glue: Kinase anchoring and scaffold proteins. Cell 85: 9–12, 1996

35. Yokote K, Margolis B, Heldin C-H, Claesson-Welsh L: Grb7 is a downstream signalling component of platelet-derived growth factor α- and β-receptors. J Biol Chem 271: 30942–30949, 1996

36. Stein D, Wu J, Fuqua SAW, Roonprapunt C, Yajnik V, d'Eustachio P, Moskow JJ, Buchberg AM, Osborne CK, Margolis B: The SH2 domain protein GRB-7 is co-amplified, overexpressed and in a tight complex with HER2 in breast cancer. EMBO J 13: 1331–1340, 1994

37. Hansen H, Svensson U, Zhu J, Laviola L, Giorgino F, Wolf G, Smith R, Riedel H: Interaction between the Grb10 SH2 domain and the insulin receptor carboxyl terminus. J Biol Chem 271: 8882–8886, 1996

38. Pandey A, Liu X, Dixon J, Di Fiore P, Dixit V: Direct association between the Ret receptor tyrosine kinase and the Src homology 2-containing adaptor protein Grb7. J Biol Chem 271: 10607–10610, 1996

39. Dey BR, Frick K, Lopaczynski W, Nissley SP, Furlanetto RW: Evidence for the direct interaction of the insulin-like growth factor 1 receptor with IRS-1, Shc, and Grb10. Mol Endocrin 10: 631–641, 1996

40. Morrione A, Valentinis B, Li S, Ooi J, Margolis B, Baserga R: Grb10: A new substrate of the insulin-like growth factor I receptor. Cancer Res 56: 3165–3167, 1996

41. Stein E, Cerretti D, Daniel T: Ligand activation of ELK receptor tyrosine kinase promotes its association with Grb10 and Grb2 in vascular endothelial cells. J Biol Chem 271: 23588–23593, 1996

42. Pandey A, Duan H, Di Fiore P, Dixit V: The Ret receptor protein tyrosine kinase associates with the SH2-containing adaptor protein Grb10. J Biol Chem 270: 21461–21463, 1995

Molecular and Cellular Biochemistry **182**: 79–89, 1998.
© 1998 *Kluwer Academic Publishers. Printed in the Netherlands.*

Insulin receptor-associated protein tyrosine phosphatase(s): Role in insulin action

Paul G. Drake and Barry I. Posner

Polypeptide Hormone Laboratory, McGill University, Montreal, Quebec, Canada

Abstract

Protein tyrosine phosphatases (PTPs) play a critical role in regulating insulin action in part through dephosphorylation of the active (autophosphorylated) form of the insulin receptor (IRK) and attenuation of its tyrosine kinase activity. Following insulin binding the activated IRK is rapidly internalized into the endosomal apparatus, a major site at which the IRK is dephosphorylated *in vivo*. Studies in rat liver suggest a complex regulatory process whereby PTPs may act, via selective IRK tyrosine dephosphorylation, to modulate IRK activity in both a positive and negative manner. Use of peroxovanadium (pV) compounds, shown to be powerful PTP inhibitors, has been critical in delineating a close relationship between the IRK and its associated PTP(s) *in vivo*. Indeed the *in vivo* administration of pV compounds effected activation of IRK in parallel with an inhibition of IRK-associated PTP activity. This process was accompanied by a lowering of blood glucose levels in both normal and diabetic rats thus implicating the IRK-associated PTP(s) as a suitable target for defining a novel class of insulin mimetic agents. Identification of the physiologically relevant IRK-associated PTP(s) should facilitate the development of drugs suitable for managing diabetes mellitus. Mol Cell Biochem **182**: 79–89, 1998)

Key words: insulin receptor, tyrosine kinase, protein tyrosine phosphatase, endosome, peroxovanadium, diabetes mellitus

Introduction

Tyrosine phosphorylation of specific intracellular proteins, controlled by the actions of protein tyrosine kinases (PTKs) and protein tyrosine phosphatases (PTPs), is recognized as a key process by which a number of polypeptide hormones and growth factors transduce and coordinate their biological effects *in vivo*. Although early work established a critical role of PTKs in effecting protein tyrosine phosphorylation, it has become increasingly apparent that PTPs play an equally important and complex role. Accordingly, tyrosine phosphorylation of intracellular proteins is considered to reflect a balance between the competing activities of PTKs and PTPs.

Insulin initiates its biological effects through interaction with a specific cell surface receptor. The insulin receptor (IRK) is a disulfide-linked heterotetrameric glycoprotein composed of two extracellular α-subunits containing insulin-binding sites and two transmembrane β-subunits possessing tyrosine kinase activity in their cytosolic domains [1]. Upon insulin binding, a conformational change in the IRK leads to the rapid autophosphorylation of multiple tyrosine residues on the intracellular portion of the β-subunit, and the activation of the IRK towards a number of intracellular substrates including insulin receptor substrates-1 and -2 (IRS-1 and IRS-2). Evidence suggests that these molecules, rather than the IRK itself, are the predominant proteins which couple to and activate downstream signalling pathways by providing specific phosphotyrosine binding sites for a number of Src homology 2 (SH2) domain-containing signalling proteins including phosphatidylinositol 3-kinase (PI3K) and Grb2 [2].

Attenuation of insulin action is mediated to a large extent via PTPs which have the potential to control insulin signal transduction at three levels by; (1) regulating the level of phosphorylation of IRK tyrosines that control IRK activity towards endogenous substrates; (2) controlling the level of phosphorylation of IRK tyrosines that mediate interaction

Footnotes: [1]The full structure for this compound can be found in reference [54]; [2]Contreres, J.O., Band, C., Dumas, V.M. and Posner, B.I. manuscript submitted; [3]Drake, P.G., Bevan, A.P. and Posner, B.I. unpublished observations

Address for offprints: B.I. Posner, Polypeptide Hormone Laboratory, Strathcona Anatomy and Dentistry Building, 3640 University Street, Room W3. 15, Montreal, Quebec, H3A 2B2, Canada

of the IRK with other signalling proteins; and (3) modulating the tyrosine phosphorylation of downstream signalling proteins. Regulation of IRK auto phosphorylation involves both IRK activity and PTP activity. This article will address the role of PTP(s) that directly dephosphorylate and hence regulate IRK activity *in vivo* and outline evidence indicating that such PTP activity represents a therapeutic target for the development of a novel class of drugs with which to manage diabetes mellitus.

Regulation of insulin receptor function

IRK activation
Studies with kinase inhibitory antibodies and kinase inactive IRK mutants have demonstrated that IRK activity is essential for insulin action *in vivo* [3]. There are several IRK β-subunit tyrosine phosphorylation sites that occur in three clusters or domains: the juxtamembrane domain which contains tyrosines 953, 960 and 972 (Tyr^{953}, Tyr^{960}, Tyr^{972}), the kinase regulatory domain which contains tyrosines at positions 1146, 1150 and 1151 (Tyr^{1146}, Tyr^{1150} and Tyr^{1151}) and the carboxyterminus which has tyrosines at positions 1316 and 1322 (Tyr^{1316} and Tyr^{1322}) [11].

Phosphorylation at Tyr^{960} appears to be involved in mediating the tyrosine phosphorylation of IRS-1 [4, 5] and a group of 60 kDa proteins [6]. Phosphorylation of Tyr^{1146}, Tyr^{1150} and Tyr^{1151} increases the activity of the IRK toward exogenous substrates [7–9] with maximal kinase activity expressed by the tris-phosphorylated form of the kinase regulatory domain. Recent studies indicate that autophosphorylation is initiated predominantly at Tyr^{1150}, immediately followed by Tyr^{1146}, and finally Tyr^{1151} [10]. Analysis of the crystal structure of the IRK kinase domain in its unphosphorylated form has revealed a novel autoinhibitory mechanism whereby Tyr^{1150} in the activation loop is bound to the active site, blocking both substrate peptide and ATP binding. Stimulation of kinase activity by autophosphorylation is believed to result from stabilization of the activation loop in a non-inhibiting conformation, mediated via phosphotyrosine interactions with positively charged residues [11].

The role of carboxyterminal tyrosine residues is less clear. A number of SH2-containing signalling proteins including the p85 regulatory subunit of PI3K [12], Grb10 [13], and the PTP, SH-PTP2 (also known as PTP1D, PTP-2C or Syp) [14–16], have been shown to bind to this region of the activated IRK. However, the physiological significance of these interactions is unclear. Several studies have indicated a role for the phosphorylation of Tyr^{1316} and Tyr^{1322} in restraining mitogenic signalling [17–21]. In contrast, some studies support a role for phosphorylation of these tyrosine residues in mediating metabolic responses [17, 22] whereas others do not show this relationship [18, 20].

Following IRK activation there is an increase in phosphorylation of the β-subunit serine and threonine residues [23]. The functional significance of IRK serine/threonine phosphorylation is unclear but may serve to inhibit or dampen IRK activity and hence insulin signalling [23, 24]. The nature of the insulin-stimulated serine/threonine kinase(s) that phosphorylate the IRK *in vivo* remains controversial. A recent study reported the purification of a closely associated enzyme [25] whereas others have suggested that the IRK contains intrinsic serine/threonine autophosphorylating activity [26, 27].

IRK internalization
Following insulin binding, the activated insulin-IRK complex undergoes rapid internalization into a heterogeneous population of tubulovesicular structures collectively known as endosomes (ENs; see Fig. 1). ENs mediate the sorting and processing of a number of ligand-receptor complexes [28] and have been implicated in both signal transduction and signal termination [29, 30]. Once internalized, the ligand-receptor complex is exposed to an increasingly acidic lumen due to the action of an ATP-dependent proton pump. Many ligands dissociate from their receptors at lower pHs, allowing delivery of ligand and receptor to different intracellular locations where they may be recycled or degraded (Fig. 1). ENs have been shown to be a major site at which internalized insulin is degraded *in vivo* [31] with recent work identifying a novel insulin-specific endosomal protease in rat liver. This protease, termed endosomal acidic insulinase (EAI), has an acidic pH optimum and is distinct from insulin degrading enzyme (IDE), now recognized to be a peroxisomal protease, and hence unlikely to metabolize insulin *in vivo* [32].

Studies in rat liver [33, 34], hepatoma cells [35] and isolated adipocytes [36–38] have established that insulin treatment results in the accumulation and concentration within ENs of IRKs that are both tyrosine phosphorylated and active towards exogenous substrates. Indeed, kinetic analyses have revealed that internalized IRKs exhibit transiently elevated autophosphorylating and exogenous kinase activity compared to receptors located at the plasma membrane [34, 39]. Use of an *in vivo* model in which endosomally-located IRKs are exclusively activated has revealed that IRKs in this intracellular compartment can phosphorylate IRS-1 and induce hypoglycemia [40]. Moreover, studies in adipocytes have implicated ENs as the principle, if not exclusive, site of insulin-stimulated IRS-1 tyrosine phosphorylation and associated PI3K activation [37, 41]. Thus mounting evidence has demonstrated internalization of activated IRKs to be an integral and essential component of insulin-signal transduction *in vivo*.

IRK dephosphorylation
Attenuation of IRK activity is thought to involve two main processes: (1) dissociation and degradation of insulin and;

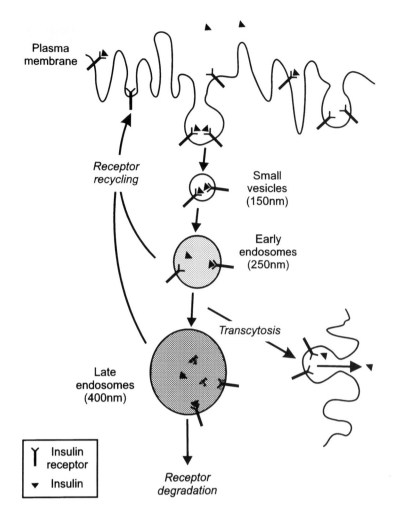

Fig. 1. Receptor-mediated endocytosis of the insulin-IRK complex. Insulin binds to its specific receptor (IRK) on the cell surface (t = 0) whereupon the insulin/IRK complex is rapidly internalized and appears sequentially in small vesicles (t = 2 min), early ENs (t = 5 min) and late ENs (t = 10 min). The insulin/IRK complex remains largely intact in the near-neutral pH interior of early ENs, but insulin dissociates from the IRK and is degraded upon reaching the acidic interior of late ENs. Insulin receptors may be recycled to the plasma membrane or delivered to secondary lysosomes for degradation. In some circumstances (i.e. insulin transport into the CNS [87]) ENs serve to transport insulin via its receptor across endothelial cells (transcytosis). Figure adapted from references [28] and [30].

(2) dephosphorylation of specific β-subunit tyrosine residues required to deactivate the IRK. Dissociation and degradation of insulin is necessary to prevent continuous stimulation of IRK β-subunit autophosphorylation. Since the IRK retains its phosphorylation state and tyrosine kinase activity after insulin is removed from the α-subunit [42], dephosphorylation of specific β-subunit tyrosine residues is necessary to deactivate the intrinsic kinase activity of the receptor and down-regulate insulin-signalling [24]. Studies of rat hepatoma cells demonstrated that internalized IRKs were dephosphorylated and inactivated prior to recycling back to the plasma membrane [35]. Other work in rat liver [34], permeabilized adipocytes [43, 44], and Chinese hamster ovary (CHO) cells [45] showed that the activated IRK is rapidly dephosphorylated *in vivo*. Thus tyrosine

phosphorylation of the IRK is a dynamic, rapidly reversible process.

In vitro studies have indicated that the tris-phosphorylated regulatory domain of IRK is the most sensitive target for PTPs in rat liver extracts [24]. However, it has not yet been established which of the three-phosphotyrosyl residues in the catalytic domain is the primary target of the physiologically relevant PTP(s).

Characterization of IRK-associated PTP activity

Cellular location of IRK-associated PTP activity
Although several PTPs have been identified in the major insulin-sensitive tissues [46], the specific PTP(s) that

dephosphorylate the IRK β-subunit *in vivo* remain unclear. Since the majority of PTPs show broad substrate specificity *in vitro*, it has been suggested that their role and selectivity is defined by their cellular localization *in vivo* [47–49]. In rat liver, the majority (~70%) of PTP activity measured against a ^{32}P-autophosphorylated IRK substrate is located in the particulate fraction [50]. A similar distribution of IRK-associated PTP activity is observed in rat skeletal muscle [51] and hepatoma cells [52] where the highest specific activity towards a phosphopeptide substrate modelled on the IRK kinase domain is detected in isolated membrane fractions. Work in our laboratory has focused on establishing the precise cellular location(s) of this membrane-bound PTP activity. Since activated IRKs are concentrated in ENs, and because IRKs are dephosphorylated prior to recycling to the cell surface, the endosomal compartment represents a logical site in which to find major dephosphorylating activity. This hypothesis was tested by Faure *et al.* [53] who identified substantial PTP(s) activity towards the IRK in rat liver endosomes. In these studies, ^{32}P-labelling of IRKs in intact endosomal membranes (*in situ* labelling) allowed for an assessment of PTP activity in a more physiological environment where the potential spatial relationship between the IRK and membrane-associated PTP(s) was not disrupted. Under these conditions the IRK was observed to dephosphorylate at 37°C with a $t_{1/2}$ of 1.6 min (Fig. 2). Dephosphorylation was unaffected by either KCl treatment, designed to remove peripheral membrane proteins, or the addition of rat liver cytosol. By contrast, treatment with Triton X-100 abolished IRK dephosphorylation (Fig. 2) suggesting that the observed PTP(s) activity resulted from the action of an intrinsic endosomal enzyme closely associated with the IRK [53]. Whether this PTP(s) represents a transmembrane protein and/or an intracellular enzyme tightly associated with the endosomal membrane remains unclear. Although these studies did not investigate IRK-dephosphorylating activity in plasma membrane fractions, more recent work with peroxovanadium PTP inhibitors has suggested that plasma membrane-located PTPs play a relatively minor role in IRK dephosphorylation *in vivo* (see below). Therefore it appears that IRK-associated PTP(s), either present or recruited to the endosomal compartment with internalized IRKs, are responsible for most IRK dephosphorylation *in vivo*.

Inhibition of IRK-associated PTPs by peroxovanadium compounds

We have recently described the insulin-mimetic properties of a number of crystallizable (> 95% pure) peroxovanadium (pV) compounds, each containing one or two peroxo anions, an oxo anion, and an ancillary ligand which confers stability on the complex [54]. When administered *in vivo* these agents activate the hepatic IRK, promote hypoglycaemia [40,

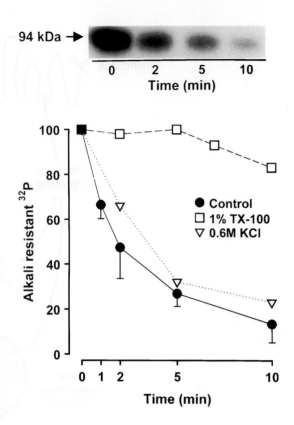

Fig. 2. Timecourse of dephosphorylation of ^{32}P-labelled endosomal insulin receptors. Rat liver ENs (25 μg), isolated 2 min following *in vivo* (intrajugular) injection of insulin (1.5 μg/100 g body weight), were autophosphorylated with 25 μM [γ-^{32}P]ATP. EDTA (10 μM) and ATP (500 μM) were then added to terminate ^{32}P-labelling of the IRK (*zero time*) hence unmasking the dephosphorylation process. To measure tyrosine dephosphorylation, fractions were solubilized at the noted times and IRKs immunoprecipitated and subjected to 7.5% SDS-PAGE followed by treatment with KOH treatment to remove ^{32}P-serine and threonine. Following autoradiography, the 94 kDa IRK band was quantitated by laser densitometry and the data plotted as a percent of the initial (*zero time*) level of phosphorylation. To study the effect of Triton X-100 (TX-100) dephosphorylation was initiated by adding ATP/EDTA (as above) containing TX-100 at a final concentration of 1.0%. To study the effect of removing peripheral membrane proteins, the autophosphorylation reaction was stopped by the addition of ice cold 0.6 M KCl. Following 30 min incubation at 4°C, the suspension was centrifuged at 100,000 g_{av} for 60 min. The pellet was resuspended in assay buffer (37°C) and dephosphorylation initiated by adding DTT (1 mM final concentration). The *Top Panel* shows a representative autoradiograph of endosomal IRK dephosphorylation under control conditions. The *Bottom panel* shows densitometric analyses of endosomal IRK dephosphorylation under control conditions (± S.E.) or following treatment with TX-100 or KCl. Figure adapted from reference [53].

54] and, depending on the compound, stimulate skeletal muscle glycogen synthesis [55]. The pVs have also been shown to activate the IRK in hepatoma cells and inhibit *in situ* dephosphorylation of rat liver endosomal IRKs with a potency > 10^3 times that of vanadate [56]. Indeed, synthetic pV compounds appear to be the most potent inhibitors of PTP activity yet described. The mechanism by which PTPs are inactivated by pV compounds appears to involve oxidation of the essential cysteine residue in the catalytic cleft of the PTP [57].

Since the potencies of different pV compounds as inhibitors of IRK dephosphorylation correlated well with their capacities to activate the IRK in hepatoma cells, it was inferred that IRK activation was linked to the inhibition of IRK dephosphorylation. Furthermore, following the administration to rats of one such compound, bpV(phen),[1] the time course and extent of inhibition of hepatic endosomal IRK-specific PTP activity closely paralleled the activation of IRK within ENs [40] (Fig. 3). In addition pV did not activate partially purified IRKs *in vitro* [58] indicating that pV-induced activation of the IRK was not mediated by a direct effect of pV on the IRK but rather via the inhibition of IRK-associated PTP(s). Cultured cells display a low level of IRK activity even in the absence of insulin [54, 59]. It is therefore possible that, in this basal state, a low-level futile cycle operates in which tyrosine phosphorylation and dephosphorylation of the IRK occur in the absence of net autophosphorylation. By inhibiting IRK-specific PTP(s), pV compounds disturb this equilibrium, leading to net IRK autophosphorylation and subsequent kinase activation (Fig. 4). This view is further supported by our recent observation that pV increases both IRK autophosphorylation and exogenous kinase activity in rat hepatoma (HTC) cells overexpressing normal but not mutant kinase-inactive IRKs.[2]

Following bpV(phen) administration *in vivo* there was a more rapid and marked stimulation of IRK activity in hepatic ENs compared to that of IRKs located at the cell surface. This is the opposite of what is observed following insulin administration [39]. When bpV(phen) was administered to colchicine-treated rats the activation of IRK at the plasma membrane was completely blocked without affecting either the extent or time course of IRK activation in ENs. Since colchicine is an inhibitor of receptor recycling this observation suggested that, following pV administration, autophosphorylated and activated cell surface IRKs were derived from the recycling of activated endosomal IRKs to the cell surface [55]. Thus endosomally-located PTP(s) appear to be the primary site of action of pV compounds in rat liver. This view has received additional support from the observation that bpV(phen) administration *in vivo* effected virtually complete inhibition of hepatic endosomal IRK dephosphorylation [60] whereas the activity of cytosolic PTP(s) [40] and the dephosphorylation of plasma membrane IRKs [60] were unaffected by this treatment. The mechanism by which pV compounds preferentially target endosomal PTP(s) is unclear. However, the potent insulin-mimetic effects of these compounds support earlier evidence [53] that endosomally-located PTP(s) play a key, if not principal, role in mediating IRK dephosphorylation *in vivo*.

Role of IRK-associated PTP(s) in IRK activation
In addition to a role in insulin-signal termination PTP(s) also appear to act to augment IRK activity. Thus it has been shown that the internalization of IRKs into rat hepatic ENs is associated with a transient increase in both IRK auto-phosphorylating and exogenous tyrosine kinase activity [34, 39]. Surprisingly this increase in endosomal IRK activity was accompanied by a 2–3 fold reduction in the phosphotyrosine content of the IRK β-subunit, compared to IRKs located at the plasma membrane. These observations suggested that IRK internalization in liver was associated with a partial dephosphorylation of phosphotyrosine residues. Thus full activation of the hepatic IRK appears to require limited and probably specific dephosphorylation of the internalized IRK β-subunit [39]. Recent work, utilizing bpV(phen) to specifically block endosomal IRK dephosphorylation *in vivo* has strengthened this hypothesis [60]. Thus by inhibiting dephosphorylation of the internalized IRK, augmentation of IRK activity within this intracellular compartment was

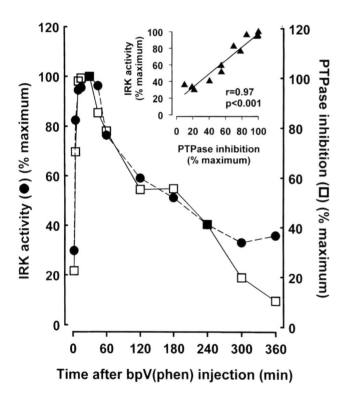

Fig. 3. Correlation between IRK activation and PTP inhibition in ENs after treatment with bpV(phen) and insulin. Rats were fasted overnight, given an intrajugular (i.j.) injection of 0.6 μmol/100 g body weight bpV(phen) and killed at the indicated times thereafter. Two minutes prior to sacrifice animals received an injection (i.j.) of 1.5 μg/100 g body weight insulin. Hepatic ENs were prepared and IRK and PTP activities measured as described in [40]. *Main panel*: Time course of activation of the IRK (●) and the corresponding inhibition of PTP activity (□). *Inset*: Linear correlation between IRK activation and PTP inhibition. Each observation is the mean of determinations on 2–5 animals. Figure adapted from reference [40].

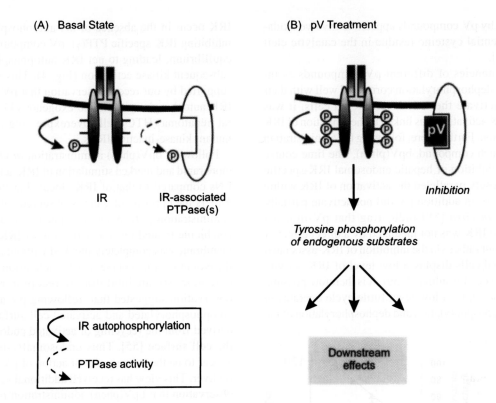

(A) Basal State

IR   IR-associated
PTPase(s)

IR autophosphorylation

PTPase activity

(B) pV Treatment

pV

Inhibition

*Tyrosine phosphorylation
of endogenous substrates*

Downstream
effects

Fig. 4. Proposed model for the activation of the IRK by pV compounds. (A) *Basal State*: Under basal conditions in the absence of insulin a low-level cycle operates in which IRK autophosphorylation and dephosphorylation occurs in the absence of net IRK autophosphorylation. (B) *pV Treatment*: by inhibiting IRK-specific PTP(s), pV compounds disturb this equilibrium, leading to IRK autophosphorylation, the tyrosine phosphorylation of endogenous substrates and the activation of downstream insulin-signalling pathways. IR – insulin receptor; PTP – protein tyrosine phosphatase; pV – peroxovanadium compound.

prevented [60]. The mechanism by which IRK activity is enhanced in ENs is unclear. Several groups have shown that carboxyterminal truncations of the IRK β-subunit [22, 45, 61] or point mutations of the β-subunit carboxyterminal tyrosines [62] increase both the autophosphorylating capacity [45] and the exogenous tyrosine kinase activity [22, 45, 62] of the receptor. These data are consistent with a restraining role for the two carboxyterminal tyrosines on the level of kinase activation attained following phosphorylation of the kinase regulatory domain of the IRK β-subunit. Thus an endosomal PTP(s) may act to dephosphorylate these carboxyterminal tyrosines promoting increased IRK activity (compared to that observed at the plasma membrane) before the continuation of IRK β-subunit dephosphorylation leads to eventual inactivation.

In contrast to the situation in rat liver, studies of insulin-stimulated receptor internalization in isolated adipocytes reveal that, although intracellular IRK tyrosine kinase activity is elevated compared to IRKs at the cell surface, a direct correlation exists between endosomal IRK tyrosine phosphorylation and exogenous kinase activity [36, 37]. Whether these differences reflect tissue-specific regulatory mechanisms related to the different physiological roles of

fat and liver remain to be established. Nonetheless, the observations in rat liver raise the intriguing possibility that, in addition to attenuating insulin action, PTP(s) may act to transiently increase the IRK signal. Moreover, in addition to modulating the level of IRK activity, IRK-associated PTP(s) may specifically and temporally regulate the phosphotyrosine-mediated interaction of the IRK with various signalling proteins (viz. IRS-1, IRS-2, PI3K, etc.) as the IRK proceeds through its intracellular itinerary.

IRK-associated PTPs: a key therapeutic target

In recent years agents that inhibit IRK-associated PTPs, and hence increase IRK activity, have received increasing attention as potential oral adjuvants or alternatives to insulin treatment in diabetic patients. Vanadium salts (vanadyl sulfate, sodium metavanadate, orthovanadate) are well known insulin-mimetics with *in vitro* studies demonstrating their ability to stimulate glucose uptake and metabolism, antilipolysis and glycogen synthase activation [63]. Moreover, oral administration of vanadate reduces hyperglycemia in a number of animal models of diabetes or insulin-resistance

[64]. Although the mode of action of vanadate remains to be fully elucidated, it is thought, by acting as a phosphate analogue, to be partly attributable to a competitive inhibition of PTP activity [57, 65]. However, concerns about the specificity and associated *in vivo* toxicity of vanadium salts which have been raised in recent years [66] may limit future clinical application.

Peroxovanadium compounds represent a new class of PTP inhibitor that may provide greater potential in this respect. When administered to normal rats *in vivo*, these agents promote acute hypoglycemia [40, 54] and, depending on the compound, stimulate skeletal muscle glucose uptake and glycogen synthesis [55]. Moreover, studies have shown that treatment with pV normalizes blood glucose levels in hypo-insulinemic diabetic BB rats completely removed from insulin treatment [67]. Studies in these rats indicate that parentally administered pV compounds produce hypoglycemia at doses which are substantially less than those inducing acute mortality (Fig. 5). By contrast, alternative insulin-mimetic agents, vanadate and vanadyl sulfate, fail to induce hypo-glycemia in BB rats before mortality occurs [67]. This enhanced 'therapeutic window' for pV compounds offers promise that therapeutically effective doses of suitable compounds will not induce toxic side effects.

Another potential advantage of pV treatment stems from observations with isolated human adipocytes. Whereas either pV [54] or vanadate [68] treatment induce an insulin-like response in rat adipocytes, only pV is able to mimic insulin action in human adipocytes [69, 70]. Although the reasons for these differences are unclear, the data suggest that pV compounds may have greater efficacy in human application.

Interestingly, a number of pV compounds with similar potencies as PTP inhibitors show markedly different potencies as hypoglycemic agents. These differences may relate to varying abilities of the compounds to stimulate both liver and skeletal muscle IRK kinase activity and skeletal muscle glycogen synthesis [56]. Thus tissue specificity may reflect differing abilities of these com-pounds to traverse the plasma membrane of skeletal muscle and/or different sensitivities of skeletal muscle PTPs to the inhibitory activities of these compounds. It is possible that a given ancillary ligand in the pV complex may target one tissue in preference to another, a feature that could be exploited clinically. Evidence suggests that the ancillary ligand may also define the specificity of PTP inhibition [56] allowing the design of pV compounds that specifically target the enzyme(s) responsible for IRK dephosphorylation while leaving other PTPs unaffected.

Studies *in vitro* have shown that PTPs display a certain degree of selectivity against different phosphotyrosine-containing proteins and synthetic peptides [71–74], suggest-ing that PTPs may have a preference for a particular amino acid sequence surrounding the phosphotyrosine residue of

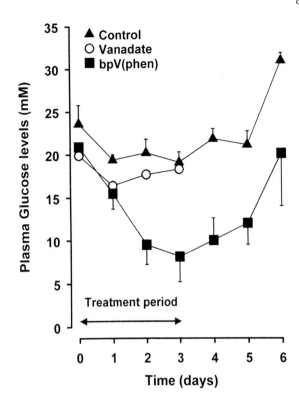

Fig. 5. Effect of bpV(phen) and, vanadate on fasting plasma glucose levels in insulin-deprived BB rats. Diabetic BB rats were treated with subcutaneous insulin to maintain hyperglycemia between 15 and 25 mM in the absence of ketonuria. Insulin was then withdrawn. Sixteen hours after the last subcutaneous insulin injection, animals received subcutaneous injections of either bpV(phen) (36 μmol/kg body wt b.i.d.) vanadate (36 μmol/kg body wt b.i.d.) or PBS (control) at 8:00 am and 4:00 pm for three days (*Treatment period*). During the three days of treatment and subsequent three days, rats were fed *ad libitum* from 4:00 pm to 8:00 am and fasted from 8:00 am to 4:00 pm. Data are presented as means for 6 animals ± S.E. Figure adapted from reference [67].

a protein. This possibility has prompted the development of a number of non-hydolyzable phosphotyrosine peptide analogs with the aim of selectively inhibiting PTPs that act on specific protein substrates. Using this approach, a synthetic tris-sulfotyrosyl dodecapeptide modelled on the IRK kinase domain was observed to potently inhibit IRK dephosphorylation [75] and enhance insulin-stimulated PI3K and mitogen-activated protein kinase (MAPK) activation [76] in permeabilized CHO cells. By contrast, this agent was without effect on epidermal growth factor (EGF) receptor activation or signalling [76] suggesting that selective inhibition of IRK dephosphorylation allowed specific activation of downstream components of insulin signal transduction pathways. However, although promising results have been obtained with permeabilized cells, it remains to be established whether such a PTP inhibitor can act to mimic or enhance the effects of insulin in animal models of insulin resistance and/or deficiency.

The nature of IRK-associated PTP(s)

The design of more specific PTP inhibitors for the management of diabetes should be facilitated by identification of the physiologically relevant PTP(s) which dephosphorylate(s) the IRK *in vivo*. To date, approximately 75 PTP(s) have been described that can be classified into three main groups: (1) receptor-like PTPs – composed of a variable extracellular domain, a single hydrophobic transmembrane domain and one or two intracellular catalytic domains; (2) intracellular PTPs – which contain a single catalytic domain and possess carboxyl and amino-terminal extensions thought to be responsible for subcellular localization and/or enzymatic regulation [48] and; (3) dual-specificity PTPs – a more recently identified class of PTP able to dephosphorylate both phosphotyrosine and phosphoserine/threonine residues [77]. Although the dual-specific PTPs have been shown to regulate the activity of the MAPK enzyme family [77], the physiological substrates and *in vivo* functions for most PTPs remain elusive.

Studies with purified enzymes and recombinant catalytic domains have revealed several PTPs that are active against the autophosphorylated IRK *in vitro* including the receptor-like PTPs, CD45, leukocyte antigen-related PTP (LAR) and leukocyte common antigen-related PTP (LRP, R-PTPα or PTPα), and the cytosolic PTPs, PTP1B and TC-PTP [78]. Using phosphopeptides modelled on the kinase regulatory domain of the IRK, differences have been observed in the *in vitro* specificity and rates by which particular PTPs dephosphorylate the three phosphotyrosines in the IRK kinase domain [78]. However, the lack of a defined three-dimensional structure may represent a limitation in using phosphopeptides to assess PTP specificity. In addition, PTP specificity appears to be determined, in part, by subcellular localization where associated proteins, targeted to specific domains of the cell, could influence enzyme activity. As a consequence, in recent years, work has focused on assessing PTP activity towards cellular substrates in intact cells.

The use of insulin-responsive cells transfected with and overexpressing selected PTPs has been one approach for assessing the role of a particular PTP *in vivo*. With this technique it has been shown that overexpression of LAR in CHO cells reduces insulin-stimulated IRK autophosphorylation [79]. Since overexpression of a cytosolic, truncated form of LAR was without effect on insulin-stimulated IRK autophosphorylation this suggested that LAR required a transmembrane localization to directly interact with the IRK *in situ* [79]. Although these observations support a role for LAR in IRK dephosphorylation, overexpression of this PTP in hamster kidney (BHK-IRK) cells had no effect on insulin-stimulated IRK autophosphorylation [80]. In this latter study a number of PTPs were assessed for their ability to inhibit IRK autophosphorylation and prevent insulin-stimulated growth inhibition of adherent cells. Of the PTPs tested, only overexpression of the receptor-like PTPs, PTPα (R-PTPα or LRP) and PTPε, significantly inhibited these insulin responses. Interestingly, although PTP1B expression was without effect in BHK-IRK cells [80], overexpression of PTP1B in kidney 293 cells [49] or Rat-1 fibroblasts [81] inhibited insulin-stimulated IRK autophosphorylation. The reasons for these conflicting observations are unclear but could reflect the use of different cells lines where relative levels of overexpressed PTP and IRK may vary.

An alternative approach to defining an *in vivo* role for a particular PTP has entailed 'knocking out' the PTP of interest. In this way, possible non-physiological interactions between highly overexpressed PTPs and IRKs, perhaps created through errant subcellular localization, are avoided. Use of antisense RNA, one strategy for selectively reducing PTP levels *in vivo*, has revealed that suppression of LAR levels by ~60% in rat hepatoma cells, increased both insulin-stimulated IRK autophosphorylation and exogenous kinase activity [82]. In another study, reduction in PTP1B levels by osmotically loading rat hepatoma cells with PTP1B antibodies was also observed to augment insulin-stimulated IRK tyrosine phosphorylation and activity [83].

A number of studies have described the direct interaction of the SH2 domain-containing PTP, SH-PTP2, with the carboxy-terminal region of the IRK β-subunit [14–16] via the interaction of the SH2-domains of this enzyme with both the kinase regulatory domain and Tyr^{1332} of the IRK β-subunit [84]. However, insulin-stimulated IRK phosphorylation was unaffected in hemizygous knockout mice expressing 50% of normal SH-PTP2 levels suggesting that this phosphatase does not play a role in IRK dephosphorylation *in vivo* [85].

As significant IRK dephosphorylating activity has been described in hepatic ENs [53, 60], this subcellular compartment represents a key site in which to identify the physiologically relevant PTP(s). Although immunoblotting failed to identify PTP1B or TC-PTP in hepatic ENs [53], LAR, PTPα and PTPε were detected in this intracellular compartment[3] raising the possibility of their involvement in IRK dephosphorylation. Studies of LAR have revealed an insulin-stimulated recruitment of this enzyme to rat hepatic ENs that was associated with a decrease in levels at the cell surface [86]. However, these studies did not directly compare the temporal relationship between endosomal IRK and LAR content in rat liver. In this regard it is important to note that previous *in vivo* studies have established that, following insulin administration, activated IRK is rapidly internalized into ENs, reaching maximal levels at 2–5 min post injection before returning to basal levels by ~15 min [34, 39]. In contrast the kinetics of LAR internalization were very different with LAR being maximally recruited to ENs at 30 min following insulin administration [86]. Nonetheless, incubation of endosomal fractions with neutralizing antibodies to LAR was reported to reduce *in situ* IRK dephosphorylating activity in this intracellular compartment by ~25% [86]. However, the fact that endosomal IRK dephosphorylation was reduced but not prevented by LAR inhibition raises the possibility that multiple PTPs may be involved in effecting the dephosphorylation of activated IRKs.

In summary, the identity of the physiologically relevant PTP(s) effecting IRK dephosphorylation remains uncertain, although certain interesting candidates have been identified. As *in vitro* studies suggest that different PTPs may dephosphorylate specific tyrosine residues on the IRK [78] this raises the possibility of a complex regulatory process whereby IRK dephosphorylation may involve several PTPs acting in tandem or at different stages of the IRKs activation 'lifespan' to regulate IRK-mediated signalling.

Conclusions

Evidence from various sources has demonstrated the critical role that PTP(s) play in regulating IRK autophosphorylation and kinase activity *in vivo*. Studies suggest that the majority of cellular PTP activity involved in dephosphorylating the IRK *in vivo* is membrane-bound and located in the endosomal apparatus. IRK dephosphorylation at this intracellular locus appears to be tightly regulated. Studies in rat liver suggest a complex regulatory process whereby PTPs may act, via selective IRK β-subunit tyrosine dephosphorylation, to modulate IRK activity in both a positive and negative fashion. Studies with a novel class of insulin-mimetic, the peroxovanadium compounds, has helped define the key role of ENs in IRK dephosphorylation *in vivo* and has indicated that selectively inhibiting IRK-associated PTP(s) offers great potential as a therapeutic approach to treating insulin-resistant and insulin-deficient forms of diabetes mellitus. As yet the identity of the physiologically relevant PTP(s) remains unclear. However, current evidence has raised the possibility that a number of PTPs may act in concert or at different stages of the 'lifespan' of the activated IRK to regulate IRK-mediated signalling.

Acknowledgements

Studies performed at the Polypeptide Hormone Laboratory (McGill University, Montreal, Quebec) are supported by grants from the Medical Research Council of Canada, Novo Nordisk in Denmark, and the Maurice Pollack Foundation, Montreal, Quebec, Canada. Paul G. Drake was previously supported by a Fellowship from the Royal Victoria Hospital Research Institute (Montreal, Quebec). We wish to thank Gerry Baquiran, Barbara Foster and Celyne Brule for their excellent technical assistance; Professor Alan Shaver and Dr. Jesse N. Ng (Department of Chemistry, McGill University) for advice on vanadium chemistry and the synthesis of the peroxovanadium compounds; and Drs. John J.M. Bergeron, A. Paul Bevan, James W. Burgess, Robert Faure and Jean-Francois Yale for their valuable contributions. Finally we thank Dr. Pascal Hingcamp for proofreading the manuscript.

References

1. Kahn CR: Insulin action, diabetogenes, and the cause of type 2 diabetes. Diabetes 43: 1066–1084, 1994
2. Waters SB, Pessin JE: Insulin receptor substrate 1 and 2 (IRS1 and IRS2): What a tangled web we weave. Trends Cell Biol 6: 1–4, 1996
3. Lee J, Pilch PF: The insulin receptor: Structure, function and signalling. Am J Physiol 266: C319–C334, 1994
4. White MF, Livingston JN, Backer JM, Lauris V, Dull TJ, Ullrich A: Mutation of the insulin receptor at tyrosine 960 inhibits signal transmission but does not effect its tyrosine kinase activity. Cell 54: 641–649, 1988
5. Backer JM, Schroeder GG, Kahn CR, Myers J, Wilden PA, Cahill DA, White MF: Insulin stimulation of phosphatidylinositol 3-kinase maps to insulin receptor regions required for endogenous substrate phosphorylation. J Biol Chem 267: 1367–1374, 1992
6. Danielsen G, Roth RA: Role of the juxtamembrane tyrosine in insulin receptor-mediated tyrosine phosphorylation of p60 endogenous substrates. Endocrinology 137: 5326–5331, 1996
7. White MF, Shoelson SE, Keutmann H, Kahn CR: A cascade of tyrosine autophosphorylation in the β-subunit activates the phosphotransferase of the insulin receptor. J Biol Chem 263: 2969–2980, 1988
8. Herrera R, Rosen O: Autophosphorylation of the insulin receptor *in vitro*: Designation of phosphorylation sites and correlation with receptor kinase activation. J Biol Chem 261: 11980–11985, 1986
9. Kwok YC, Nemenoff RA, Powers AC, Avruch J: Kinetic properties of the insulin receptor tyrosine protein kinase: Activation through an insulin-stimulated tyrosine specific, intramolecular autophosphorylation. Arch Biochem Biophys 244: 102–113, 1986
10. Wei L, Hubbard SR, Hendrickson WA, Ellis L: Expression, characterization, and crystallization of the catalytic core of the human insulin receptor protein-tyrosine kinase domain. J Biol Chem 270: 8122–8130, 1995
11. Hubbard SR, Wei L, Ellis L, Hendrickson WA: Crystal structure of the tyrosine kinase domain of the human insulin receptor. Nature 372: 746–754, 1994
12. Levy-Toledano R, Taouis M, Blaettler DH, Gorden P, Taylor SI: Insulin-induced activation of phosphatidyl inositol 3-kinase. Demonstration that the p85 subunit binds directly to the COOH terminus of the insulin receptor in intact cells. J Biol Chem 269: 31178–31182, 1994
13. Hansen H, Svensson U, Zhu J, Laviola L, Giorgino F, Wolf G, Smith RJ, Riedel H: Interaction between the Grb10 SH2 domain and the insulin receptor carboxyl terminus. J Biol Chem 271: 8882–8886, 1996
14. Ugi S, Maegawa A, Olefsky JM, Shigeta Y, Kashiwagi A: Src homology 2 domains of protein tyrosine phosphatase are associated *in vitro* with both the insulin receptor and insulin receptor substrate-1 via different phosphotyrosine motifs. FEBS Lett 340: 216–220, 1994
15. Staubs PA, Reichart DR, Saltiel AR, Milarsky KM, Maegawa H, Berhanu P, Olefsky JM, Seely BL: Localization of the insulin receptor binding sites for the SH2 domain proteins p85, Syp, and GTPase activating protein. J Biol Chem 269: 27186–27192, 1994

16. Kharintonenov A, Schnekenburger J, Chen Z, Knyazev P, Ali S, Zwick E, White M, Ullrich A: Adapter function of protein phosphatase 1D in insulin receptor/insulin receptor substrate-1 interaction. J Biol Chem 270: 29189–29193, 1995

17. Theis RS, Ullrich A, McClain DA: Augmented mitogenesis and impaired metabolic signalling mediated by a truncated insulin receptor. J Biol Chem 264: 12820–12825, 1989

18. Takata Y, Webster NJG, Olefsky JM: Mutation of the two carboxyl-terminal tyrosines results in an insulin receptor with normal metabolic signalling but enhanced mitogenic signalling properties. J Biol Chem 266: 9135–9139, 1991

19. Baron V, Gautier N, Kalliman P, Dolais-Kitabgi J, Van Obberghen E: The carboxy-terminal domain of the insulin receptor: Its potential role in growth promoting effects. Biochemistry 30: 9365–9370, 1991

20. Takata Y, Webster NJG, Olefsky JM: Intracellular signalling by a mutant human insulin receptor lacking the carboxy-terminal tyrosine. J Biol Chem 267: 9065–9070, 1992

21. Ando A, Momomura K, Tobe K, Yamamoto-Honda R, Sakura H, Tamori Y, Kaburagi Y, Koshio O, Akunuma Y, Yazaki Y, Kasuga M, Kadowaki T: Enhanced insulin-induced mitogenesis and mitogen-activated protein kinase activities in mutant insulin receptors with substitution of two COOH-terminal tyrosine autophosphorylation sites by phenylalanine. J Biol Chem 267: 12788–12796, 1992

22. Maegawa H, McClain DA, Freidenberg G, Olefsky JM, Napier M, Lipari T, Dull TJ, Lee J, Ullich A: Properties of a human insulin receptor with a COOH-terminal truncation. 2. Truncated receptors have normal kinase activity but are defective in signalling metabolic effects. J Biol Chem 263: 8912–8917, 1988

23. Cheatham B, Kahn CR: Insulin action and the insulin signalling network. Endocrine Rev 16: 117–142, 1995

24. Sale G: Serine/threonine kinases and tyrosine phosphatases that act on the insulin receptor. Biochem Soc Trans 20: 664–670, 1992

25. Carter WG, Sullivan AC, Asamoah KA, Sale GJ: Purification and characterization of an insulin-stimulated insulin receptor serine kinase. Biochemistry 35: 14340–14351, 1996

26. Baltensperger K, Lewis RE, Woon CW, Vissavajjhala AH, Ross AH, Czech MP: Catalysis of serine and tyrosine autophosphorylation by the human insulin receptor. Proc Natl Acad Sci USA 89: 7885–7889, 1992

27. Tauer TJ, Volle DJ, Rhode SL, Lewis RE: Expression of the insulin receptor with a recombinant vaccinia virus. Biochemical evidence that the insulin receptor has intrinsic serine kinase activity. J Biol Chem 271: 331–336, 1996

28. Burgess JW, Bevan AP, Bergeron JJM, Posner BI: Intracellular trafficking and processing of ligand receptor complexes in the endosomal system. Exp Clin Endocrinol 11: 67–78, 1992

29. Baass PC, Di Guglielmo GM, Authier F, Posner BI, Bergeron JJM: Compartmentalized signal transduction by receptor tyrosine kinases. Trends Cell Biol 5: 465–470, 1995

30. Bevan AP, Drake PG, Bergeron JJM, Posner BI: Intracellular signal transduction: The role of endosomes. TEM 7: 13–21, 1996

31. Authier F, Posner BI, Bergeron JJM: Hepatic endosomes are the major physiological locus of insulin and glucagon degradation in vivo. Cell Prot Sys: 89–113, 1994

32. Authier F, Posner BI, Bergeron JJM: Insulin-degrading enzyme. Clin Invest Med 19: 149–160, 1996

33. Khan MN, Savoie S, Bergeron JJM, Posner BI: Characterization of rat liver endosomal structures. In vivo activation of insulin-stimulable receptor kinase in these structures. J Biol Chem 261: 8462–8472, 1986

34. Khan MN, Baquiran G, Brule C, Burgess J, Foster B, Bergeron JJM, Posner BI: Internalization and activation of the rat liver insulin receptor kinase in vivo. J Biol Chem 264: 12931–12940, 1989

35. Backer JM, Kahn CR, White MF: Tyrosine phosphorylation of the insulin receptor during insulin-stimulated internalization in rat hepatoma cells. J Biol Chem 294: 1694–1701, 1989

36. Klein HH, Freidenberg GR, Matthaei S, Olefsky JM: Insulin receptor kinase following internalization in isolated rat adipocytes. J Biol Chem 262: 10557–10564, 1987

37. Kublaoui B, Lee J, Pilch PF: Dynamics of signalling during insulin-stimulated endocytosis of its receptor in adipocytes. J Biol Chem 270: 59–65, 1995

38. Wang B, Balba Y, Knutson VP: Insulin-induced in situ phosphorylation of the insulin receptor located in the plasma membrane versus endosomes. BBRC 227: 27–34, 1996

39. Burgess JW, Wada I, Ling N, Khan MN, Bergeron JJM, Posner BI: Decrease in β-subunit phosphotyrosine correlates with internalization and activation of the endosomal insulin receptor kinase. J Biol Chem 267: 10077–10086, 1992

40. Bevan AP, Burgess JW, Drake PG, Shaver A, Bergeron JJM, Posner BI: Selective activation of the rat hepatic endosomal insulin receptor kinase. Role for the endosome in insulin signalling. J Biol Chem 270: 10784–10791, 1995

41. Kelly KL, Ruderman NB: Insulin-stimulated phosphatidylinositol 3-kinase: Association with a 185 kDa tyrosine phosphorylated protein (IRS-1) and localization in a low density membrane vesicle. J Biol Chem 268: 4391–4398, 1993

42. Rosen OM, Herrera R, Olowe Y, Petruzzelli LM, Cobb MH: Phosphorylation activates the IR tyrosine protein kinase. Proc Natl Acad Sci USA 80: 3237–3240, 1983

43. Mooney RA, Anderson DL: Phosphorylation of the insulin receptor in permeablized adipocytes is coupled to a rapid dephosphorylation reaction. J Biol Chem 264: 6850–6857, 1989

44. Mooney RA, Bordwell KL: Differential dephosphorylation of the insulin receptor and its 160 kDa substrate (pp160) in rat adipocytes. J Biol Chem 267: 14054–14060, 1992

45. Bernier M, Liotta AS, Kole HK, Shock DD, Roth J: Dynamic regulation of intact and C-terminal truncated insulin receptor phosphorylation in permeablized cells. Biochemistry 33: 4343–4351, 1994

46. Goldstein BJ: Protein-tyrosine phosphatases and the regulation of insulin action. J Cell Biochem 48: 33–42, 1992

47. Faure R, Posner BI: Differential intracellular compartmentalization of phosphotyrosine phosphatases in a glial cell line: TC-PTP Versus PTP-1B. GLIA 9: 311–314, 1993

48. Mauro LJ, Dixon JE: 'Zip codes' direct intracellular protein tyrosine phosphatases to the correct cellular 'address'. TIBS 19: 151–155, 1994

49. Lammers R, Bossenmaier B, Cool DE, Tonks NK, Schlessinger J, Fischer EH, Ullrich A: Differential activities of protein tyrosine phosphatases in intact cells. J Biol Chem 268: 22456–22462, 1993

50. King MJ, Sale GJ: Insulin-receptor phosphotyrosyl-protein phosphatases. Biochem J 256: 893–902, 1988

51. Ahmad F, Goldstein BJ: Purification, identification and subcellular distribution of three predominant protein-tyrosine phosphatase enzymes in skeletal muscle tissue. Biochim Biophys Acta 1248: 57–69, 1995

52. Meyerovitch J, Backer JM, Csermely P, Shoelson SE, Kahn CR: Insulin differentially regulates protein phosphotyrosine phosphatase activity in rat hepatoma cells. Biochemistry 31: 10338–10344, 1992

53. Faure R, Baquiran G, Bergeron JJM, Posner BI: The dephosphorylation of insulin and epidermal growth factor receptors. The role of endosome-associated phosphotyrosine phosphatase(s). J Biol Chem 267: 11215–11221, 1992

54. Posner BI, Faure R, Burgess JW, Bevan AP, Lachance D, Zhang-Sun G, Fantus IG, Ng JN, Hall DA, Soo Lum B, Shaver A: Peroxovanadium compounds. A new class of potent phosphotyrosine phosphatase inhibitors which are insulin mimetics. J Biol Chem 269: 4596–4604, 1994

55. Bevan AP, Burgess JW, Yale J-F, Drake PG, Lachance D, Baquiran G, Shaver A, Posner BI: In vivo insulin mimetic effects of pV compounds: Role for tissue targeting in determining potency. Am J Physiol 268: E60–E66, 1995

56. Bevan AP, Drake PG, Yale J-F, Shaver A, Posner BI: Peroxovanadium compounds: Biological actions and mechanism of insulin-mimesis. Mol Cell Biochem 153: 49–58, 1995

57. Huyer G, Liu S, Kelly J, Moffat J, Payette P, Kennedy B, Tsaprailis G, Gresser MJ, Ramachandran C: Mechanism of inhibition of protein-tyrosine phosphatases by vanadate and pervanadate. J Biol Chem 272: 843–851, 1997

58. Fantus IG, Kadota S, Deragon G, Foster B, Posner BI: Pervanadate [peroxides of vanadate] mimics insulin action in rat adipocytes via activation of the insulin receptor kinase. Biochemistry 28: 8864–8871, 1989

59. Kadota S, Fantus IG, Deragon G, Guyada HJ, Hersh B, Posner BI: Stimulation of insulin-like growth factor 2 receptor binding and insulin receptor kinase activity in rat adipocytes. Effects of vanadate and H_2O_2. J Biol Chem 262: 8252–8256, 1987

60. Drake PG, Bevan AP, Burgess JW, Bergeron JJM, Posner BI: A role for tyrosine phosphorylation in both activation and inhibition of the insulin receptor tyrosine kinase in vivo. Endocrinology 137: 4960–4968, 1996

61. Tavare JM, Ramos P, Ellis L: An assessment of human insulin receptor phosphorylation and exogenous kinase activity following deletion of 69 residues from the carboxy-terminus of the receptor β-subunit. Biochem Biophys Res Comm 188: 86–93, 1992

62. Kaliman P, Baron V, Alengrin F, Takata Y, Webster NJG, Olefsky JM, Van Obberghen E: The insulin receptor C-terminus is involved in the regulation of the receptor kinase activity. Biochemistry 32: 9539–9544, 1993

63. Shechter Y: Insulin-mimetic effects of vanadate: Possible implications for future treatment of diabetes. Diabetes 39: 1–5, 1990

64. Shechter Y, Li J, Meyerovitch J, Gefel D, Bruck R, Elberg G, Miller DS, Shisheva A: Insulin-like actions of vanadate are mediated in an insulin-receptor-independent manner via non-receptor protein tyrosine kinases and protein phosphotyrosine phosphatases. Mol Cell Biochem 153: 39–47, 1995

65. Tracey AS, Gressner MJ: Interaction of vanadate with phenol and tyrosine: Implications for the effects of vanadate on systems regulated by tyrosine phosphorylation. Proc Natl Acad Sci USA 83: 609–613, 1986

66. Domingo JL, Gomez M, Sanchez DJ, Llobet JM, Keen CL: Toxicology of vanadium compounds in diabetic rats: The action of chelating agents on vanadium accumulation. Mol Cell Biochem 153: 233–240, 1995

67. Yale J-F, Lachance D, Bevan AP, Vigeant C, Shaver A, Posner BI: Hypoglycemic effects of peroxovanadium compounds in Sprague-Dawley and diabetic BB rats. Diabetes 44: 1274–1279, 1995

68. Eriksson JW, Lonnroth P, Smith U: Vanadate increases cell surface insulin binding and improves insulin sensitivity in both normal and insulin-resistant rat adipocytes. Diabetologia 35: 510–516, 1992

69. Lonnroth P, Eriksson JW, Posner BI, Smith U: Peroxovanadate but not vanadate exerts insulin-like effects in human adipocytes. Diabetologia 36: 113–116, 1993

70. Eriksson JW, Lonnroth P, Posner BI, Shaver A, Wesslau C, Smith UPG: A stable peroxovanadium compound with insulin-like action in human fat cells. Diabetologia 39: 235–242, 1996

71. Hashimoto N, Feener EP, Zhang W-R, Goldstein BJ: Insulin receptor protein tyrosine phosphatases. Leukocyte common antigen-related phosphatase rapidly deactivates the insulin receptor kinase by preferential dephosphorylation of the receptor regulatory domain. J Biol Chem 267: 13811–13814, 1992

72. Ramachandran C, Aebersold R, Tonks NK, Pot DA: Sequential dephosphorylation of a multiply phosphorylated insulin receptor peptide by protein tyrosine phosphatases. Biochemistry 31: 4232–4238, 1992

73. Zhang ZY, Thieme-Sefler AM, Maclean D, McNamara DJ, Dobrusin EM, Sawyer TK, Dixon JE: Substrate specificity of the protein tyrosine phosphatases. Proc Natl Acad Sci USA 90: 4446–4450, 1993

74. Cho H, Krishnaraj R, ltoh M, Kitas E, Bannwarth W, Saito H, Walsh CT: Substrate specificity's of catalytic fragments of protein tyrosine phosphatases (HPTP beta, LAR, and CD45) toward phosphotyrosyl-peptide substrates and thiophosphotyrosylated peptides as inhibitors. Protein Science 2: 977–984, 1993

75. Liotta AS, Kole HK, Fales HM, Roth R, Bernier M: A synthetic tris-sulfotyrosyl dodecapeptide analogue of the insulin receptor 1146-kinase domain inhibits tyrosine dephosphorylation of the insulin receptor in situ. J Biol Chem 269: 22996–23001, 1994

76. Kole HK, Garant MJ, Kole S, Bernier M: A peptide based protein-tyrosine phosphatase inhibitor specifically enhances insulin receptor function in intact cells. J Biol Chem 271: 14302–14307, 1996

77. Tonks NK, Neel BJ: From form to function: Signalling by protein tyrosine phosphatases. Cell 87: 365–368, 1996

78. Goldstein BJ: Regulation of insulin receptor signalling by protein-tyrosine dephosphorylation. Receptor 3: 1–15, 1993

79. Zhang W-R, Li P-M, Oswald MA, Goldstein BJ: Modulation of insulin signal transduction by eutopic overexpression of the receptor-type protein tyrosine phosphatase LAR. Mol Endocrinol 10: 575–584, 1996

80. Moller NPH, Moller KB, Lammers R, Kharitonenkov A, Hoppe E, Wiberg FC, Sures I, Ullrich A: Selective down-regulation of the insulin receptor signal by protein-tyrosine phosphatases α and ε. J Biol Chem 270: 23126–23131, 1995

81. Kenner KA, Anyanwu E, Olefsky JM, Kusari J: Protein-tyrosine phosphatase 1B is a negative regulator of insulin- and insulin-like growth factor-1-stimulated signalling. J Biol Chem 271: 19810–19816, 1996

82. Kulas DT, Zhang W-R, Goldstein BJ, Furianetto RW, Mooney RA: Insulin receptor signalling is augmented by antisense inhibition of the protein tyrosine phosphatase LAR. J Biol Chem 270: 2435–2438, 1995

83. Ahmad F, Li P-M, Meyerovitch J, Goldstein BJ: Osmotic loading of neutralizing antibodies demonstrates a role for protein-tyrosine phosphatase 1B in negative regulation of insulin action pathway. J Biol Chem 270: 20503–20508, 1995

84. Rocchi S, Tartare-Deckert S, Swaka-Verhelle D, Gamha A, Van Obberghen E: Interaction of SH2-containing protein tyrosine phosphatase -2 with the insulin receptor and the insulin-like growth factor-1 receptor: Studies of the domains involved using the yeast two-hybrid system. Endocrinology 137: 4944–4952, 1996

85. Arrandale JM, Gore-Wilise A, Rocks S, Ren J-M, Zhu J, Davis A, Livingston JN, Rabin DU: Insulin signalling in mice expressing reduced levels of Syp. J Biol Chem 271: 21353–21358, 1996

86. Ahmad F, Goldstein BJ: Functional association between the insulin receptor and the transmembrane protein-tyrosine phosphatase LAR in intact cells. J Biol Chem 272: 448–457, 1997

87. King GL, Johnson SM: Receptor-mediated transport of insulin across endothelial cells. Science 227: 1583–1586, 1985

Molecular and Cellular Biochemistry **182**: 91–99, 1998.
© 1998 *Kluwer Academic Publishers. Printed in the Netherlands.*

Regulation of the insulin signalling pathway by cellular protein-tyrosine phosphatases

Barry J. Goldstein, Faiyaz Ahmad, Wendi Ding, Pei-Ming Li and Wei-Ren Zhang
Dorrance H. Hamilton Research Laboratories, Division of Endocrinology, Diabetes and Metabolic Diseases, Department of Medicine, Jefferson Medical College of Thomas Jefferson University, Philadelphia, USA

Abstract

Protein-tyrosine phosphatases (PTPases) have been implicated in the physiological regulation of the insulin signalling pathway. In cellular and molecular studies, the transmembrane, receptor-type PTPase LAR and the intracellular, non-receptor enzyme PTP1B have been shown to have a direct impact on insulin action in intact cell models. Since insulin signalling can be enhanced by reducing the abundance or activity of specific PTPases, pharmaceutical agents directed at blocking the interaction between individual PTPases and the insulin receptor may have potential clinical relevance to the treatment of insulin-resistant states such as obesity and Type II diabetes mellitus. (Mol Cell Biochem **182**: 91–99, 1998)

Key words: insulin action, transmembrane signalling, insulin resistance, tyrosine kinase, insulin receptor, receptor internalization

Introduction

Over the past several years, evidence has supported the hypothesis that specific cellular protein-tyrosine phosphatases (PTPases) are expressed in insulin-sensitive tissues and have a physiological role in the regulation of reversible tyrosine phosphorylation of the insulin receptor and post-receptor substrates in the insulin action pathway. Since the purified insulin receptor will retain its autophosphorylation state *in vitro* even after insulin is removed from the ligand binding site, cellular PTPase enzymes play an important role in the regulation of insulin action *in vivo* by catalyzing the rapid dephosphorylation and deactivation of the receptor kinase in a manner that determines the steady-state insulin receptor kinase activity. This brief review will discuss recent cellular and molecular evidence for involvement of PTPases in the negative regulation of insulin signalling, notably the transmembrane, receptor-type PTPase LAR and the intracellular, non-receptor enzyme PTP1B. Since insulin signalling can be enhanced by reducing the abundance or activity of specific PTPases, this work has potential clinical relevance to the treatment of insulin-resistant states with pharmaceutical agents directed at individual PTPases.

Reversible tyrosine phosphorylation in insulin signalling

One of the major advances in our understanding of insulin action has been the characterization of the central role of reversible tyrosine phosphorylation of the insulin receptor and its cellular substrate proteins in the mechanism of insulin action [1]. Insulin initiates its cellular effects by binding to a specific heterotetrameric plasma membrane receptor which encodes a tyrosyl-specific protein kinase. The kinase activity of the insulin receptor is essential for downstream activation of virtually all of insulin's growth-promoting and metabolic effects, as shown in studies using cells from patients with naturally occurring insulin receptor mutations and studies employing site-directed mutagenesis of the receptor [1]. Insulin binding elicits the rapid autophosphorylation of specific tyrosine residues as a cascade involving the receptor kinase domain, the C-terminus and the

Address for offprints: B.J. Goldstein, Division of Endocrinology, Diabetes and Metabolic Diseases, Jefferson Medical College, Room 349, Alumni Hall, 1020 Locust Street, Philadelphia, PA 19107, USA

juxtamembrane domain. Detailed studies on the activation of the insulin receptor kinase have shown that phosphorylation of two tyrosines in the kinase domain, involving Tyr-1158 and either Tyr-1162 or 1163 occurs first and that the resulting receptors with mono- or bis-phosphorylated kinase domains exhibit minimal activation of the β-subunit kinase activity. Phosphorylation of the third tyrosyl residue in the receptor 'regulatory domain' follows the bis-phosphorylation stage rapidly and leads to full activation of the receptor kinase towards exogenous substrates [2–4].

In intact cells, the relative amount of tris-phosphorylated receptors present after insulin stimulation is less than that seen after activation of partially purified receptors *in vitro* [5], suggesting that the tris-phosphorylated, activated insulin receptor is a preferential substrate for one or more cellular PTPases that can attenuate receptor activation. The tris-phosphorylated insulin receptor is also more rapidly dephosphorylated than the bis-phosphorylated receptor *in vitro* by subcellular fractions of rat liver [6]. Since only the tris-phosphorylated insulin receptor has full kinase activity, the PTPase-catalyzed conversion of tris- to bis-phosphorylation of the regulatory domain can be considered to be a discrete molecular 'switch' which determines the overall degree of insulin receptor kinase activation.

Tyrosyl phosphorylation of the receptor juxtamembrane region at Tyr-972 is involved in recognition and phosphorylation of substrates by the receptor kinase, including the ~185–190 kDa phosphoprotein insulin receptor substrates-1 and 2 (IRS-1 and 2) and Shc and the subsequent activation of glycogen and DNA synthesis by insulin [7]. In the receptor C-terminus, the autophosphorylation of Tyr-1328 and Tyr-1334 affects the catalytic efficiency and stability of the receptor β-subunit [8]. C-terminal tyrosine phosphorylation of the insulin receptor may affect signalling to distal effects of insulin, including MAP kinase, glycogen synthesis and mitogenesis, although the magnitude of these effects varies in different cell types [9]. IRS-1 is phosphorylated by the insulin receptor kinase on at least 8 tyrosines, including residues 460, 608, 628, 895, 939, 987, 1172 and 1222 [7]. When phosphorylated on these tyrosines, IRS-1 is thought to act as an adapter or 'docking' protein for the binding and activation of a variety of src-homology 2 (SH2) domain-containing signalling proteins, which form a tight but non-covalent association with the phosphotyrosyl domains of IRS-1. Included in the variety of proteins known to interact with IRS-1 in this way are the p85 subunit of phosphatidyl-inositol (PI)-3′ kinase, the SH2/SH3 adaptor proteins Grb-2 and Nck, and the intracellular protein-tyrosine phosphatase, SH-PTP2 [7]. The reversible phosphorylation of various domains of the insulin receptor and its substrate proteins in cells may differentially regulate insulin action and serve as a mechanism for sorting of some of the pleiotropic insulin responses.

Role of protein-tyrosine phosphatases in the regulation of insulin action

The current view of reversible tyrosine phosphorylation in the regulation of the insulin receptor has lead to the hypothesis that the steady-state balance of these events is determined by the opposing actions of receptor auto-phosphorylation, which activates the kinase activity, and cellular PTPases, which deactivate the receptor kinase. An essential observation regarding insulin receptor regulation is that purified heterotetrameric insulin receptors do not self-dephosphorylate and they retain their autophosphoryl-ation state *in vitro* even after insulin is removed from the ligand binding site [10, 11]. In contrast, when studied *in situ*, in intact [10] or permeabilized cells [12, 13], dissociation of insulin from the receptor is followed by a rapid dephos-phorylation of the β-subunit and a concomitant deactivation of the receptor kinase. Thus, cellular PTPases play a central role in the regulation of insulin action by dephosphorylating and inactivating the receptor kinase in a manner that balances and terminates the insulin receptor signal [14].

Internalization and dephosphorylation of the insulin receptor

Following autophosphorylation of the insulin receptor in the plasma membrane and receptor clustering at the surface of liver cells and adipocytes, the insulin receptor is internalized through an endosomal compartment by a process that is associated with dynamic changes in the receptor phos-phorylation state [15–19]. Posner and his colleagues have shown that internalization of the insulin receptor leads to a transient increase in the insulin receptor kinase activity (by as much as 3–5 fold) by a mechanism involving partial insulin receptor dephosphorylation [20, 21], prior to insulin receptor recycling back to the plasma membrane in the basal state [22]. The nature of the limited receptor dephos-phorylation has not yet been determined. In adipocytes, Kublaoui *et al.* [16] demonstrated that after a brief exposure to insulin the internalized insulin receptor is initially more highly phosphorylated and has a higher kinase activity than the insulin receptor in the plasma membrane, which subsequently declines, suggesting that insulin receptor inactivation and its interaction with cellular PTPases may occur in the endosomal fraction of adipocytes.

The PTPase superfamily of enzymes

Since one of the major goals of work in this area has been to identify specific PTPases that regulate the insulin signalling pathway and determine their mechanism of action, it is essential to understand some of the properties of PTPases from recent molecular cloning and biochemical studies. The PTPases comprise an extensive family of proteins that exert both positive and negative influences on several pathways of cellular signal transduction and metabolism [23, 24]. PTPases have in common a conserved ~230 amino acid domain that contains the PTPases signature sequence motif – (I/V)HCXAGXGR(S/T)G – which includes the cysteine residue that catalyzes the hydrolysis of protein-phosphotyrosine residues by the formation of a cysteinyl-phosphate intermediate [25].

PTPases have been divided into two broad categories: *receptor-type*, which have a general structure like a membrane receptor with an extracellular domain, a single transmembrane segment and one or two tandemly conserved PTPase catalytic domains; and *nonreceptor*-type, which have a single PTPase domain and additional functional protein segments (Fig. 1). Unfortunately, this classification fails to convey the exact subcellular localization of many PTPase homologs, since 'non-receptor' or intracellular PTPases are frequently found in particulate fractions as protein-complexes or due to functional segments that target the enzymes to specific intracellular sites, such as the endoplasmic reticulum. Many of the receptor-type PTPases also have conserved Ig-like and fibronectin-III structural repeats in the extracellular segments, as well as functional domains such as the MAM motif which may confer homotypic interactions between PTPases on neighboring cells. Only in a few instances have putative intracellular targets been identified for PTPase enzymes [25]. The best characterized examples include the activation of p561[lck] by the dephosphorylation of a negative regulatory residue (Tyr-505) near the C-terminus of lck by the lymphocyte-specific PTPase CD45 (Leukocyte Common Antigen), the regulation

of the p34[cdc2] kinase by cdc25, a PTPase that dephosphorylates Tyr-15 and triggers cellular entry through the $G_2 \rightarrow$ M phase of the cell cycle, and MKP-1, an intracellular dual-specificity (Tyr/Thr) PTPase that has been shown to have specificity towards the deactivation of MAP kinase both *in vivo* and *in vitro*.

Identification of PTPases that regulate the insulin action pathway

In order to identify candidate PTPases that act on the insulin signalling pathway, the tissue expression of various PTPase homologs, their subcellular localization and catalytic specificity for the insulin receptor should be considered. These criteria, as well as recent data that has addressed the modulation of insulin signalling by PTPases in intact cells, will be discussed below.

Tissue distribution

Since the restricted tissue distribution of some PTPases can be an important factor in determining their specialized cellular roles, identification of PTPases expressed in insulin-sensitive tissues, including liver, skeletal muscle and adipose tissue, has been an important criterion for identifying candidate enzymes for the physiological regulation of insulin signalling. Liver is a rich source of PTPase activities in crude particulate and soluble tissue fractions that can dephosphorylate the insulin receptor [26]. RNA and immunoblot analysis as well as cDNA library screening has revealed relatively abundant expression of PTP1B, SH-PTP2, LAR (Leukocyte Antigen Related, named

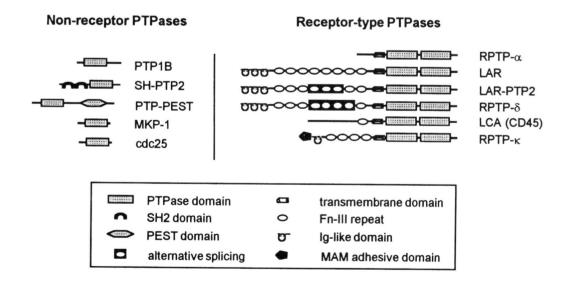

Fig. 1. Representative structures of protein-tyrosine phosphatases. Non-receptor and receptor-type protein tyrosine phosphatases are shown schematically with protein motifs as indicated in the legend. References to sequence citations and biochemical data on these enzymes can be found in reference [25].

94

by its homology to CD45), LRP (RPTP-α), and RPTP-κ in normal liver [25]. Our molecular cloning studies using an LAR cDNA probe (gift of Dr. Haruo Saito) demonstrated a similar abundance of 7 per 10^6 cDNA inserts for LAR and the insulin receptor in a rat liver cDNA library [27].

The identification of PTPases expressed in skeletal muscle is of particular importance, since the clinical insulin resistance associated with Type II diabetes mellitus and obesity is predominantly due to defects in insulin action in this tissue [28]. Amplification of rat skeletal muscle mRNA by a reverse transcription/polymerase chain reaction technique with degenerate primers to conserved PTPase domains revealed a limited set of PTPase cDNA transcripts which included LAR and LRP (RPTP-α) [29]. However, subsequent library screening using the LRP cDNA as a hybridization probe at reduced stringency demonstrated that this PTPase has particularly low expression in muscle tissue. Northern analysis of skeletal muscle RNA has also confirmed the expression of PTP1B and SH-PTP2 expression in muscle [30, 31]. In related studies, purification of the major peaks of PTPase activity from rat skeletal muscle particulate and cytosol fractions by serial ion exchange and molecular sieving FPLC followed by affinity chromatography, enabled us to identify major candidate skeletal muscle PTPases by immunoblotting [32]. These confirmed the relatively abundant expression of LAR in the particulate fraction and PTP1B and SH-PTP2, which were distributed between the cytosol and particulate fractions. By immunodepletion, these 3 enzymes accounted for more than 70% of the total PTPase activity in the muscle homogenates [32]. Immunoblotting also failed to detect significant amounts of LRP in muscle subcellular fractions, suggesting that it may be less likely to play a significant role in the regulation of insulin action in this tissue.

Screening of a rat cDNA library from isolated fat cells at reduced stringency with a panel of candidate PTPase probes was used to identify PTPases in adipocytes [33]. Inserts for LRP, PTP1B, SH-PTP2, and LAR were found at 16, 7, 6 and 3 per million, respectively. Furthermore, a 5 amino acid sequence variant of SH-PTP2 (positions 409–413) was identified within the catalytic domain (GenBank #U09307) that may affect the tissue-specific activity of this enzyme, since it can reduce the enzyme V_{max} by 8–20 fold [34].

Subcellular localization
In studies performed with insulin receptors in situ in permeabilized adipocytes or CHO cells essentially devoid of cytoplasmic contents, dissociation of insulin was followed by a rapid dephosphorylation of the insulin receptor, suggesting that a major tyrosine phosphatase for the insulin receptor is an integral membrane protein or one otherwise closely linked to membrane proteins or to the receptor itself [12, 13]. The bulk of PTPase activity towards the insulin receptor is also recovered in a particulate fraction with the highest

specific PTPase activity in a glycoprotein extract of the solubilized plasma membrane [26]. In skeletal muscle, the PTPase activity towards the insulin receptor is ~6 fold higher in the particulate fraction than in the cytosol [32]. PTP1B is expressed on the cytoplasmic face of the endoplasmic reticulum [35]; however, a significant portion of PTP1B is also found in the cytosol of fibroblasts [35] as well as in the cytosol of normal rat skeletal muscle and liver tissue [32, 36]. The transmembrane PTPase LAR is localized to both plasma membranes and the microsomal fraction of liver and skeletal muscle and absent from the cell cytosol. SH-PTP2 is distributed between the cytosol and particulate fractions, perhaps associated as SH2 domain/protein-phosphotyrosyl complexes with membrane-bound proteins or receptors [32, 37, 38]. Both LAR and PTP1B have been identified in liver endosomes, where PTPases appear to have a role in the dynamic dephosphorylation of the insulin receptor that occurs in this subcellular fraction (see below).

Substrate specificity
PTPases will generally dephosphorylate a variety of substrates with different kinetic parameters, and several PTPases are active against the autophosphorylated insulin receptor *in vitro*, including PTP1B, LAR, LRP, SH-PTP2, and CD45 [25]. *In vitro*, the recombinant catalytic domains of LAR, PTP1B and LRP can efficiently dephosphorylate insulin and EGF receptors, and LRP appeared to have a catalytic preference for the autophosphorylated EGF receptor which it dephosphorylated ~2 fold more rapidly than the insulin receptor [30]. The regional dephosphorylation of tyrosine residues in the insulin receptor regulatory domain by LAR, LRP or PTP1B *in vitro* was also investigated [39]. Using recombinant PTPase catalytic domains from an *E. coli* expression system, relative to the level of overall receptor dephosphorylation, LAR was found to deactivate the receptor kinase 2–3 times more rapidly than either PTP1B or LRP. Furthermore, tryptic mapping of the insulin receptor β-subunit after dephosphorylation by PTPases showed that LAR dephosphorylated the tris-phosphorylated (tyr-1150) receptor kinase domain 3–4 times more rapidly than either PTP1B or LRP, indicating that the effect of LAR to inactivate the receptor kinase was due to a preferential dephosphorylation of the receptor regulatory domain by this enzyme [39]. These biochemical data have supported a potential role for the LAR PTPase in the regulation of the insulin receptor kinase.

Candidate PTPases for the insulin action pathway – cellular studies

In order to establish the potential physiological relevance of specific PTPase homologs, cellular studies have provided important *in vivo* data, as summarized below.

LAR

LAR is one of the PTPases expressed in insulin-sensitive tissues that has been implicated in the physiological regulation of insulin action. LAR has an interesting post-translational itinerary that may have important consequences for regulation of its PTPase activity and/or its association with physiological substrates in target cells [40, 41]. It is initially synthesized as a ~200 kDa proprotein that is processed at a pentabasic site (aa 1148–1152) by a subtilisin-like protease into a complex of two non-covalently associated subunits: the extracellular or E-subunit (150 kDa) contains the cell adhesion molecule domains, and the phosphatase or P-subunit (85 kDa) contains an 82 amino acid extracellular region and the transmembrane and cytoplasmic domains (Fig. 2). Interestingly, a portion of the E-subunit is apparently shed during growth of HeLa cells and pre-B lymphocytes transfected with the LAR cDNA by proteolytic cleavage at a second site within the P-subunit ectodomain near the transmembrane domain [41]. As with other transmembrane PTPase homologs with a similar structure, the role of the tandem PTPases of LAR is unclear. The catalytic cysteine residue of LAR (Cys-1539) was identified in studies using site-directed mutagenesis of this residue; changing it to serine which prevents the formation of the catalytic phosphoenzyme intermediate [42, 43]. Since this single amino acid change ablates the activity of the cytoplasmic domain, the second (D2) domain is thought to be catalytically inactive. The close homology between PTPase domains D1 and D2 (each ~40% identical to the PTPase domain of PTP1B) suggests that each domain might bind to potential phosphotyrosyl-protein substrates, although only domain D1 appears to have catalytic activity.

LAR has emerged as an important candidate enzyme for the regulation of the insulin receptor for several reasons. LAR is widely expressed in insulin-sensitive tissues [26, 44],

it is a transmembrane protein that is localized to the membrane fraction of the cell [32, 38], and its cytoplasmic domain has a catalytic preference for the regulatory phosphotyrosines of the insulin receptor kinase domain *in vitro* [30, 39]. To further support the hypothesis that LAR has a role in the physiological regulation of insulin receptor phosphorylation in intact cells, in collaboration with Dr. Robert Mooney, we transfected LAR antisense mRNA in hepatoma cells which reduced LAR mass by 63% and resulted in an amplification of insulin-stimulated PI3′-kinase activity to 350% over the level observed in cells transfected with the null expression vector. In addition, decreased LAR expression resulted in an augmentation of additional post-receptor events including IRS-1 tyrosine phosphorylation, IRS-1 complexing with p85 subunit of PI3′-kinase, IRS-1 associated PI3′-kinase activity and the activation of both MAP kinase kinase (MEK) and MAP kinase [45, 46]. After insulin stimulation, the reduced LAR mass was associated with a 3 fold increase in IRS-1 phosphorylation by the insulin receptor kinase *in vitro*, and an increase in insulin receptor autophosphorylation by 150%, indicating a proximal site of action of LAR directly on the insulin receptor *in situ*.

Additional recent data has indicated that LAR must be expressed eutopically, at the plasma membrane, to have its effects on insulin receptor signalling [47, 48]. In transfected clonal lines of CHO-HIR cells overexpressing ~5 fold more LAR than control cells, the LAR P-subunit was localized to the cell membrane where it led to a reduction of insulin-stimulated tyrosine phosphorylation of the insulin receptor by 31–42% (p < 0.01), and that of IRS-1 was decreased by 34–56% (p < 0.01), compared to empty vector transfectants. Studies in stable or transiently transfected cells gave quantitatively similar results. Full-length LAR over-expression also blocked insulin-stimulated receptor kinase activation as well as thymidine incorporation into DNA. In

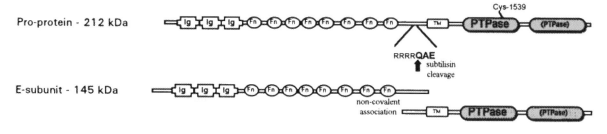

Fig. 2. Proteolytic processing of the LAR protein-tyrosine phosphatase pro-protein. The LAR PTPase has been shown to be initially synthesized as a ~212 kDa proprotein that is proteolytically processed at an external domain near the transmembrane segment. The catalytic cysteine residue in the D1 domain (Cys-1539) is indicated. The E-subunit and P-subunit exist as a non-covalently associated complex on the cell surface as well as in internal membrane fractions of the cell. In some settings, the E-subunit is shed from the cell, leaving the transmembrane P-subunit at the cell surface. The proportion of E/P-complexed and E-shed LAR in various types of cultured cells or in tissues, is not known. See text for references and further discussion.

96

contrast, overexpression of the free LAR cytoplasmic domain, detected as a catalytically active 72 kDa protein in the cell cytosol, did not significantly affect the insulin-stimulated tyrosine phosphorylation of the insulin receptor or IRS-1. These studies provided some of the first evidence that increased expression of LAR has negative regulatory effects at a proximal site in the insulin signalling pathway. Since this effect occurs only when LAR is eutopically expressed at the cell membrane, these data further suggested that in intact cells, LAR requires a transmembrane localization to directly interact with the insulin receptor.

A physical association between LAR and the insulin receptor was recently demonstrated by immunoprecipitation of cell lysates with LAR antibody and immunoblotting with antibody to the insulin receptor, or vice-versa [49]. Up to 11.8% of the LAR protein in the lysates of CHO-hIR cells transfected with LAR co-immunoprecipitated with the insulin receptor. In CHO cells, the LAR/insulin receptor association was increased 6.5 fold by chemical cross-linking of the cell surface and was related to the level of expression of LAR and the insulin receptor in transfected cells. Interestingly, LAR/insulin receptor crosslinking was also increased 3.9 fold by treatment with insulin, suggesting a dynamic role for ligand-induced conformational changes in the insulin receptor or in its phosphorylation state that enhances its association with LAR as a negative-regulatory PTPase. Treatment of the cells with the PTPase inhibitor vanadate, which dramatically enhances the phosphorylation state of the insulin receptor, had no demonstrable effect on the association of LAR with the insulin receptor, further suggesting that the interaction between LAR and the insulin receptor may not be primarily between the PTPase catalytic domain and the phosphotyrosyl residues of the activated insulin receptor β-subunit. Also, since substrate dephosphorylation by LAR is extraordinarily rapid (the turnover number for LAR has been estimated to be 8990 nmol Pi released/min per nmol of enzyme [30], additional domains of protein-protein interaction may enhance and stabilize the association between these two proteins.

Some degree of specificity was also evident in the LAR interaction with the insulin receptor in the KRC-7 cells, since immunoprecipitation with an antibody to the EGF receptor failed to demonstrate chemical crosslinking between the EGF receptor and LAR. This was unexpected, since we recently demonstrated that in McA-RH7777 rat hepatoma cells, a reduction of LAR mass by overexpression of antisense mRNA to LAR resulted in enhanced phosphorylation and signalling through the EGF receptor [46], suggesting that LAR may participate to regulate multiple tyrosine kinase receptors at the cell surface. It is possible that receptor regulation varies among hepatoma cell lines, and the EGF receptor is associated with LAR to a lesser extent in the KRC-7 cells. Alternatively, the interaction

between the EGF receptor and LAR may be of lower affinity or capacity, so that it may be less evident using techniques of immunoprecipitation and immunoblotting compared with the results obtained for the LAR/insulin receptor association.

Since activated insulin receptors are rapidly internalized into an endosomal membrane compartment, an important site of receptor dephosphorylation as the receptors are recycled back to the plasma membrane in the basal state [15, 19, 20, 22, 50], it seems likely that the PTPases responsible for the physiological dephosphorylation of the insulin receptor in the endosomal compartment would be coupled to the movement of the receptor and perhaps interact with the receptors in the endosomal fraction. In support of this hypothesis, insulin stimulation of liver cells leads to internalization of LAR into an endosomal fraction in a close temporal relationship with the insulin receptor itself [49]. An insulin-stimulated translocation of LAR at 15 min of incubation was suggested by a 52% decrease in the mass of LAR in the plasma membrane fraction with a concomitant 2.2 fold increase of LAR in the endosomal fraction. Furthermore, incubation of endosomes with neutralizing antibody to the LAR catalytic domain decreased insulin receptor dephosphorylation in situ by 28% (p = 0.01 vs. control), while incubation with inhibitory PTP1B antibodies only diminished insulin receptor dephosphorylation by 9%. Thus, LAR appears to follow the itinerary of insulin receptor internalization and movement into an endosomal fraction where it can act in situ as a receptor PTPase, and have an integral role in the dynamic regulation of reversible insulin receptor tyrosine phosphorylation during its movement into the cell [49].

PTP1B

PTP1B is a widely expressed enzyme that was first identified as a prominent PTPase in the cytosol fraction of placenta [51]. The full-length enzyme is ~50 kDa, with a cleavable C-terminal segment downstream from the PTPase domain that directs its association with the endoplasmic reticulum either through a hydrophobic interaction or by attachment to a non-catalytic subunit [35]. We and others have found that a substantial portion of the uncleaved form of PTP1B is also present in the cytosol of rat tissues [32, 52]. In early studies, microinjection of PTP1B into *Xenopus* oocytes blocked insulin-stimulated S6 kinase activation and retarded insulin-induced oocyte maturation [53, 54]. Lammers *et al.* [55] showed that overexpression of PTP1B almost completely dephosphorylated insulin proreceptors and β-subunits in the basal state, and reduced the phosphotyrosine content of the ligand-activated receptor β-subunits to less than 50% of the control level.

In order to gain insight into the potential role of PTP 1 B in the regulation of insulin signalling in intact cells, we used an osmotic shock technique to load rat KRC-7 hepatoma cells with affinity-purified antibodies that immunoprecipitate and neutralize the enzymatic activity of recombinant rat PTP1B [56]. PTP1B antibody loading increased insulin-stimulated DNA synthesis and PI3'-kinase activity by 42 and 38%, respectively, compared to control cells loaded with pre-immune IgG (p < 0.005). We also determined that insulin-stimulated receptor kinase activity towards an exogenous peptide substrate was increased by 57% in the PTP1B antibody-loaded cells and that insulin-stimulated receptor autophosphorylation and IRS-1 tyrosine phosphorylation were increased 2.2 and 2.0 fold, respectively.

Kenner et al. [57] have reported data to also suggest that PTP1B acts as a negative regulator of insulin action in cellular models, including an enhancement of insulin signalling by transfection of a catalytically inactive mutant of PTP1B, which appears to act in a dominant negative fashion. Seely, et al. [58] have recently demonstrated a physical interaction between PTP1B and the activated insulin receptor in intact cells. An interesting study in a gene knock-out model for the G-protein subunit $G_{i\alpha2}$ suggested that the observed insulin resistance might result from increased expression of PTP1B [59]. Overall, these studies indicate that PTP1B can negatively regulate insulin signalling and appears to act, at least in part, directly at the level of the insulin receptor.

SH-PTP2
This widely expressed PTPase with two SH2 domains and a single catalytic PTPase domain [31, 33] associates with autophosphorylated PDGF and EGF receptors as well as with tyrosine-phosphorylated IRS-1 by its SH2 domains in a process that activates its catalytic domain [60–63]. In recombinant *in vitro* systems, SH-PTP2 can dephosphorylate the insulin receptor and IRS-1 [63–66], although studies in intact cells have failed to demonstrate a direct interaction between SH-PTP2 and insulin receptors, or any effect of overexpression of catalytically active SH-PTP2 on insulin signalling [67–69]. Recent studies have demonstrated a positive role for SH-PTP2 in insulin-induced mitogenesis [64, 68], consistent with the recognition of SH-PTP2 as the mammalian homolog of the *Drosophila csw* gene product, which positively transrnits signals downstream of the *torso* receptor tyrosine kinase [70]. Thus, SH-PTP2 appears to play a positive role in downstream post-receptor signalling by insulin, but not involving the insulin receptor itself.

LRP (RPTP-α) and RPTP-ε
LRP is a receptor-type PTPase expressed in insulin-sensitive tissues that we have shown has catalytic activity *in vitro* towards the general dephosphorylation of the insulin receptor [39]. Using a novel transfection assay for identifying PTPases that negatively affect insulin action, Moller *et al.* [71] found that LRP and the closely related transmembrane enzyme RPTP-ε can act as negative regulators

of the insulin receptor tyrosine kinase. LRP has also been shown to activate pp60[c-src] by dephosphorylation of the negative-regulatory tyrosine at position 527 which leads to transformation of rat embryo fibroblasts [72], and it may be involved in triggering a neuronal differentiation pathway [73]. LRP has also been suggested to play a potential role in attenuation of GRB-2 mediated signalling, thus possibly influencing a variety of interrelated signalling pathways [74]. Additional work will be required to further substantiate the potential role of LRP and RPTP-ε in the regulation of the insulin signalling pathway.

Conclusion

From the variety of PTPases found in insulin-sensitive tissues, the available data have provided considerable support for a physiological role for LAR and PTP1B in the negative regulation of the insulin action pathway. Furthermore, the finding that insulin signalling can be enhanced by the specific inhibition of LAR or PTP1B has potential clinical relevance to the treatment of insulin-resistant disease states and Type II diabetes mellitus with pharmaceutical agents. While significant progress has been made in this area, much remains to be learned, including the relative specificity and site(s) of action of LAR and PTP1B on insulin signalling, the molecular mechanism of their inhibition of the insulin receptor kinase, and the role of PTPases in the regulation of tyrosine dephosphorylation of post-insulin receptor cellular substrate proteins.

Acknowledgements

The authors are grateful for the valuable collaboration we have enjoyed with Dr. Robert Mooney collaborative and for helpful advice regarding the liver endosome preparation that was provided by Drs. Barry Posner and Paul Bevan. Work in the author's laboratory has been supported by NIH grant R01-DK43396 to Dr. Goldstein.

References

1. Cheatham B, Kahn CR: Insulin action and the insulin signalling network. Endocrinol Rev 16: 117–142, 1995
2. Ebina Y, Ellis L, Jamagin K *et al.*: Human insulin receptor cDNA: The structural basis for hormone activated transmembrane signalling. Cell 40: 747–758, 1985
3. White MF, Shoelson SE, Keutmann H, Kahn CR: A cascade of tyrosine autophosphorylation in the β-subunit activates the phosphotransferase of the insulin receptor. J Biol Chem 263: 2969–2980, 1988

4. Flores-Riveros JR, Sibley E, Kastelic T, Lane MD: Substrate phosphorylation catalyzed by the insulin receptor tyrosine kinase. Kinetic correlation to autophosphorylation of specific sites in the beta subunit. J Biol Chem 264: 21557–21572, 1989

5. White W, Kahn CR: The cascade of autophosphorylation in the β-subunit of the insulin receptor. J Cell Biochem 39: 429–441, 1989

6. King MJ, Sharma RP, Sale GJ: Site-specific dephosphorylation and deactivation of the human insulin receptor tyrosine kinase by particulate and soluble phosphotyrosyl protein phosphatases. Biochem J 275: 413–418, 1991

7. Myers MG, White MF: Insulin signal transduction and the IRS proteins. Ann Rev Pharmacol Toxicol 36: 615–658, 1996

8. Yan PF, Li SL, Liang SJ, Giannini S, Fujita-Yamaguchi Y: The Role of COOH-terminal and acidic domains in the activity and stability of human insulin receptor protein tyrosine kinase studied by purified deletion mutants of the beta-subunit domain. J Biol Chem 268: 22444–22449, 1993

9. Tavaré JM, Siddle K: Mutational analysis of insulin receptor function – consensus and controversy. Biochim Biophys Acta 1178: 21–39, 1993

10. Haring HU, Kasuga M, White MF, Crettaz M, Kahn CR: Phosphorylation and dephosphorylation of the insulin receptor: Evidence against an intrinsic phosphatase activity. Biochemistry 23: 3298–3306, 1984

11. Kowalski A, Gazzano H, Fehlmann M, Van Obberghen E: Dephosphorylation of the hepatic insulin receptor: Absence of intrinsic phosphatase activity in purified receptors. Biochem Biophys Res Comm 117: 885–893, 1983

12. Mooney RA, Anderson DL: Phosphorylation of the insulin receptor in permeabilized adipocytes is coupled to a rapid dephosphorylation reaction. J Biol Chem 264: 6850–6857, 1989

13. Bernier M, Liotta AS, Kole HK, Shock DD, Roth J: Dynamic regulation of intact and C-terminal truncated insulin receptor phosphorylation in permeabilized cells. Biochemistry 33: 4343–4351, 1994

14. Goldstein BJ: Regulation of insulin receptor signalling by protein-tyrosine dephosphorylation. Receptor 3: 1–15, 1993

15. Backer JM, Kahn CR, White MF: Tyrosine phosphorylation of the insulin receptor during insulin-stimulated internalization in rat hepatoma cells. J Biol Chem 264: 1694–1701, 1989

16. Kublaoui B, Lee J, Pilch PF: Dynamics of signalling during insulin-stimulated endocytosis of its receptor in adipocytes. J Biol Chem 270: 59–65, 1995

17. Klein HH, Freidenberg GR, Matthaei S, Olefsky JM: Insulin receptor kinase following internalization in isolated rat adipocytes. J Biol Chem 262: 10557–10564, 1987

18. Bevan AP, Drake PG, Bergeron JJM, Posner BI: Intracellular signal transduction: The role of endosomes. Trends Endocrinol Metab 7: 13–21, 1996

19. Bevan AP, Burgess JW, Drake PG, Shaver A, Bergeron JJ, Posner BI: Selective activation of the rat hepatic endosomal insulin receptor kinase. Role for the endosome in insulin signalling. J Biol Chem 270: 10784–10791, 1995

20. Burgess JW, Wada I, Ling N, Khan MN, Bergeron JJM, Posner BI: Decrease in β-subunit phosphotyrosine correlates with internalization and activation of the endosomal insulin receptor kinase. J Biol Chem 267: 10077–10086, 1992

21. Khan MN, Baquiran G, Brule C et al.: Internalization and activation of the rat liver insulin receptor kinase in vivo. J Biol Chem 264: 12931–12940, 1989

22. Faure R, Baquiran G, Bergeron JJM, Posner BI: The dephosphorylation of insulin and epidermal growth factor receptors. Role of endosome-associated phosphotyrosine phosphatase(s). J Biol Chem 267: 11215–11221, 1992

23. Walton KM, Dixon JE: Protein tyrosine phosphatases. Ann Rev Biochem 62: 101–120, 1993

24. Fischer EH, Charbonneau H, Tonks NK: Protein tyrosine phosphatases – A diverse family of intracellular and transmembrane enzymes. Science. 253: 401–406, 1991

25. Goldstein BJ: Phospho-protein phosphatases 1: Tyrosine phosphatases. In: P Sheterline (ed). Protein Profile. Academic Press, London, 1995, pp 1425–1585

26. Goldstein BJ, Meyerovitch J, Zhang WR et al.: Hepatic protein-tyrosine phosphatases and their regulation in diabetes. Adv Prot Phosphatases 6: 1–17, 1991

27. Zhang WR, Hashimoto N, Ahmad F, Ding W, Goldstein BJ: Molecular cloning and expression of a unique receptor-like protein-tyrosine phosphatase in the leukocyte common-antigen-related phosphatase family. Biochem J 302: 39–47, 1994

28. DeFronzo RA, Bonadonna RC, Ferrannini E: Pathogenesis of NIDDM. A balanced overview. Diabetes Care 15: 318–368, 1992

29. Zhang WR, Goldstein BJ: Identification of skeletal muscle protein-tyrosine phosphatases by amplification of conserved cDNA sequences. Biochem Biophys Res Commun 178: 1291–1297, 1991

30. Hashimoto N, Zhang WP, Goldstein BJ: Insulin receptor and epidermal growth factor receptor dephosphorylation by three major rat liver protein-tyrosine phosphatases expressed in a recombinant bacterial system. Biochem J 284: 569–576, 1992

31. Freeman RM, Plutzky J, Neel BG: Identification of a human src homology 2-containing protein-tyrosine-phosphatase – A putative homolog of drosophila corkscrew. Proc Natl Acad Sci USA 89: 11239–11243, 1992

32. Ahmad F, Goldstein BJ: Purification, identification and subcellular distribution of three predominant protein-tyrosine phosphatase enzymes in skeletal muscle tissue. Biochim Biophys Acta 1248: 57–69, 1995

33. Ding W, Zhang WR, Sullivan K, Hashimoto N, Goldstein BJ: Identification of protein-tyrosine phosphatases prevalent in adipocytes by molecular cloning. Biochem Biophys Res Commun 202: 902–907, 1994

34. Mei L, Doherty CA, Huganir RL: RNA splicing regulates the activity of a SH2 domain-containing protein tyrosine phosphatase. J Biol Chem 269: 12254–12262, 1994

35. Frangioni JV, Beahm PH, Shifiin V, Jost CA, Neel BG: The nontransmembrane tyrosine phosphatase PTP-1B localizes to the endoplasmic reticulum via its 35 amino acid C-terminal sequence. Cell 68: 545–560, 1992

36. Goldstein BJ: Protein-tyrosine phosphatases and the regulation of insulin action. In: D LeRoith, JM Olefsky, SI Taylor (eds). Diabetes Mellitus: A Fundamental and Clinical Text. Lippincott, Philadelphia, 1996, pp 174–186

37. Ahmad F, Goldstein BJ: Increased abundance of specific skeletal muscle protein tyrosine phosphatases in a genetic model of obesity and insulin resistance. Metabolism 44: 1175–1184, 1995

38. Ahmad F, Goldstein BJ: Alterations in specific protein-tyrosine phosphatases accompany the insulin resistance of streptozotocin-diabetes. Am J Physiol 268: E932–E940, 1995

39. Hashimoto N, Feener EP, Zhang WR, Goldstein BJ: Insulin receptor protein tyrosine phosphatases – Leukocyte common antigen-related phosphatase rapidly deactivates the insulin receptor kinase by preferential dephosphorylation of the receptor regulatory domain. J Biol Chem 267: 13811–13814, 1992

40. Streuli M, Krueger NY, Ariniello PD et al.: Expression of the receptor-linked protein tyrosine phosphatase LAR: Proteolytic cleavage and shedding of the CAM-like extracellular region. EMBO J 11: 897–907, 1992

41. Serra-Pages C, Saito H, Streuli M: Mutational analysis of proprotein processing, subunit association, and shedding of the LAR transmembrane protein tyrosine phosphatase. J Biol Chem 269: 23632–23641, 1994

42. Pot DA, Woodford TA, Remboutsika E, Haun RS, Dixon JE: Cloning, bacterial expression, purification, and characterization of the cytoplasmic domain of rat LAP, a receptor-like protein tyrosine phosphatase. J Biol Chem 266: 19688–19696, 1991

43. Streuli M, Krueger NX, Thai T, Tang M, Saito H: Distinct functional roles of the two intracellular phosphatase like domains of the receptor-linked protein tyrosine phosphatases LCA and LAR. EMBO J 9: 2399–2407, 1990.

44. Longo FM, Martignetti JA, Le Beau JM, Zhang JS, Bames JP, Brosius J: Leukocyte common antigen-related receptor-linked tyrosine phosphatase. Regulation of mRNA expression. J Biol Chem 268: 26503–26511, 1993

45. Kulas DT, Zhang WR, Goldstein BJ, Furlanetto RW, Mooney RA: Insulin receptor signalling is augmented by antisense inhibition of the protein-tyrosine phosphatase LAR. J Biol Chem 270: 2435–2438, 1995

46. Kulas DT, Goldstein BJ, Mooney RA: The transmembrane protein-tyrosine phosphatase LAR modulates signalling by multiple receptor tyrosine kinases. J Biol Chem 271: 748–754, 1996

47. Zhang WR, Li PM, Oswald MA, Goldstein BJ: Modulation of insulin signal transduction by eutopic overexpression of the receptor-type protein-tyrosine phosphatase LAR. Mol Endocrinol 10: 575–584, 1996

48. Li PM, Zhang WR, Goldstein BJ: Suppression of insulin receptor activation by overexpression of the protein-tyrosine phosphatase LAR in hepatoma cells. Cell Signal 1996

49. Ahmad F, Goldstein BJ: Functional association between the insulin receptor and the transmembrane protein-tyrosine phosphatase LAR in intact cells. J Biol Chem 1996

50. Baass PC, Diguglielmo GM, Authier F, Posner BI, Bergeron JJM: Compartmentalized signal transduction by receptor tyrosine kinases. Trends Cell Biol 5: 465–470, 1995

51. Charbonneau H, Tonks NK, Kumar S et al.: Human placenta protein-tyrosine phosphatase: Amino acid sequence and relationship to a family of receptor-like proteins. Proc Natl Acad Sci USA 86: 5252–5256, 1989

52. Ide R, Maegawa H, Kikkawa R, Shigeta Y, Kashiwagi A: High glucose condition activates protein tyrosine phosphatases and deactivates insulin receptor function in insulin sensitive rat 1 fibroblasts. Biochem Biophys Res Commun 201: 71–77, 1994

53. Cicirelli MF, Tonks NK, Diltz CD, Weiel JE, Fischer EH, Krebs EG: Microinjection of a protein-tyrosine-phosphatase inhibits insulin action in Xenopus oocytes. Proc Natl Acad Sci USA 87: 5514–5518, 1990

54. Tonks NK, Cicirelli MF, Diltz CD, Krebs EG, Fischer EH: Effect of microinjection of a low-M$_r$ human placenta protein tyrosine phosphatase on induction of meiotic cell division in Xenopus oocytes. Mol Cell Biol 10: 458–463, 1990

55. Lammers R, Bossenmaier B, Cool DE et al.: Differential activities of protein tyrosine phosphatases in intact cells. J Biol Chem 268: 22456–22462, 1993

56. Ahmad F, Li PM, Meyerovitch J, Goldstein BJ: Osmotic loading of neutralizing antibodies defines a role for protein-tyrosine phosphatase 1B in negative regulation of the insulin action pathway. J Biol Chem 270: 20503–20508, 1995

57. Kenner KA, Anyanwu E, Olefsky JM, Kusari J: Protein-tyrosine phosphatase 1B is a negative regulator of insulin- and insulin-like growth factor-1 stimulated signalling. J Biol Chem 271: 19810–19816, 1996

58. Seely BL, Staubs PA, Reichart DR et al.: Protein tyrosine phosphatase 1B interacts with the activated insulin receptor. Diabetes. 45: 1379–1385, 1996

59. Moxham CM, Malbon CC. Insulin action impaired by deficiency of the g-protein subunit g(i-alpha-2). Nature 379: 840–844, 1996

60. Kuhné MP, Pawson T, Lienhard GE, Feng GS: The insulin receptor substrate-1 associates with the SH2-containing phosphotyrosine phosphatase Syp. J Biol Chem 268: 11479–11481, 1993

61. Case RD, Piccione E, Wolf G et al.: SH-PTP2/Syp SH2 domain binding specificity is defined by direct interactions with platelet-derived growth factor beta-receptor, epidermal growth factor receptor, and insulin receptor substrate-1-derived phosphopeptides. J Biol Chem 269: 10467–10474, 1994

62. Sugimoto S, Wandless TJ, Shoelson SE, Neel BG, Walsh CT: Activation of the SH2-containing protein tyrosine phosphatase, SH-PTP2, by phosphotyrosine-containing peptides derived from insulin receptor substrate-1. J Biol Chem 269: 13614–13622, 1994

63. Kuhné MR, Zhao ZZ, Rowles J et al.: Dephosphorylation of insulin receptor substrate 1 by the tyrosine phosphatase PTP2C. J Biol Chem 269: 15833–15837, 1994

64. Xiao S, Rose DW, Sasaoka T et al.: Syp (SH-PTP2) is a positive mediator of growth factor-stimulated mitogenic signal transduction. J Biol Chem 269: 21244–21248, 1994

65. Maegawa H, Ugi S, Ishibashi O et al.: Src homology-2 domains of protein tyrosine phosphatase are phosphorylated by insulin receptor kinase and bind to the COOH-terminus of insulin receptors in vitro. Biochem Biophys Res Commun 194: 208–214, 1993

66. Ugi S, Maegawa H, Olefsky JM, Shigeta Y, Kashiwagi A: Src homology 2 domains of protein tyrosine phosphatase are associated in vitro with both the insulin receptor and insulin receptor substrate-1 via different phosphotyrosine motifs. FEBS Lett 340: 216–220, 1994

67. Vogel W, Lammers R, Huang JT, Ulrich A: Activation of a phosphotyrosine phosphatase by tyrosine phosphorylation. Science 259: 1611–1614, 1993

68. Miarski KL, Saltiel AR: Expression of catalytically inactive Syp phosphatase in 3T3 cells blocks stimulation of mitogen-activated protein kinase by insulin. J Biol Chem 269: 21239–21243, 1994

69. Yamaguchi K, Miarski KL, Saltiel AR, Pessin JE: Protein-tyrosine-phosphatase SHPTP2 is a required positive effector for insulin downstream signalling. Proc Natl Acad Sci USA 92: 664–668, 1995

70. Perkins LA, Larsen I, Perrimon N: corkscrew encodes a putative protein tyrosine phosphatase that functions to transduce the terminal signal from the receptor tyrosine kinase torso. Cell 70: 225–236, 1992

71. Moller NPH, Moller KB, Lammers R et al.: Selective down-regulation of the insulin receptor signal by protein-tyrosine phosphatases alpha and epsilon. J Biol Chem 270: 23126–23131, 1995

72. Zheng XM, Wang Y, Pallen CJ: Cell transformation and activation of pp60c-src by overexpression and activation of a protein tyrosine phosphatase. Nature 359: 336–339, 1992

73. den Hertog J, Pals CEGM, Peppelenbosch MP, Tertoolen LGJ, Delaat SW, Kruijer W: Receptor protein tyrosine phosphatase-α activates pp60(c-src) and is involved in neuronal differentiation. EMBO J 12: 3789–3798, 1993

74. den Hertog J, Tracy S, Hunter T: Phosphorylation of receptor protein-tyrosine phosphatase α on Tyr789, a binding site for the SH3-SH2-SH3 adaptor protein GRB- 2 in vivo. EMBO J 13: 3020–3032, 1994

Molecular and Cellular Biochemistry **182**: 101–108, 1998.

Protein-tyrosine phosphatase-1B acts as a negative regulator of insulin signal transduction

John C.H. Byon,[1] Anasua B. Kusari[1] and Jyotirmoy Kusari[1,2]
Departments of [1]Physiology and [2]Molecular and Cellular Biology Program, Tulane University Medical Center, New Orleans, LA, USA

Abstract

Insulin signaling involves a dynamic cascade of protein tyrosine phosphorylation and dephosphorylation. Most of our understanding of this process comes from studies focusing on tyrosine kinases, which are signal activators. Our knowledge of the role of protein-tyrosine phosphatases (PTPases), signal attenuators, in regulating insulin signal transduction remains rather limited. Protein-tyrosine phosphatase 1B (PTP-1B), the prototypical PTPase, is ubiquitously and abundantly expressed. Work from several laboratories, including our own, has implicated PTP-1B as a negative regulator of insulin action and as a potentially important mediator in the pathogenesis of insulin-resistance and non-insulin dependent diabetes mellitus (NIDDM). (Mol Cell Biochem **182**: 101–108, 1998)

Key words: PTP-1B, insulin action, diabetes

Introduction

Insulin is an important regulator of cell growth and metabolism. The pleiotropic actions of insulin are initiated by the binding of insulin to its receptor, which is a heterotetrameric protein consisting of two α and two β subunits linked by disulfide bonds to form a β-α-α-β structure. The α subunit is entirely extracellular and contains the insulin binding site. The β subunit is a transmembrane protein and contains the tyrosine kinase domain on its intracellular segment. Upon ligand binding, the insulin receptor undergoes autophosphorylation on tyrosine residues, which increases the receptor's intrinsic tyrosine kinase activity [1]. This in turn causes phosphorylation of one or more cellular substrates leading to a cascade of secondary phosphorylation and dephosphorylation [2–7].

Considerable progress has been made in defining the molecular mechanism of insulin action and the role of protein tyrosine phosphorylation within this context. The regulation of tyrosine phosphorylation represents a balance of tyrosine kinase and phosphatase activities. To date, most attempts to assess the role of protein tyrosine phosphorylation have focused mainly on the role of protein tyrosine kinases, and thus offer an incomplete picture of this dynamic process.

Protein-tyrosine phosphatase 1B (PTP-1B) was one of the first PTPases to be cloned, sequenced, and characterized [8–10]. It was originally purified as the major PTPase of human placenta [8]. PTP-1B is a member of the non-receptor family of PTPases and contains the conserved catalytic domain characterized by the amino-acid sequence (I/V)HCXAGXXR(S/T)G [10–12]. Cys 215 and Arg 221 within this motif are crucial for the catalytic activity of this enzyme [11–13]. Although, PTP-1B was one of the first PTPases identified, only recently insight into the potential physiological function of this enzyme has come to light. Over the years our laboratory has focused on the potential role of PTP-1B in regulating insulin signal transduction.

Insulin regulates PTP-1B expression and activity

We have examined the effects of insulin on the activity and expression of PTP-1B in L6 myotubes. PTPase activity was assessed using a synthetic phosphopeptide substrate (IRP), which is identical to the major site of insulin receptor

Address for offprints: J. Kusari, Tulane University School of Medicine, Department of Physiology SL39, 1430 Tulane Avenue, New Orleans, LA 70112-2699, USA

102

autophosphorylation. In unstimulated L6 myotubes, almost all of the PTPase activity resided in the Triton X-100 soluble fraction (P1) [14]. Incubation with insulin resulted in a dramatic increase in PTPase activity within the P1 fraction over basal levels, which did not result from a redistribution of PTPase activities from other subcellular fractions to P1. Insulin stimulation caused a biphasic increase in PTPase activity, with a small increase seen at 30 min, followed by a decline to basal levels within 2 h, and a subsequent increase in PTPase activity after 4 h, which gradually reached a maximum at 32 h [14]. Total PTPase activity was mirrored by increased levels of PTP-1B-specific activity. PTP-1B activity rose concomitantly with increasing PTP-1B protein levels, following increased PTP-1B mRNA levels, which were maximal after 12 h of stimulation with insulin. Insulin caused a dose-dependent increase in PTP-1B. The low concentrations required to generate these effects (EC_{50} = 5 nM) suggested that the hormone acted through its own receptor [14].

PTP-1B is a negative regulator of insulin action

The direct effects of PTP-1B on insulin signaling have been shown by microinjecting purified PTP-1B into *Xenopus* oocytes, which blocked insulin-stimulated S6 peptide phosphorylation and retarded insulin-induced oocyte maturation [15, 16]. Based on these previous studies, it appears that PTP-1B acts as a negative regulator of insulin action. To further explore this hypothesis, we established clonal cell lines overexpressing wild type or catalytically inactive PTP-1B (Cys^{215} → Ser, CS) in cells overexpressing insulin (Hirc B) receptors [17]. Cysteine 215 is an essential residue within the catalytic domain of PTP-1B, and its mutation to serine abolishes the PTPase's enzymatic capability [13]. PTPase activity in cells overexpressing wild type PTP-1B (Hirc A) was strikingly higher than that found in parental cells, while in cells expressing the CS PTP-1B (Hirc M) it remained unaltered. In comparison with the parental cell line, insulin-stimulated insulin receptor autophosphorylation and IRS-1 tyrosine phosphorylation were significantly reduced in Hirc A cells, and was higher in Hirc M cells. Basal glucose incorporation rates of Hirc-A and Hirc M did not significantly differ from that seen in Hirc B. Ligand stimulation increased [^{14}C]glucose incorporation in all cell lines in a dose-dependent manner. However, the relative [^{14}C]glucose incorporation was lowest in cells expressing wild type PTP-1B and was greatest in cells containing the CS PTP-1B.

Thus, catalytically active PTP-1B could act as a negative regulator of insulin action by dephosphorylating the insulin receptor [17]. An excess of PTP-1B could completely dephosphorylate and inactivate a large subset of receptors. The remaining pool of fully active receptors may be unable to compensate for the loss, resulting In a reduction of

receptor signaling. Another possible scenario is that active PTP-1B could dephosphorylate only specific, critical receptor tyrosine residues necessary for receptor activation without completely dephosphorylating all tyrosine residues in a non-specific manner. As a final possibility, PTP-1B could dephosphorylate both the receptors and IRS proteins, with the IRS proteins as the preferred substrates. The enhanced receptor autophosphorylation and IRS phosphorylation observed during overexpression of CS PTP-1B [17] could be due to binding of the inactive PTPase to phospho-tyrosine residues, thereby protecting them from dephosphorylation by the endogenous PTP-1B.

The role of PTP-1B as a negative regulator of insulin signal transduction pathways has been independently confirmed by other groups [18–20]. Osmotic loading of rat KRC-7 hepatoma cells with affinity-purified neutralizing PTP-1B antibodies increased insulin stimulated DNA synthesis and PI-3-kinase activity. Concurrently, insulin-stimulated receptor autophosphorylation and IRS-1 tyrosine phosphorylation were also significantly augmented, as well as insulin receptor tyrosine kinase activity towards an exogenous peptide substrate [18]. Osmotic loading did not decrease the intracellular content of PTP-1B, suggesting that the antibodies blocked PTP-1B activity by sterically inhibiting the interaction between PTP-1B and its substrates [18]. In another study, synthetic tris-sulfotyrosyl dodecapeptide (TRDIY-(S)ETDY(S)Y(S)RK-amide), identical in primary sequence to amino acids 1142–1153 of the insulin proreceptor, inhibited insulin receptor dephosphorylation in solubilized membranes, and digitonin-permeabilized Chinese hamster ovary cells, expressing high levels of the human insulin receptor (CHO/HIRc) [19]. It also inhibited recombinant PTP-1B dephosphorylation of a synthetic tyrosine phosphorylated substrate [19]. A N-stearyl derivative of the peptide increased insulin-stimulated receptor autophosphorylation in intact CHO/HIRc cells. The peptide showed specificity towards tyrosine-class phosphatases; alkaline phosphatase and serine/threonine phosphatases were unaffected [19]. The sulfotyrosyl peptide functions as a nonhydrolyzable phosphotyrosyl peptide analogue capable of direct interaction with PTP-1B's phosphatase domain, which may result in blocking PTP-1B activity [19]. A recent report shows that overexpression of wild-type PTP-1B in rat adipocytes greatly inhibited insulin stimulated translocation of epitope-tagged GLUT4 [20].

PTP-1B interacts directly with the activated insulin receptor

Two methods were utilized to investigate whether PTP-1B directly interacts with the insulin receptor. The first involved using a PTP-1B glutathione S-transferase fusion protein, containing a point mutation in its catalytic domain, which rendered the enzyme inactive (PTPC^{215S}-GST). This mutant retained the phosphotyrosine binding site, thus providing a PTP-1B fusion protein which could bind to, but not dephosphorylate phosphotyrosine substrates. Hirc B whole cell lysates were isolated and incubated with insulin and PTPC^{215S}-GST. Precipitation with glutathione sepharose pulled down the insulin receptor (IR) as well as an unidentified 120 kDa phosphoprotein [21]. In a complementary *in vitro* study, purified insulin receptors were stimulated with insulin and activated in the presence of PTPC^{215S}-GST and recombinant IRS-1. The IR β-subunit, the PTP-1B fusion protein, and IRS-1 all showed a marked increase in phosphotyrosine content after hormone stimulation, suggesting that PTP-1B becomes tyrosine phosphorylated upon insulin stimulation [21]. The interaction between PTP-1B and the insulin receptor was also evaluated by incubating the labeled PTP-1B fusion protein ([S^{35}]PTPC^{215S}-GST) with autophosphorylated insulin receptor kinase domain peptides immobilized on amylose-agarose beads in the presence or absence of unlabelled fusion protein under equilibrium binding conditions. The radioactive PTP-1B fusion protein bound to the activated IR peptides and was displaced in a concentration dependent manner, indicating that the interaction of the two proteins occurs in a specific, competitive manner [21].

The second method used to investigate interactions between PTP-1B and the insulin receptor, employed the overexpression of the catalytically inactive CS mutant of PTP-1B in Hirc cells (Hirc M) [22]. Lysates from control and insulin treated Hirc M cells were immunoprecipitated with antibody to PTP-1B. Subsequent immunoblotting was performed sequentially using anti-phosphotyrosine, anti-insulin receptor, and anti-PTP-1B antibodies. In a parallel experiment, insulin receptors were immunoprecipitated from control and insulin-treated Hirc M cells. These lysates were also immunoblotted with the same series of antibodies. A 95 kDa protein co-precipitated with anti-PTP-1B after insulin stimulation, and was identified as the β-subunit of the insulin receptor. Only a residual amount of IR β-subunit was found in precipitates of control cells. Immunoprecipitation with the IR antibody revealed the coprecipitation of a 50 kDa protein identified as PTP-1B in insulin-treated cells. Again, little PTP-1B was coprecipitated by anti-IR in unstimulated cells. The presence of small quantities of PTP-1B in immunoprecipitates from unstimulated cells could be due to a low level of insulin receptor tyrosine phosphorylation under basal conditions. These results indicated that PTP-1B complexes with the intact insulin receptor in an insulin-dependent manner and confirms our earlier *in vitro* results [21]. The insulin stimulated interaction between CS PTP-1B and the insulin receptor was accompanied by an increase in CS PTP-1B tyrosine phosphorylation, similar to that observed *in vitro* [21]. A presently unidentified protein(s) of 180 kDa

was also co-precipitated by both PTP-1B and insulin receptor antibodies. This protein could represent IRS-1, IRS-2 or a combination of both [22].

Most protein tyrosine kinases phosphorylate substrates at tyrosine residues preceded by acidic amino acids [23]. The coding sequence for PTP-1B possesses three such candidate sites, Tyr 66 (QEDDNDY) and Tyrs 152 and 153 (EDIKSYY). To determine whether these residues become phosphorylated upon insulin stimulation, either Tyr 66 or Tyrs 152 and 153 were mutated to phenylalanine. Expression vectors containing CS PTP-1B, CS Y66F (YF), or CS Y152/153F (YYFF) mutants were coexpressed along with wild-type insulin receptor in COS cells. Cells transfected with CS PTP-1B showed increased phosphorylation of PTP-1B after ligand-stimulation. However, in neither YF nor YYFF transfected cells was PTP-1B phosphorylation evident upon insulin treatment [22], suggesting that tyrosine residues 66, 152, and 153 are all necessary for the insulin-induced phosphorylation of PTP-1B. In order to determine if these residues are also essential for the interaction of PTP-1B with the activated insulin receptor, CS PTP-1B, YF and YYFF transfectants were treated with insulin and then immunoprecipitated with anti-PTP-1B. As compared to CS PTP-1B cells, YF and YYFF cells coprecipitated significantly less of the insulin receptor [22]. These data suggest that PTP-1B tyrosines 66, 152, and 153 play some role in the interaction with the insulin receptor, although the extent of their importance is unclear. The sequence surrounding Y66 (pYINA) conforms to the consensus binding site for the SH2 domain of Grb2 (pYXNX), suggesting that Grb2 might act as an adaptor molecule for IR:PTP-1B interaction. Thus, PTP-1B may interact with the insulin receptor via an intermediary Grb2 association.

To identify insulin-receptor phosphotyrosines required for complex formation with PTP-1B, cell lines overexpressing different insulin receptor mutants were utilized. The insulin receptor mutant, ΔCT lacks the C-terminal 30 amino-acids, which includes tyrosines 1316 and 1322. YFF contains tyrosine to phenylalanine mutations at tyrosines 1150 and 1151. FYY is mutated at tyrosine 1146, and YF' possesses a point mutation at tyrosine 960. Anti-PTP-1B antibody coprecipitated relatively equivalent quantities of insulin receptor from insulin treated cells expressing either the wild-type, ΔCT, or YF' receptors. On the other hand, very little insulin receptor was seen in anti-PTP-1B precipitates from insulin treated cell expressing either the FYY or YFF receptors. Thus, PTP-1B interaction *in vivo* is directly related to the level of insulin receptor autophosphorylation. Insulin receptor tyrosine residues essential for efficient complex formation minimally include residues 1146, 1150 and/or 1151 [22].

The requirement for the phosphorylation of insulin receptor tyrosine residues 1146, 1150, and 1151 may be

104

to create a binding site for PTP- I B. Alternatively, the phosphorylation of these residues might be necessary for the subsequent tyrosine phosphorylation of alternate PTP-1B binding sites, or in order to expose a conformationally-dependent binding site. Although, PTP-1B does not contain a recognizable SH2 or PTB domain, it does contain the sequence FKVRES, located 19 residues N-terminal of the catalytically essential Cys 215. This sequence is very similar to the highly conserved FLVRES motif of the SH2 domains [24], which has been shown to be essential for the binding of SH2 domains to tyrosine phosphorylated proteins [25]. It is possible that PTP-1B interacts with the insulin receptor through this FKVRES sequence. The alternative possibility that other accessory proteins indirectly mediate this interaction cannot be ruled out.

Phosphorylation of PTP-1B by the insulin receptor regulates its activity

Demonstration of a direct interaction between the insulin receptor and PTP-1B raises an interesting question. Does the insulin receptor regulate PTP-1B activity via tyrosine phosphorylation, and thereby its own downregulation? To test this hypothesis, the phosphorylation of catalytically inactive mutant PTP-1B (CS) was measured in the presence of partially purified insulin receptor [26]. The insulin receptor tyrosine kinase catalyzed the tyrosine phosphorylation of CS in an insulin-dependent manner. Phosphorylation of CS by the IR was absolutely dependent upon insulin-stimulated receptor autophosphorylation and required an intact kinase domain, which contains tyrosines 1146, 1150, and 1151. The tyrosine phosphorylation of wild-type PTP-1B by the IR kinase increased phosphatase activity *in vitro*. When incubated with wild-type PTP-1B *in vitro*, the tyrosine phosphorylation of CS was significantly reduced, suggesting that PTP-1B might modulate its own tyrosine phosphorylation state. These results suggest that in response to insulin, PTP-1B becomes tyrosine phosphorylated by the IR kinase. The phosphorylation of PTP-1B increases its phosphatase activity, which could in turn, dephosphorylate the insulin receptor, thereby inhibiting receptor tyrosine kinase activity. Eventually, PTP-1B attenuates its own enzymatic activity by autodephosphorylation (Fig. 1). This is similar to the mechanism previously shown for PTP-1D [27].

PTPase activity and PTP-1B content are associated with insulin resistance

Insulin resistance is the major pathophysiological abnormality in non-insulin dependent diabetes mellitus (NIDDM), resulting in reduced rates of insulin-mediated glucose

uptake, primarily in the skeletal muscle tissue [28]. While the exact cellular basis of insulin resistance in NIDDM remains unknown, it is now clear that in most cases post-insulin binding defects are responsible for decreased insulin action [29, 30]: in other words, a dysregulation of the insulin signaling network.

Current evidence indicates that insulin receptor auto-phosphorylation and tyrosine kinase activity are severely impaired in skeletal muscle and other tissues in NIDDM [31–34], suggesting that changes in PTPase activity may contribute to NIDDM-related insulin resistance. Support for this hypothesis comes from data collected from various animal models of insulin-resistance [35–37], insulin-resistant, non-diabetic Pima Indians [38], and from patients with NIDDM [39, 40].

Several studies have examined PTPase enzyme activity and PTPase protein levels in various models of insulin resistance. The streptozotocin-diabetic rat has been primarily used as a disease model to study diabetes associated with pancreatic β-cell destruction. However, it also serves as a model for exploring the mechanism of insulin resistance of insulinopenic diabetes. When PTPase activity was measured in the cytosolic and particulate fractions of skeletal muscle and liver from these animals, cytosolic PTPase activity in both tissues was increased to 120–125% of control in diabetic animals, and particulate enzyme activity was decreased to 65–70% of control in both liver and skeletal muscle [35]. PTP-1B levels were shown to be significantly increased in diabetes, along with LAR and SHP-2 (SH-PTP2/Syp) [35]. Therefore, changes in abundance and distribution of PTPases may be involved in the pathogenesis of insulin resistance in insulinopenic diabetes.

Similar findings were observed in genetic models of insulin-resistance and NIDDM. Ahmad and Goldstein examined PTPase activity and protein levels in subcellular fractions of lean (+/?), insulin-resistant obese (fa/fa), and diabetic (ZDF/Drt – fa/fa) Zucker rats. Using a phospho-tyrosyl-myelin basic protein substrate, PTPase activity in the solubilized-particulate fraction was found to be increased by 65% in obese animals and by 74% in diabetic animals [36]. *In vitro* dephosphorylation of a recombinant rat insulin receptor kinase domain was also increased by 104 and 114% in obese and diabetic animals, respectively [36]. Immunoblotting revealed an increase in LAR, PTP-1B, and SHP-2 (SH-PTP2/Syp) protein levels in the solubilized-particulate fraction of obese and diabetic animals, with PTP-1B levels showing the greatest increase [36]. Thus, at least in genetic animal models of insulin-resistant obesity and NIDDM, there appears to be an important role played by PTPases.

In the Pima Indian population, NIDDM is extremely prevalent. In order to understand the molecular mechanisms underlying the insulin resistance associated with NIDDM, skeletal muscle PTPase activity of various cellular fractions

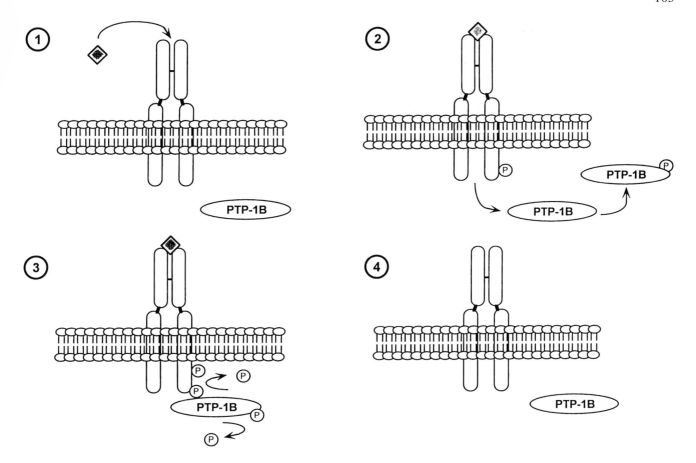

Fig. 1. The insulin receptor regulates PTP-1B tyrosine phosphorylation and activity. Binding of insulin to its receptor leads to tyrosine autophosphorylation and activation of the insulin receptor's tyrosine kinase activity. The activated insulin receptor can then posphorylate PTP-1B on tyrosine residues, which results in activation of PTP-1B enzymatic activity. Activated PTP-1B removes insulin receptor phosphotyrosines and autodephosphorylates itself, and thus inactivates the insulin receptor.

was measured under basal and insulin stimulated conditions in insulin-sensitive and insulin-resistant Pima Indians [38]. Basal PTPase activities in the soluble fraction were comparable between insulin-sensitive and resistant subjects. After insulin infusion, the soluble-PTPase activity in insulin-sensitive subjects decreased by 25%, with no change in the insulin-resistant subjects. These results suggest that either abnormal regulation of PTPase activity by insulin or excessive PTPase activity associated with the particulate fraction of muscle, or possibly both characteristics may be central to the cause of insulin resistance associated with NIDDM in Pima Indians.

When particulate and cytosolic protein tyrosine phosphatase activity was measured in skeletal muscle from 15 insulin-sensitive subjects and 5 insulin-resistant, non-diabetic subjects, as well as 18 subjects with the common form of NIDDM, approximately 90% of total PTPase activity was found in the particulate fraction [39]. In comparison with lean non-diabetic subjects, particulate PTPase activity was reduced by 21% ($p < 0.05$) and 22% ($p < 0.005$) in obese

non-diabetic and NIDDM subjects, respectively [39]. PTP-1B levels were then measured in the skeletal muscle of these subjects, in order to determine if the reduced PTPase activity in NIDDM subjects was due to a decrease in PTP-1B expression. Results from these experiments revealed a 38% decrease in PTP-1B protein content in NIDDM subjects [39].

The discrepancy between the PTPase activities observed in Pima Indians and our later study involving subjects with the common form of NIDDM could be due to several reasons. The Pima Indians represent a population with a highly homogeneous genetic background. Thus, the cause of insulin resistance in this population may not be similar to the cause in other more genetically diverse ethnic populations. However, the reduced PTPase activity and PTP-1B protein levels observed in skeletal muscle of NIDDM subjects could lead to insulin resistance. After insulin binding, the insulin receptor is internalized into an endosomal compartment and is then either recycled to the plasma membrane or degraded. However, it seems that all recycled receptors are not capable

106

of tyrosine kinase activity. It has been reported that patients with NIDDM have reduced insulin-receptor tyrosine kinase activity which appears to be attributable to changes in the ratio of two pools of receptors, both of which bind insulin but only one (the 'active' pool) of which is capable of tyrosine autophosphorylation and subsequent kinase activation [41]. Patients with NIDDM have a low ratio of active:inactive receptor pools [41]. Since PTP-1B has been localized to the endoplasmic reticulum, it is possible that PTP-1B can dephosphorylate internalized receptors, and that this process is required for maintenance of the active receptor pool. A decrease in PTP-1B levels might lead to a decrease in the 'activated' receptor population and eventually lead to impaired insulin signal transduction and thus cause insulin resistance.

Collectively, the data accumulated from both animal models of insulin resistance and human subjects suffering from NIDDM clearly indicate that tight regulation of PTPase activity and levels is necessary for maintenance of the insulin-sensitive state. Furthermore, fluctuations in PTPase activity and/or levels may lead to insulin-resistance. PTP-1B's ability to bind directly to the activated insulin receptor *in vivo* [22], and its subsequent dephosphorylation of the insulin receptor [22], in combination with its differential expression in NIDDM and normal subjects [38], mark this enzyme as a potential candidate in mediating the pathophysiology of NIDDM.

Perspectives

Although the work done in recent years has added greatly to our biochemical and physiological understanding of PTP-1B, one very important unresolved issue is how does PTP-1B, which is localized to the endoplasmic reticulum [42], interact with the insulin receptor in the plasma membrane. One possibility is that a fraction of the PTP-1B is released into the cytosol [43]. However, our studies have shown no observable translocation of PTPase activity between subcellular fractions following hormone stimulation. A fraction of PTP-1B might be associated with the plasma membrane instead of the ER, thus bringing the enzyme into a more advantageous location for interaction with its substrates. A final possibility is that activated endosomal receptors could be more relevant in insulin signaling than plasma membrane receptors [44, 45]. Endosome-associated, activated receptors could be rapidly brought into close proximity with PTP-1B at the ER. Subsequent dephosphorylation could inactivate the receptors, preventing further kinase activity.

Another challenge is to identify the exact domains by which PTP-1B interacts with the insulin receptor. Besides the FKVRES sequence, PTP-1B contains two proline-rich regions in the carboxy-terminus, which correspond to the canonical class II SH3 domain binding motif, PXXPXR [46, 47]. A recent study has shown that PTP-1B interacts with Crk, Grb2, and p130Cas *in vitro* and with p130Cas *in vivo*, via its proline-rich domains [48]. The *in vivo* binding of PTP-1B to p130Cas was independent of p130Cas phosphorylation [48]. Furthermore, a substantial amount of p130Cas cosedimented with the endoplasmic reticulum membranes [48], suggesting that the proline-rich domains on PTP-1B may direct it to some of its intracellular targets. It remains to be seen whether PTP-1B employs these domains to interact with the insulin receptor.

Identification of all the intracellular substrates of PTP-1B remains a major hurdle to full understanding of its physiological function. We and others have attempted to identify PTP-1B substrates by mutating the nucleophilic Cys → Ser or Ala in order to generate a catalytically inactive enzyme that retains the ability to bind substrate. This is effective for identification of certain substrates, but not for all. Recently, Flint *et al.* have developed a novel 'substrate-trapping' mutation to identify physiological substrates of PTP-1B [49]. Their procedure was based on the crystal structure of PTP-1B [50, 51]. Asp-181 functions as a general acid in protonating the tyrosyl leaving group of the substrate [49] and is essential for enzyme activity. Mutation of this invariant residue to Ala seems to stabilize the interaction of the active site Cys with the substrate, thus retaining specificity of enzyme binding but inhibiting catalytic function. Thus, the enzyme-substrate complex is stabilized sufficiently to permit isolation. Using this Asp-181 → Ala mutation, it was shown that the EGF receptor is another bona fide *in vivo* substrate of PTP-1B [49]. This technique should prove to be a valuable tool in the quest to fully identify the PTP-1B substrates involved in insulin signal transduction.

Finally, the evidence from genetic animal models of insulin resistance and human studies suggest a role for PTPases and PTP-1B in particular in mediating the pathophysiology of obesity-related insulin resistance and NIDDM. It remains to be seen if modulating PTP-1B levels via gene transfer can ameliorate the symptoms of insulin-resistance and NIDDM in animal models of these diseases. If so, this may prove to be a potential future therapeutic for human patients.

References

1. Kasuga M, Karlsson FA, Kahn CR: Insulin stimulates the phosphorylation of the 95,000-dalton subunit of its own receptor. Science 215: 185–187, 1982
2. Kahn CR, White MF, Shoelson SE, Backer JM, Araki E, Cheatham B, Csermely P, Folli F, Goldstein BJ, Huertas P, Rothenberg PL, Saad MJA, Siddle K, Sun X-J, Wilden PA, Yamada K, Kahn SA: The insulin receptor and its substrate: molecular determinants of early events in insulin action. Rec Prog Horm Res 48: 291–339, 1993

3. Rosen OM: After insulin binds. Science 237: 1452–1458, 1987
4. Cheatham B, Kahn CR: Insulin action and the insulin signaling network. Endo Rev 16: 117–142, 1995
5. Kahn CR, White MF: The insulin receptor and the molecular mechanism of insulin action. J Clin Invest 82: 1151–1156, 1988
6. Olefsky JM: The insulin receptor. A multifunctional protein. Diabetes 39: 1009–1016, 1990
7. Goldfine ID: The insulin receptor: Molecular biology and trans-membrane signaling. Endo Rev 8: 235–255, 1987
8. Tonks NK, Diltz CD, Fischer EH: Purification of the major protein-tyrosine phosphatase of human placenta. J Biol Chem 263: 6722–6730, 1988
9. Tonks NK, Diltz CD, Fischer EH: Characterization of the major protein-tyrosine phosphatase of human placenta. J Biol Chem 263: 6731–6737, 1988
10. Charbonneau H, Tonks NK, Kumar S, Diltz CD, Harrylock M, Cool DE, Krebs EG, Fisher EH: Human placenta protein-tyrosine phosphatase: Amino-acid sequence and relationship to a family of receptor-like proteins. Proc Natl Acad Sci USA 86: 5152–5256, 1990
11. Streuli M, Drueger NX, Thai T, Tang M, Saito H: Distinct functional roles of the two intracellular phosphatase like domains of the receptor-linked protein-tyrosine phosphatases LCA and LAR. EMBO J 9: 2399–2407, 1990
12. Guan KL, Huan RS, Watson SJ, Geahlen RL, Dixon JE: Cloning and expression of a protein-tyrosine phosphatase. Proc Natl Acad Sci USA 87: 1501–1505, 1991
13. Guan KL, Dixon JE: Evidence for protein-tyrosine-phosphatase catalysis proceeding via a cysteine-phosphate intermediate. J Biol Chem 266: 17026–17030, 1991
14. Kenner KA, Hill DE, Olefsky JM, Kusari J: Regulation of protein tyrosine phosphatases by insulin and insulin-like growth factor 1. J Biol Chem 268: 25455–25462, 1993
15. Cicirelli MF, Tonks NK, Diltz CD, Weiel JE, Fischer EH, Krebs EG: Microinjection of a protein-tyrosine phosphatase inhibits insulin action in Xenopus oocytes. Proc Natl Acad Sci USA 87: 5514–5518, 1990
16. Tonks NK, Cicirelli MF, Diltz CD, Krebs EG, Fisher EH: Microinjection of a low M$_r$-human placenta protein tyrosine phosphatase on induction of meiotic cell division in Xenopus oocytes. Mol Cell Biol 10: 458–463, 1990
17. Kenner KA, Hill DE, Olefsky JM, Kusari J: Protein-tyrosine phosphatase 1B is a negative regulator of insulin and insulin-like growth factor-1-stimulated signaling. J Biol Chem 271: 19810–19816, 1996
18. Ahmad F, Li PM, Meyerovitch J, Goldstein BJ: Osmotic loading of neutralizing anitbodies demonstrates a role for protein tyrosine phosphatase 1B in negative regulation of the insulin action pathway. J Biol Chem 270: 20503–20508, 1996
19. Liotta AS, Kole HK, Fales HM, Roth J, Bemier M: A synthetic tris-sulfotyrosyl dodecapeptide analogue of the insulin receptor 1146-kinase domain inhibits tyrosine dephosphorylation of the insulin receptor in situ. J Biol Chem 269: 22996–23001, 1994
20. Chen H, Wertheimer SJ, Lin CH, Katz SL, Amrein KE, Bum P, Quon MJ: Protein-tyrosine phosphatases PTP-1B and Syp are modulators of insulin-stimulated translocation of GLUT4 in transfected rat adipose cells. J Biol Chem 272: 8026–8031, 1997
21. Seely BL, Staubs PA, Reichart DR, Berhanu P, Milarski KL, Saltiel AR, Kusari J, Olefsky JM: Protein tyrosine phosphatase 1B interacts with the activated insulin receptor. Diabetes 45: 1379–1385, 1996
22. Bandyopadhyay D, Kusari A, Kenner KA, Liu F, Chernoff J, Gustafson TA, Kusari J: Protein tyrosine phosphatase 1B complexes with the insulin receptor in vivo and is tyrosine phosphorylated in the presence of insulin. J Biol Chem 272: 1639–1645, 1997
23. Pearson RB, Kemp BE: Design and use of peptide substrates for protein kinases. In: T Hunter, BM Setton (eds). Methods in Enzymology. Academic Press, New York, Vol. 200 (Part A), 1991, pp 62–81
24. Mayer BJ, Jackson PK, Van Etten RA, Baltimore D: Point mutations in the abl SH2 domain coordinately impair phosphotyrosine binding in vitro and transforming activity in vivo. Mol Cell Biol 12: 609–618, 1992
25. Koch CA, Anderson D, Moran MF, Ellis C, Pawson T: SH2 and SH3 domains: Elements that control interactions of cytoplasmic signaling proteins. Science 252: 668–674, 1991
26. Kusari J, Bandhyopadhay D, Kenner K, Kusari A: Activated insulin receptor phosphorlyates and increases the catalytic activity of protein tyrosine phosphatase 1B (PTPase 1B). Diabetes 46(Suppl. I): (abstr) 204A, 1997
27. Stein-Gerlach M, Kharitonenkov A, Vogel W, Ali S, Ullrich A: Protein-tyrosine phosphatase 1D modulates its own state of tyrosine phosphorylation. J Biol Chem 270: 24635–24637, 1995
28. Olefsky JM: Pathogenesis of non-insulin-dependent diabetes (type II). In: LJ Degroot, GM Besser, GF Cahill, JC Marshall, DH Nelson, WD Odell, JT Potts, AH Rubenstein Jr, E Steinberger (eds). Endocrinology. 2nd ed. W.B. Saunders Co., Philadelphia, USA, 1989, pp 1369–1388
29. Kolertman OG, Gray RS, Griffin J, Burstein P, Insel J, Scarlett JA, Olefsky JM: Receptor and post-receptor defects contribute to the insulin resistance in non-insulin-dependent diabetes mellitus. J Clin Invest 68: 957–969, 1981
30. Truglia JA, Livingston JN, Lockwood DH: Insulin resistance: Receptor and post-binding defects in human obesity and non-insulin-dependent diabetes mellitus. Am J Med 79 (Suppl 2B): 13–21, 1985
31. Freidenberg GR, Henry RR, Klein HH, Reichart HH, Olefsky JM: Decreased kinase activity of insulin receptors from adipocytes of non-insulin-dependent diabetic subjects. J Clin Invest 79: 240–250, 1987
32. Sinha MK, Poires WJ, Flickinger EG, Meelheim D, Caro JF: Insulin-receptor kinase activity of adipose tissue from morbidly obese humans with and without NIDDM. Diabetes 36: 620–625, 1987
33. Caro JF, Ittoop O, Pories WJ, Meelheim D, Flickinger EG, Thomas F, Jenquin JF, Silverman JF, Khazanie PG, Sinha MK: Studies on the mechanism of insulin resistance in the liver from humans with non insulin-dependent diabetes. J Clin Invest 78: 249–258, 1986
34. Caro JF, Sinha MK, Raju SM, Ittoop O, Pories WJ, Flickinger EG, Meelheim D, Dohm GL: Insulin receptor kinase in human skeletal muscle from obese subjects with and without non insulin-dependent diabetes. J Clin Invest 79: 1330–1337, 1987
35. Ahmad F, Goldstein BJ: Alterations in specific protein-tyrosine phosphatases accompany insulin resistance of streptozotocin diabetes. Am J Physiol 268: E932–E940, 1995
36. Ahmad F, Goldstein BJ: Increased abundance of specific skeletal muscle protein-tyrosine phosphatases in a genetic model of insulin-resistant obesity and diabetes mellitus. Metabolism 44: 1175–1184, 1995
37. Olichon-Berthe C, Hauguel-De Mouzon S, Peraldi P, Van Obberghen E, Le Marchand-Brustel Y: Insulin receptor dephosphorylation by phosphotyrosine phosphatases obtained from insulin-resistant obese mice. Diabetologia 37: 56–60, 1994
38. McGuire MC, Fields RM, Nyomba BL, Raz I, Bogardus C, Tonks NK, Sommercorn J: Abnormal regulation of protein tyrosine phosphatase activities in skeletal muscle of insulin resistant humans. Diabetes 40: 939–942, 1991
39. Kusari J, Kenner KA, Suh K-Y, Hill DE, Henry RR: Skeletal muscle protein-tyrosine phosphatase activity and tyrosine phosphatase 1B protein content are associated with insulin action and resistance. J Clin Invest 93: 1156–1162, 1994
40. Worm D, Vinten J, Staehr P, Henriksen JE, Handberg A, Beck-Nielsen H: Altered basal and insulin-stimulated phosphotyrosine phosphatase (PTPase) activity in skeletal muscle from NIDDM patients compared with control subjects. Diabetologia 39: 1208–1214, 1996

108

41. Olefsky JM, Garvey WT, Henry RR, Brillan D, Matthaei S, Freidenberg GR: Cellular mechanisms of insulin resistance in non-insulin-dependent (type II) diabetes. Am J Med 85: 86–105, 1988

42. Frangloni JV, Beahm PH, Shifrin V, Jost CA, Neel BG: The non-transmembrane tyrosine phosphatase PTP-1B localizes to the endoplasmic reticulum via its 35 amino acid C-terminal sequence. Cell 68: 545–560, 1992

43. Maegawa H, Ide R, Hasegawa M, Ugi S, Egawa D, Iwanishi M, Kikkawa R, Shigeta Y, Kashiwagi A: Thiazolidine derivatives ameliorate high glucose-induced insulin resistance via the normalization of protein-tyrosine phosphatase activities. J Biol Chem 270: 7724–7730, 1995

44. Faure R, Baquiran G, Bergeron JJ, Posner BI: The dephosphorylation of the insulin and epidermal growth factor receptors. Role of endosome-associated phosphotyrosine phosphatases. J Biol Chem 267: 11215–11221, 1992

45. Burgess JW, Wada I, Ling N, Khan MN, Bergeron JJ, Posner BI: Decrease in beta-subunit phosphotyrosine correlates with internalization and activation of the endosomal insulin receptor kinase. J Biol Chem 267: 10077–10086, 1992

46. Yu H, Chen JK, Feng S, Dalgamo DC, Brauer AW, Schreiber S: Structural basis for the binding of proline-rich peptides to SH3 domains. Cell 76: 933–945, 1994

47. Feng S, Chen JK, Yu H, Simon JA, Schreiber S: Two binding orientations for peptides to the Src SH3 domain: development of a general model for SH3-ligand interactions. Science 266: 1241–1247, 1994

48. Liu F, Hill DE, Chernoff J: Direct binding of the proline-rich of protein tyrosine phosphatase 1B to the Src homology 3 domain of p130[Cas]. J Biol Chem 271: 31290–31295, 1996

49. Flint AJ, Tiganis T, Barford D, Tonks NK: Development of 'substrate-trapping' mutants to identify physiological substrates of protein tyrosine phosphatases. Proc Natl Acad Sci USA 94: 1680–1685, 1997

50. Barford D, Flint AJ, Tonks NK: Crystal structure of human protein tyrosine phosphatase 1B. Science 263: 1397–1404, 1994

51. Jia Z, Barford D, Flint AJ, Tonks NK: Structural basis for phosphotyrosine peptide recognition by protein tyrosine phosphatase 1B. Science 268: 1754–1758, 1995

Molecular and Cellular Biochemistry **182**: 109–119, 1998.
© 1998 *Kluwer Academic Publishers. Printed in the Netherlands.*

Multifunctional actions of vanadium compounds on insulin signaling pathways: Evidence for preferential enhancement of metabolic versus mitogenic effects

I. George Fantus and Evangelia Tsiani

Department of Medicine, Mount Sinai Hospital, Department of Physiology and Banting and Best Diabetes Centre, University of Toronto, Toronto, Canada

Abstract

The pathophysiologic importance of insulin resistance in diseases such as obesity and diabetes mellitus has led to great interest in defining the mechanism of insulin action as well as the means to overcome the biochemical defects responsible for the resistance. Vanadium compounds have been discovered to mimic many of the metabolic actions of insulin both *in vitro* and *in vivo* and improve glycemic control in human subjects with diabetes mellitus. Apart from its direct insulinmimetic actions, we found that vanadate modulates insulin metabolic effects by enhancing insulin sensitivity and prolonging insulin action. All of these actions appear to be related to protein tyrosine phosphatase (PTP) inhibition. However, in contrast to its stimulatory effects, vanadate inhibits basal and insulin-stimulated system A amino acid uptake and cell proliferation. The mechanism of these actions also appears to be related to PTP inhibition, consistent with the multiple roles of PTPs in regulating signal transduction. While the precise biochemical pathway of vanadate action is not yet known, it is clearly different from that of insulin in that the insulin receptor and phosphatidylinositol 3′-kinase do not seem to be essential for vanadate stimulation of glucose uptake and metabolism. The ability of vanadium compounds to 'bypass' defects in insulin action in diseases characterized by insulin resistance and their apparent preferential metabolic versus mitogenic signaling profile make them attractive as potential pharmacological agents. (Mol Cell Biochem **182**: 109–119, 1998)

Key words: vanadium compounds, insulin action, tyrosine kinase/phosphatase, L6 myocytes

Introduction

Growth factors such as PDGF, EGF and insulin which bind to and activate receptors with instrinsic protein tyrosine kinase activity are well documented to exert their effects by tyrosine (tyr) phosphorylation [1–3]. Moreover a number of actions of cytokines [4, 5] integrins [6], as well as ligands which signal via the seven transmembrane domain family of G-protein (GTP-binding protein) coupled receptors [7] have also been found to require tyr phosphorylation. This signal transduction mechanism is balanced by tyr dephosphorylation catalyzed by a family of protein tyrosine phosphatases (PTPs) [8–10].

Insulin binding to its receptor stimulates tyr autophosphorylation of three regulatory tyr residues in the cytoplasmic domain of the β-subunit [11]. This trans or cross-phosphorylation between β-subunits within the receptor dimer results in an augmentation of receptor tyr kinase activity. Subsequently, further phosphorylation of the receptor itself as well as substrates such as IRS-1, IRS-2 (insulin receptor substrate- 1,2) and shc (src homology and α-collagen-like) [12] is required to signal downstream events. These protein substrates contain PTB (protein tyr binding) domains which facilitate their interactions with the IR via a specific tyr phosphorylated region (NPEpY) in the juxtamembrane region [13, 14]. Once tyr phosphorylated, IRS-1 and/or

Address for offprints: I.G. Fantus, Department of Medicine, Mount Sinai Hospital, 600 University Avenue, Suite 780, Toronto, Ontario, M5G 1X5, Canada

IRS-2 serve as 'docking' proteins for various intracellular signaling molecules containing SH2 (src homology 2) domains. The SH2 domains serve as binding partners for phosphorylated tyr residues followed by specific C-terminal amino acid sequences which determine specificity [3]. Examples of SH2-domain containing proteins which bind to IRS-1 include the enzymes PI3-kinase (phosphatidylinositol 3-kinase), SHP-2/SYP (a SH2 containing PTP) and fyn (a src-like cytosolic-tyr kinase) as well as adaptor proteins such as Grb2 (growth factor receptor bound protein 2) and Nck [15, 16]. It is these proteins which then serve to transmit the pleiotropic effects of insulin on growth, differentiation and metabolism [15]. Thus the balance between tyr phosphorylation and dephosphorylation appears to be critical in the regulation of the earliest steps in insulin action.

The element vanadium was discovered by the Swedish Chemist Neil Sefstrom in 1830. Although a common component of the earth's crust, it was not until the 1970's that biologists became more interested in this element [17]. In 1977 Cantley discovered that vanadate was an inhibitor of Na^+, K^+, ATPase [18]. Subsequently it was found to inhibit all 'P' type phosphorylated ATPases [19, 20] and in 1982 to inhibit PTPs [21, 22]. The interest in vanadate was greatly augmented by the findings that vanadate (V_2O_5) and vanadyl sulfate ($VOSO_4$) could mimic insulin to stimulate glucose uptake and metabolism in insulin target tissues *in vitro* [23–25] and *in vivo* [26]. The impact of these critical observations on research with vanadium compounds is shown in Table 1.

Using vanadate as a probe we and others have demonstrated the various biological consequences of PTP inhibition on insulin signaling. In this paper we review the evidence that PTP inhibition may alter the physiology of insulin signaling by altering insulin sensitivity as well as the duration of the biological response. In addition we show that vanadate may paradoxically inhibit rather than mimic some of the actions of insulin implicating the requirement of PTP activity in signal propagation. Finally, we provide evidence that although cell signaling by vanadium compounds is dependent on tyr kinase activity the signal transduction pathway is different from that of insulin. For more detailed information on the chemistry and biological actions of vanadium and its derivatives, the reader is referred to several reviews [27–32].

Vanadate mimics and modulates metabolic actions of insulin

Essentially all of the actions of insulin on glucose uptake and metabolism as well as lipid metabolism have been found to be stimulated by vanadate *in vitro* and *in vivo*, e.g. glucose transport, glycolysis, glycogen synthesis, lipogenesis and inhibition of lipolysis (reviewed in [29–32]). Thus it appears that stimulation of tyr phosphorylation is a critical component of the signaling pathways for these metabolic actions. In addition to these rapid metabolic effects recent studies also suggest that some changes in gene expression caused by insulin are also mimicked by vanadium compounds [31–34].

Apart from directly stimulating these actions of insulin, we reasoned that inhibition of PTPs would alter other aspects of insulin signaling. The first possibility was suggested by the phenomenon of 'spare' receptors; that is, in adipocytes only 10–20% of insulin receptors need be occupied and activated to achieve a maximum biological response [35]. Thus, in this model tissue, sensitivity to insulin is regulated by the number of cell surface insulin receptors. Since the magnitude of the biological effect is dependent on active hormone-receptor (H-R) complex formation and thereby net IR tyr kinase activity, inhibition of PTPs would be expected to augment tissue sensitivity. We demonstrated that preincubation of rat adipocytes with a low vanadate concentration (which did not significantly stimulate on its own) was able to enhance insulin sensitivity (Fig. 1A). Surprisingly some increase in insulin receptor binding affinity was also observed [36]. The mechanism of this latter effect is not clear. However, the enhanced sensitivity may be related to both increased insulin binding as well as IR kinase activity. In support of these results are reports by

Table 1. Impact of discoveries of the bioeffects of vanadium compounds on publications in the biomedical sciences

	Major discoveries	Publication rate	
		(years)	(articles/year)
1977	Inhibition of Na^+, K^+, ATPase	1976–1979	36/year
1979/80	Insulin-mimetic action on glucose metabolism *in vitro*	1980–1985	117/year
1982	Inhibition of phosphotyrosine phosphatases	1986–1990	167/year
1985	Insulin-mimetic action *in vivo* in diabetic rats	1990–1995	238/year
1995	Hypoglycemic action *in vivo* in humans with diabetes mellitus	1996	222/year

The years of discoveries of the major biological effects of vanadium compounds *in vitro* and *in vivo* are summarized (see text for details). The interest of the scientific community reflected by the number of publications in each time period is shown. Publication rates were determined by a Medline search using the terms vanadium, vanadates and the more recently introduced vanadium compounds. Articles indexed under more than one term were only counted once. Compared with the late 1970's there was a 6–7 fold increase throughout the 1990's.

others that vanadate is able to enhance insulin sensitivity both *in vitro* [37] and *in vivo* [38–40].

The hypothesis that tyr dephosphorylation is required to terminate the insulin signal suggested a second potential effect of vanadate. It was reasoned that PTP inhibition would prolong the action of insulin. Again in the presence of low concentrations of vanadate, we demonstrated that insulin-stimulated lipogenesis in rat adipocytes was prolonged after removal of insulin [41]. A kinetic analysis revealed that under the conditions of these experiments the $t_{1/2}$ of the IR tyr kinase was not prolonged but the amplitude of its stimulation was greatly augmented in the presence of vanadate [41]. Previous studies using insulin had suggested that increasing stimulating hormone concentrations would prolong the duration of the biological response [42]. Thus it appeared that regulation of the duration of the response was another function of activated 'spare' receptors, i.e. those IRs activated over and above that required to achieve a maximum biological response. Indeed biological response duration was prolonged by vanadate as long as IR tyr kinase activity was increased. At the maximum amplitude of the kinase there was no longer any prolongation by vanadate (Fig. 1B) [41]. In summary, the PTP inhibitor vanadate can interact with the insulin metabolic signaling pathway in at least 3 ways: (i) to mimic insulin metabolic actions (ii) to augment insulin sensitivity and (iii) to prolong the duration of the biological response (Table 2).

Vanadate inhibits some actions of insulin

System A amino acid transport
Apart from regulating glucose and lipid metabolism, insulin has important effects on protein metabolism and cell growth. Although the effector signaling pathways have not been completely delineated, both the PI3-kinase and ras-raf-MAPK enzyme cascades appear to play a role. When the effects of vanadate and pervanadate (aqueous peroxo-vanadium) on protein synthesis were first examined in rat adipocytes, we found that vanadate could not significantly increase protein synthesis while pervanadate had a stimulatory effect which was less than that of insulin [43]. In skeletal muscle, Clark [44] and Foot [45] reported that vanadium compounds (vanadate and pervanadate) could not mimic insulin to stimulate protein synthesis.

More recently, we examined the effect of vanadate and pervanadate on amino acid uptake by the insulin sensitive System A transporter [46]. In cultured differentiated L6 skeletal muscle cells the vanadium compounds inhibited the uptake of MeAIB (methylaminoisobutyric acid), an amino acid analogue specifically transported by System A, in a time and dose-dependent manner (Fig. 2). We found that insulin-stimulated MeAIB uptake was more sensitive to inhibition by

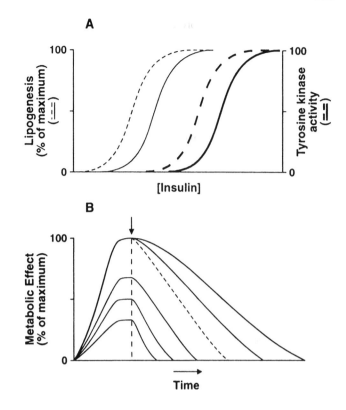

Fig. 1. Vanadate alters the physiology of insulin signaling. (A) Enhancement of insulin sensitivity by vanadate. Representative dose-response curves for insulin-stimulated lipogenesis (thin lines) and insulin receptor tyrosine kinase activity (thick lines) in the absence (solid lines) and presence (dashed lines) of vanadate are shown. It should be noted that the stimulatory effects of vanadate alone are small and have been subtracted. The concentration of insulin required to achieve maximum tyrosine kinase activity is greater than that required for maximum lipogenesis reflecting the presence of 'spare' receptors; (B) Vanadate prolongs insulin action. The duration of the biological response (lipogenesis) is determined by the stimulating concentration of insulin. This phenomenon reflects the regulation of response duration, at least in part, by the extent of activation of the IR tyrosine kinase. Thus vanadate exposure will prolong insulin action at insulin concentrations which do not achieve maximum tyrosine kinase activation. The solid lines represent stimulation by increasing concentrations of insulin. The dashed line reflects the response duration at the insulin concentration required to achieve maximum lipogenesis. The arrow depicts the time of removal of insulin.

Table 2. Modulation of insulin metabolic effects by vanadate

1.	Increased basal activity
2.	Increased insulin sensitivity
3.	Prolonged duration of response

vanadium compounds than basal uptake and was associated with an inhibition of the insulin-stimulated increase in V_{max} [46]. This response of amino acid uptake was specific since glucose uptake was stimulated by similar concentrations of the vanadium compounds in the L6 cells. Furthermore,

112

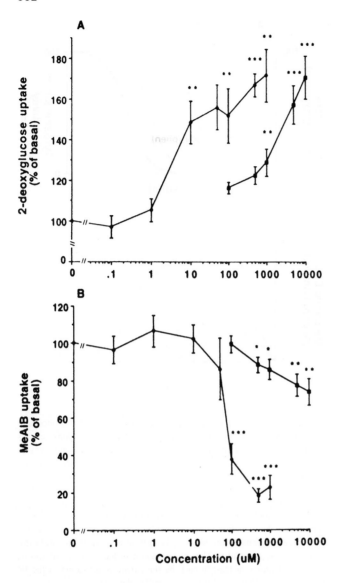

Fig. 2. Vanadate and pervanadate stimulate glucose uptake but inhibit amino acid uptake in L6 myotubes. L6 myotubes were incubated for 90 min with the indicated concentrations of vanadate (□) or pervanadate (♦). At the end of the incubation 2-deoxyglucose (A) and MeAIB (B) uptake were measured. Values are means ± S.E. of 4–9 separate experiments performed in duplicate. *p < 0.05, **p < 0.001, ***p < 0.001 compared with basal. (Adapted from Ref. [46]).

insulin-stimulated amino acid uptake was also inhibited by vanadate in rat hepatoma cells. The inhibitory effect of vanadate did not require protein synthesis and was not mimicked by the Na^+, K^+, ATPase inhibitor ouabain. The mechanism of action of insulin to stimulate system A amino acid transport is not defined while the site of inhibition by vanadate also remains unknown. The similar relative dose-response curves of vanadate and pervanadate for stimulation of glucose uptake and inhibition of amino acid uptake [46] along with their relative potencies as PTP inhibitors

[43, 47] suggest that PTP inhibition is the mechanism. Thus, one possibility is that a vanadate-inhibited PTP plays a role in the activation process of amino acid uptake. The participation of PTPs and/or dual specificity phosphatases, which are also inhibited by vanadium compounds, in processes such as ras activation, mitogenesis and cell cycle progression has been well documented (see below). However, whether amino acid uptake falls into this category requires further study.

Mitogenesis
A major action of insulin in addition to metabolic signaling is the stimulation of cell growth and differentiation. The sequence of events leading to cell proliferation has been largely defined by work in yeast and by studies of oncogenic transformation [48, 49]. Previous studies of cultured cells exposed to vanadate have yielded conflicting results in regard to cell growth. Thus, vanadate stimulation of tyr phosphorylation along with cell proliferation and the enhancement of mitogenesis in the presence of various growth factors have been reported ([50–52], reviewed in [53]). In contrast, a number of studies have demonstrated inhibition of cell proliferation by vanadate [54, 55] and peroxovanadate ([56], reviewed in [57]).

To determine whether vanadate mimicked insulin's mitogenic effect in a model insulin target tissue, we incubated L6 myoblasts, previously shown to respond to insulin and IGF-1 with a proliferative response [58, 59], with vanadate and a stable peroxovanadium compound, bpV(phen). Bisperoxovanadium 1,10-phenanthroline [46] [bpV(phen)] was employed since pervanadate, generated by mixing vanadate and H_2O_2, is unstable in the absence of excess H_2O_2 [60] and these experiments required exposure for 16 h. This compound has PTP inhibitory efficacy which is essentially equal to that of aqueous peroxovanadium [47] and we have found a similar dose-dependent stimulation of glucose uptake in L6 cells (data not shown). While exposure to insulin for 16 h stimulated ^{3}H-thymidine incorporation to 290 ± 29% of control, there was a dose-dependent inhibition by vanadate and bpV(phen) (Fig. 3). Maximum inhibition by vanadate was achieved at a concentration of 1000 μM and by bpV(phen) at 20 μM. These concentrations reflect the relative potencies of these agents as PTP inhibitors [46, 47] and suggest that PTP inhibition is the mechanism by which these agents inhibit mitogenesis.

To further explore this action of vanadate, its effect on serum and insulin stimulated mitogenesis was examined. Insulin, 10^{-7} M , and 10% FBS significantly stimulated ^{3}H-thymidine incorporation (% of basal, insulin 290 ± 29, FBS 280 ± 25). Vanadate completely inhibited the ^{3}H-thymidine incorporation stimulated by both insulin and FBS (Fig. 4).

There are at least two well documented events occurring during the course of cell proliferation in which PTP activation participates. First, similar to other growth factors insulin

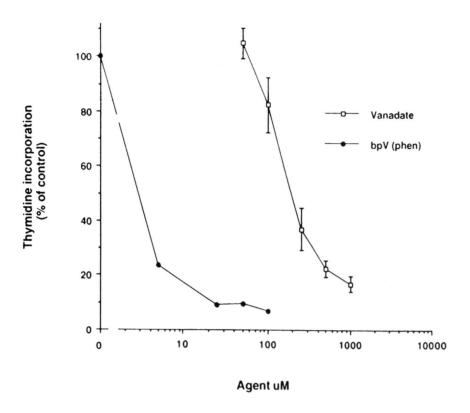

Fig. 3. Dose-dependent inhibition of ³H-thymidine incorporation by vanadate and bpV(phen). Subconfluent (40–50%) L6 myoblasts were serum deprived (0.1% FBS) for 24 h in α-MEM. The cells were subsequently incubated for 16 h in fresh medium in the presence and absence of the indicated concentrations of vanadate or bpV(phen). ³H-thymidine (2 μci/well) was added for the final 8 h. At the end of the incubation the cells were washed and solubilized in 0.2% SDS. Cell lysates were precipitated with 10% TCA and radioactivity incorporated in the pellets determined by β-scintillation counting (Packard 460). The results are means ± S.E. of 4–5 separate experiments.

activates ras, a key component of the growth response [61]. It has been demonstrated that the SH2-domain containing cytosolic PTP, SHP-2 (SHPTP2/SYP) is required for insulin-induced ras activation [62–64]. Thus inhibition of SHP-2 is one, possible mechanism by which PTP inhibitors may block the insulin stimulated growth response. Second, cell proliferation stimulated by all growth factors involves progression through a series of stages referred to as the cell cycle [49]. Progression through the cell cycle is regulated at various points by serine/threonine kinases, the CDKs (cyclin-dependent kinases) [49, 65]. These are activated by binding to various proteins termed cyclins as well as by specific ser phosphorylation and tyr and thr dephosphorylation reactions. Thus CDK1 (cdc2) which forms a complex with cyclin B in the G2 (Gap 2) phase of the cell cycle requires dephosphorylation of tyr and thr by the phosphatase cdc25 [66]. This activation of CDK1 allows progression from G2 to the M (mitosis) phase. The cdc25 phosphatases comprise one of four families within the superfamily of tyrosine phosphatases [10, 66]. They dephosphorylate both ser/thr as well as tyr residues and have an active site motif Cx_5R, characteristic of tyr phosphatases which are inhibitable by vanadate. Thus

inhibition of cdc25 would be predicted to inhibit cell proliferation.

Upon serum deprivation the majority of cultured cells are arrested in the G0/G1 phase of the cell cycle [67]. A block in the ras-raf pathway, e.g. by inhibition of SHP-2, would be likely to maintain this state. In contrast, inhibition of cdc25 would be expected to result in progression to G2 but failure to progress through M. To determine which phase of the cell cycle was most affected by vanadate, the proportion of myoblasts in the various phases after 48 h exposure to the different agents was measured by flow cytometry. Unstimulated serum-deprived cells were in G1 (80%) (Fig. 5). Insulin and serum (FBS) stimulation decreased the proportion of cells in G1 (62 and 69% respectively) and increased the proportion in S (Control 14.5%, Insulin 29%, FBS 22%) and G2/M (Control 5.5%, Insulin 9%, FBS 9%). Vanadate alone also decreased the proportion of cells in G1 to a similar extent (62.3 %). However in contrast to insulin and FBS, vanadate caused an accumulation of cells in G2/M (31.1%) with a reduction in the proportion in S phase (6.6%). Co-incubation of serum with vanadate resulted in a larger decrease in the number of cells in G1 (48%) than achieved

114

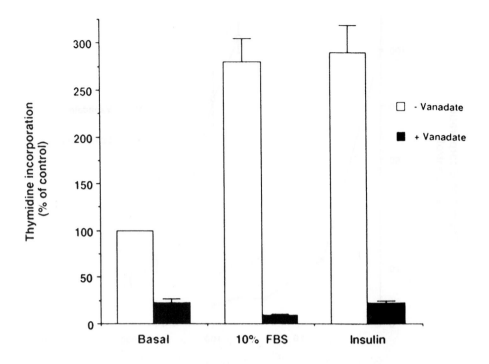

Fig. 4. Vanadate inhibits serum and insulin stimulated ^{3}H-thymidine incorporation. Subconfluent (40–50%) L6 myoblasts were serum-deprived (0.1% FBS) for 24 h in α-MEM. The cells were subsequently incubated for 16 h in fresh medium with and without either 10% FBS or 10^{-7} M insulin in the absence (open bars) and presence (filled bars) of 500 μM vanadate. ^{3}H-thymidine incorporation was determined as described in the legend to Fig. 3. The results are means ± S.E. of 4–5 experiments performed in duplicate. Both FBS (280 ± 25% of basal) and insulin (290 ± 29%) stimulated thymidine incorporation (p < 0.001 for both) while vanadate significantly inhibited incorporation under all three conditions (basal 22.4 ± 4.0%; FBS 9.6 ± 1.2%, insulin 22.2 ± 30%; p < 0.001 for all compared with basal without vanadate).

with either alone, but with a concomitant accumulation of cells in G2/M (43.5 %) (Fig. 5). In the case of stimulation with insulin however, there appeared to be an inhibition of exit from G0/G1 caused by vanadate. Thus cells remained in G1 without a change in proportion in G1 (79%), S (13 %) or G2/M (8 %) compared with control (Fig. 5). The data suggest that in the case of growth stimulation by serum, there is a block by vanadate in the G2/M transition in L6 myoblasts. This is similar to that observed previously by other investigators with pervanadate in neuroblastoma NB41 and glioma C6 cells [56] as well as vanadate in human leukemia B cells, Ball-1 [68] and is consistent with an inhibition of CDK1 dephosphorylation and function. The cause of the vanadate-induced block in G1 observed in the presence of insulin is not clear. Serum contains a number of different growth promoting factors and these may overcome a specific block of a G1-associated signaling pathway which is sensitive to vanadate. On the other hand, insulin may promote progression through G1 only via a vanadate-sensitive pathway. Inhibition of cell proliferation by another PTP inhibitor, RK-682, has also been found to be associated with a block of G1 progression [68].

As discussed above the activation of the ras-raf-MAPK pathway by insulin requires the PTP SHP-2. However,

vanadate is able to activate MAPK in the absence of insulin receptor activation [69, 70] and bpV(phen) resulted in activation of MAPK in the presence of PD98059, an inhibitor of MEK (MAP/ERK kinase) [71]. These data strongly suggest that MAPK activation by vanadate may be caused largely by inhibition of a dual specificity phosphatase, e.g. MKP-1 [72] or Pyst-1 [73]. The growth response to various mitogenic stimuli is associated with a robust but transient activation of MAPK [74]. However, in PC12 (pheochromocytoma) cells prolonged activation of MAPK was associated with differentiation rather than proliferation [75]. Thus the duration of MAPK activation may provide a signal which determines cell fate, i.e. proliferation or differentiation. The mechanism by which MAPK activation is terminated in the continued presence of growth factors is thought to be at least in part, by dephosphorylation [72, 73]. We reasoned that if MAPK activation by vanadate is caused by phosphatase inhibition, there may be prolonged activation of MAPK.

To test this hypothesis the extent of tyr phosphorylation of MAPK was assessed in lysates from L6 myoblasts stimulated for 18 h with 10^{-7} M insulin or 10% FBS in the presence and absence of 100 μM vanadate. An equal number of cells from each condition were lysed in SDS and total cell

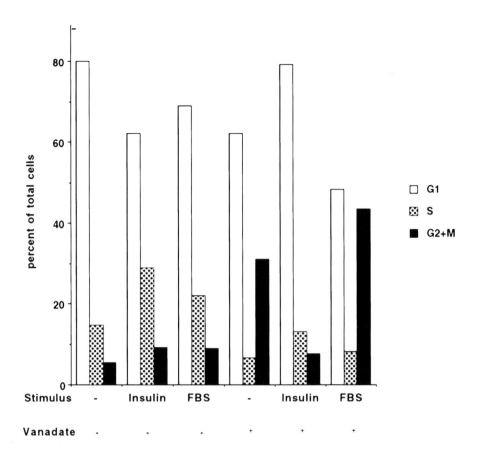

Fig. 5. Effect of vanadate on cell cycle progression. Serum-deprived (24 h, 0.1% FBS) L6 myoblasts were incubated for 48 h in (α-MEM with and without 10⁻⁷M insulin or 10% FBS in the absence and presence of 75 µM vanadate. At the end of the incubation the cells were washed, trypsinized and fixed in 80% ethanol. DNA was stained with the propidium iodide method [89] and the proportion of cells in various phases of the cell cycle determined by flow cytometry. Results are from a representative experiment. Similar results were observed in 3 separate experiments.

lysates were loaded onto gels, the proteins separated by SDS-PAGE, transferred to nitrocellulose membranes and blotted with anti-phosphotyrosine antibody as previously described [76]. Although all three agents alone, insulin, serum and vanadate activate MAPK acutely ([69, 77] and data not shown), after 18 h only vanadate exposure resulted in tyr phosphorylation of a 42 kDa band identified as MAPK (Fig. 6). Furthermore, chronic treatment with either insulin or serum in the presence of vanadate also resulted in persistent and even greater MAPK tyr phosphorylation (Fig. 6). It should be noted that MAPK tyr phosphorylation is associated with its activation. We have also found that addition of the MEK inhibitor PD98059 did not inhibit the chronic activation of MAPK by vanadate (not shown). Two other bands showed tyr phosphorylation stimulated by chronic exposure to vanadate (Fig. 6). These were at M_r 95 kDa and ~170 kDa consistent with the IR and IRS-1 which we and others [76] have confirmed to migrate at these positions on SDS-PAGE. Neither insulin nor FBS treatment alone caused tyr phosphorylation of the IR after 18 h exposure although

a very small degree of IRS-1 tyr phosphorylation was evident in the case of insulin. This is in contrast to acute exposure to insulin which is well documented to produce marked tyr phosphorylation of both IR and IRS-1. Thus, despite chronic activation of several components of the insulin action pathway by vanadate, the mitogenic response to insulin and FBS was inhibited. The potential role of chronic activation of MAPK in mediating this inhibitory action remains to be determined.

Mechanisms of the insulin-mimetic action of vanadium compounds

The mechanism of action of vanadium compounds to mimic insulin appears to be related to PTP inhibition. In this manner one or more tyr kinases may be indirectly activated. Whether the IR tyr kinase is involved remains controversial [33]. Several studies suggest that it is not and Shisheva *et al.* have suggested that a nonreceptor tyr kinase of ~54 kDa is

116

Anti-pY Immunoblot

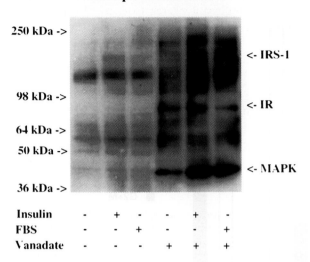

Insulin	-	+	-	-	+	-
FBS	-	-	+	-	-	+
Vanadate	-	-	-	+	+	+

Fig. 6. Chronic exposure to vanadate stimulates protein tyrosine phosphorylation. Serum-deprived L6 myoblasts were incubated in α-MEM with and without 10% FBS or 10^{-7} M insulin for 24 h in the presence and absence of 100 μM vanadate. At the end of the incubation cells (~3.6×10^5) were washed and lysed in 4% SDS, 10 mM DTT, 115 mM Tris HCL (pH 6.8), 10% glycerol, 0.24 mg/ml Bromophenol Blue, 100 μM PMSF, 10 μM E-64, 1 μM pepstatin, 1 μM leupeptin, 40 mM NaF, 7.5 mM Na pyrophosphate and 1.5 mM Na_3VO_4. The lysates were passed five times through a 27 gauge needle, boiled and 20 μg protein (Biorad) separated by SDS-PAGE (7.5%). The proteins were transferred to PVDF (polyvinylidine difluoride) membranes and immunoblotted with anti-phosphotyrosine antibodies (Transduction Laboratories). Molecular weight markers are indicated on the left and the identity of the tyr phosphorylated proteins determined by their migration indicated on the right. The IR, IRS-1 and MAPK proteins showed similar migration blotted with specific antibodies (not shown). While chronic insulin or FBS stimulation alone did not result in sustained tyr phosphorylation, chronic treatment with vanadate alone and in the presence of insulin or FBS resulted in enhanced tyr phosphorylation of IR, IRS-1 and MAPK. The band at M_r ~200 kDa shows a tyr phosphorylation pattern similar to that of the IR and likely represents the IR precursor.

activated and mediates at least some of the insulin-like effects in adipocytes [78]. More recently, a second nonreceptor tyr kinase of similar size has been found in adipocytes and suggested to be responsible for glucose transport stimulation (Shechter Y, personal communication). In the case of the more potent peroxovanadium compounds there is evidence for IR activation [47, 79]. Thus it is likely that more than one tyr kinase is involved. Furthermore, it has recently been demonstrated that the peroxovanadium compounds irreversibly oxidize the -SH group of the essential cysteine at the PTP catalytic site while there was a weaker and reversible inhibition caused by vanadate [80]. Thus the various vanadium compounds may have a somewhat different but overlapping spectrum of activities.

What is perhaps most interesting is that the signaling pathways from the tyr kinase to the final biological effects

appear to be different for the vanadium compounds compared to insulin. Band and Posner [71] found that in contrast to insulin, suppression of IGF-BP1 gene expression by bpv(phen) was not blocked by wortmannin or rapamycin, inhibitors of PI-3-kinase and p70[s6k] respectively. We [81] and others [82] have found that glucose transport stimulation by vanadate and pervanadate is not blocked by wortmannin, again in contrast to insulin. Since a number of defects in the insulin signaling pathway are evident in various insulin resistant states, e.g. PI-3kinase [83–85], the ability to bypass these would be a desirable property for a potential pharmacological agent (see [33] for review).

Summary and conclusions

Vanadate and related vanadium compounds (e.g. peroxo-vanadium) clearly have insulin-mimetic actions. These appear to be related largely, if not entirely to their actions as PTP inhibitors. The role of the IR tyr kinase remains unclear and activation of other tyr kinases may be important in mediating vanadate's biological effects. In addition to direct insulin-mimetic actions, vanadate is able to alter the physiology of insulin signaling in target tissues. Thus, in the presence of low concentrations of vanadate, insulin sensitivity is enhanced and the duration of its metabolic action prolonged. These indirect effects may have important *in vivo* consequences which would contribute to the successful treatment of Type 2 diabetes mellitus (NIDDM). Indeed two recent studies in human subjects have demonstrated that oral Na metavanadate and vanadyl sulfate both improved metabolic control and insulin sensitivity [86, 87].

In contrast to glucose and lipid metabolic effects, we and others have now found that vanadate and peroxovanadium do not mimic several other actions of insulin. Specifically, processes related to cell growth and proliferation may be paradoxically inhibited. These include protein synthesis, amino acid uptake and mitogenesis. Although the mechanism of inhibition is not clear, our increased understanding of the roles of various tyr phosphatases in cell function combined with the knowledge that PTP inhibition by these vanadium compounds is relatively nonspecific suggest that the same mechanism, namely PTP inhibition, may result in both the positive and negative effects on insulin action (Fig. 7).

The above observations raise important questions relating to the potential use of vanadate and/or its derivatives for the therapy of diabetes. Do these inhibitory effects present a problem or a benefit? One might speculate that inhibition of amino acid uptake may represent a disadvantage if present *in vivo* since protein metabolism is already disturbed in NIDDM. On the other hand the growth effects of insulin have been suggested to contribute to the macrovascular disease associated with insulin resistance and hyperinsulinemia [88].

Effect of Vanadium compounds on insulin action

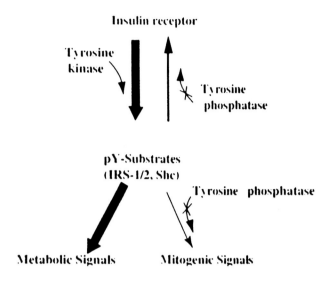

Fig. 7. Preferential metabolic versus mitogenic signaling by vanadium compounds. The early events in insulin action are mediated by tyr phosphorylation. Subsequently insulin has pleiotropic actions which can be divided into two general classes, metabolic and mitogenic. The role of tyr phosphorylation and dephosphorylation in signal transduction distal to the IR and IRS-1 appear to differ for these two classes. Thus, tyr phosphorylation is required to propagate most metabolic effects but tyr dephosphorylation is required for mitogenic signals. As vanadate is a general inhibitor of PTPS, it is proposed that vanadate and its derived compounds will turn out to be unique insulin-mimetic agents which demonstrate preferential metabolic versus mitogenic signaling.

Thus a potential advantage of vanadate treatment might be an improvement in insulin sensitivity and hyperglycemia without the risks associated with hyperinsulinemia.

Research describing the effects of vanadate has provided valuable information on the role of tyr phosphorylation and dephosphorylation in modulating insulin action. Further investigation to characterize the proteins involved in the signaling pathway of vanadate will also provide important data, since these agents may bypass several steps in the insulin signaling pathway. The identification of these proteins will point to novel molecular targets for therapy of diseases associated with insulin resistance. Further *in vivo* studies in animal models and clinical studies in people with diabetes will be required to evaluate the full potential of vanadium compounds as therapeutic agents.

Acknowledgements

The authors would like to thank Drs. B. Posner, C. Yip and A. Klip for helpful discussion and Mrs. B. Baubinas for secretarial assistance. This work was supported by a grant from the Medical Research Council of Canada to IGF (MT-7658). E. Tsiani was supported in part by a postdoctoral fellowship from the Department of Medicine, University of Toronto.

References

1. Hunter T: A thousand and one protein kinases. Cell 50: 823–829, 1987
2. Ullrich A, Schlessinger J: Signal transduction by receptors with tyrosine kinase activity. Cell 61: 203–212, 1990
3. Pawson T: Protein nodules and signalling networks. Nature 373: 573–580, 1995
4. Ihle JN, Witthuhn BA, Quelle FW, Yamamoto K, Thierfelder WE, Kreider B, Silvennoinen O: Signaling by the cytokine receptor superfamily JAKs and STATS. Trends Biochem Sci 19: 222–227, 1994
5. Roupas P, Herington AC: Postreceptor signaling mechanisms for growth hormone. Trends Endocrinol Metab 5: 154–158, 1994
6. Clark EA, Brugge JS: Integrins and signal transduction pathways: The road taken. Science 268: 233–239, 1995
7. Dikic I, Tokiwa G, Lev S, Courtneidge SA, Schlessinger J: A role for Pyk2 and Src in linking G-protein coupled receptors with MAP kinase activation. Nature 383: 547–550, 1996
8. Fischer EH, Charbonneau H, Tonks N-K: Protein tyrosine phosphatases: A diverse family of intracellular and transmembrane enzymes. Science 253: 401–406, 1991
9. Walton KM, Dixon JE: Protein tyrosine phosphatases. Annu Rev Biochem 62: 101–120, 1993
10. Fauman EB, Saper MA: Structure and function of the protein tyrosine phosphatases. Trends Biochem Sci 21: 413–417, 1996
11. White MF, Kahn CR: The insulin signaling system. J Biol Chem 269: 1–5, 1994
12. Patti M-E, Sun X-J, Bruening JC, Araki E, Lipes MA, White MF, Kahn CR: 4PS/Insulin receptor substrate (IRS)-2 is the alternative substrate of the insulin receptor in IRS-1-deficient mice. J Biol Chem 270: 24670–24673, 1995
13. Gustafson TA, He W, Craparo A, Schaub CD, O'Neill TJ: Phosphotyrosine-dependent interaction of SHC and insulin receptor substrate 1 with the NPEY motif of the insulin receptor via a novel non-SH2 domain. Mol Cell Biol 15: 2500–2508, 1995
14. van der Geer P, Wiley S, Ka-Man Lai V, Olivier JP, Gish GD, Stephens R, Kaplan D, Shoelson S, Pawson T: A conserved amino-terminal Shc domain binds to phosphotyrosine motifs in activated receptors and phosphopeptides. Curr Biol 5: 404–412, 1995
15. Cheatham B, Kahn CR: Insulin action and the insulin signaling network. Endo Rev 16: 117–142, 1995
16. Sun XJ, Crimmins DL, Myers MG, Miralpeix M, White MP: Pleiotropic insulin signals are engaged by multisite phosphorylation of IRS-1. Mol Cell Biol 13: 7418–7428, 1993
17. Macara IG: Vanadium: An element in search of a role. Trends Biochem Sci 5: 92–94, 1980
18. Cantley L, Resh M, Guidotti G: Vanadate inhibits the red cell (Na⁺, K⁺) ATPase from the cytoplasmic side. Nature Lond 272: 552–554, 1978
19. Nechay BR, Nanninga LB, Nechay PSE, Post RL, Granthan JJ, Macara IG, Kubena LF, Phillips TD, Nielsen FFH: Role of vanadium in biology. Fed Proc 45: 123–132, 1986
20. Stankiewicz PJ, Tracey AS, Crans DC: Inhibition of phosphate-metabolizing enzymes by oxovanadium comples. In: H Sigel, A Sigel (eds). Metal Ions in Biological Systems. Marcel Dekker, New York, 1995, 31: 287–324a

118

21. Swarup G, Cohen S, Garbers D: Inhibition of membrane phospho-tyrosyl-protein phosphatase activity by vanadate. Biochem Biophys Res Commun 107: 1104–1109, 1982

22. Swarup G, Speeg JKV, Coben S, Garbers DL: Phosphotyrosyl-protein phosphatase of TCRC-2 cells. J Biol Chem 257: 7298–7301, 1982

23. Tolman EL, Barris E, Bums M, Pansini A, Partridge R: Effects of vanadium on glucose metabolism in vitro. Life Sci 25: 1159–1164, 1979

24. Dubyak GR, Kleinzeller A: The insulin-mimetic effects of vanadate in isolated rat adipocytes. Dissociation from effects of vanadate as (Na⁺-K⁺)-ATPase inhibitor. J Biol Chem 255: 5306–5312, 1980

25. Shechter Y, Karlish SJD: Insulin-like stimulation of glucose oxidation in rat adipocytes by vanadyl (IV) ions. Nature 284: 556–558, 1980

26. Heyliger CE, Tahiliani AG, McNeill JH: Effect of vanadate on elevated blood glucose and depressed cardiac performance of diabetic rats. Science 227: 1474–1477, 1985

27. Crans DC: Aqueous chemistry of labile oxovanadates: Relevance to biological studies. Comm Inorganic Chem 16: 1–33, 1994

28. Shaver A, Ng JB, Hall DA, Posner BI: The chemistry of peroxovanadium compounds relevant to insulin mimesis. Mol Cell Biochem 153: 5–15, 1995

29. Shechter Y: Insulin-mimetic effects of vanadate: Possible implications for future treatment of diabetes. Diabetes 39: 1–5, 1990

30. Posner BI, Shaver A, Fantus IG: Insulin mimetic agents: Vanadium and peroxovanadium compounds. In: CJ Bailey, PR Flatt (eds). New Antidiabetic Drugs. Smith-Gordon, London, 1990, pp 107–118

31. Orvig C, Thompson KH, Battell M, McNeill JH: Vanadium compounds as insulin mimics. In: H Sigel, A Sigel (eds). Metal Ions in Biological Systems. Marcel Dekker, New York, 1995, 31: 575–594

32. Brichard SM, Ongemba LN, Henquin JC: Oral vanadate decreases muscle insulin resistance in obese fa/fa rats. Diabetologia 35: 522–527, 1992

33. Tsiani E, Fantus IG: Vanadium compounds: Biological actions and potential as pharmacological agents. Trends Endocrinol Metab 8: 51–58, 1997

34. Brichard SM, Henquin J-C: The role of vanadium in the management of diabetes. Trends Pharmacol Sci 16: 265–270, 1995

35. Kahn CR: Insulin resistance, insulin insensitivity, and insulin unresponsiveness: A necessary distinction. Metabolism 27: 1893–1902, 1978

36. Fantus IG, Ahmad F, Deragon G. Vanadate augments insulin binding and prolongs insulin action in rat adipocytes. Endocrinology 127: 2716–2725, 1990

37. Eriksson JW, Lonnrath P, Smith U: Vanadate increases cell surface insulin binding and improves insulin sensitivity in both normal and insulin-resistant rat adipocytes. Diabetologia 35: 510–516, 1992

38. Brichard SM, Pottier AM, Henquin JC: Long-term improvement of glucose homeostasis by vanadate in obese hyperinsulinemic fa/fa rats. Endocrinology. 125: 2510–2516, 1989

39. Rossetti L, Laughlin MR: Correction of chronic hyperglycemia with vanadate, but not with phlorizin, normalizes in vivo glycogen repletion and in vitro glycogen synthase activity in diabetic skeletal muscle. J Clin Invest 84: 892–899, 1989

40. Meyerovitch J, Rothenberg P, Shechter Y, Bonner-Weir S, Kahn CR: Vanadate normalizes hyperglycemia in two mouse models of non-insulin-dependent diabetes mellitus. J Clin Invest 87: 1286–1294, 1991

41. Fantus IG, Ahmad F, Deragon G: Vanadate augments insulin-stimulated insulin receptor kinase activity and prolongs insulin action in rat adipocytes: Evidence for transduction of amplitude of signaling into duration of response. Diabetes 43: 375–383, 1994

42. Haring HU, Biermann E, Kemmler W: Relation of insulin receptor occupancy and deactivation of glucose transport. Am J Physiol 242: E234–E240, 1982

43. Fantus IG, Kadota S, Deragon G, Foster B, Posner BI: Pervanadate [peroxides of vanadate] mimics insulin action in rat adipocytes via activation of the insulin receptor tyrosine kinase. Biochemistry 28: 8864–8871, 1989

44. Clark AS, Fagan JM, Mitch WE: Selectivity of insulin-like actions of vanadate on glucose and protein metabolism in skeletal muscle. Biochem J 232: 273–276, 1985

45. Foot E, Bliss T, Fernandes L, DaCosta C, Leighton B: The effects of orthovanadate, vanadyl and peroxides of vanadate on glucose metabolism in skeletal muscle preparations in vitro. Mol Cell Biochem 109: 157–162, 1992

46. Tsiani E, Abdullah N, Fantus IG: The insulin-mimetic agents vanadate and pervanadate stimulate glucose but inhibit amino acid uptake. Am J Physiol 272: C156–C162, 1997

47. Posner BI, Faure R, Burgess JW, Bevan AP, Lachance D, Zhang-Sun G, Fantus IG, Ng JB, Hall DA, Soo Lum B, Shaver A: Peroxovanadium compounds. A new class of potent phosphotyrosine phosphatase inhibitors which are insulin mimetics. J Biol Chem 269: 4596–4604, 1994

48. Nasmyth K: Viewpoint: Putting the cell cycle in order. Science 274: 1643–1645, 1996

49. Sherr CJ: Cancer cell cycles. Science 274: 1672–1677, 1996

50. Klarlund JK: Transformation of cells by an inhibitor of phosphatases acting on phosphotyrosine proteins. Cell 41: 707–717, 1985

51. Feldman RA, Lowy DR, Vass WC: Selective potentiation of C-fps/fes transforming activity by a phosphatase inhibitor. Oncogene Res 5: 187–197, 1990

52. Chen Y, Chan TM: Orthovanadate and 2,3-dimethoxy-1,4-naphtho-quinone augment growth factor-induced cell proliferation and c-fos gene expression in 3T3-L1 cells. Arch Biochem Biophys 305: 9–16, 1993

53. Wang H, Scott RE: Unique and selective mitogenic effects of vanadate on SV40-transformed cells. Mol Cell Biochem 153: 59–67, 1995

54. Hanauske U, Hanauske A-R, Marshall MH, Muggia VA, Von Hoff DD: Biphasic effect of vanadium salts on in vitro tumor colony growth. Intern J Cell Cloning 5: 170–178, 1987

55. Cruz TF, Morgan A, Min W: In vitro and in vivo antineoplastic effects of orthovanadate. Mol Cell Biochem 153: 161–166, 1995

56. Faure R, Vincent M, Dufour M, Shaver A, Posner BI: Arrest at the G2/M transition of the cell cycle by protein-tyrosine phosphatase inhibition: Studies on a neuronal and a glial cell line. J Cell Biochem 59: 389–401, 1995

57. Djordjevic C: Antitumor activity of vanadium compounds. In: H Sigel, A Sigel (eds). Metal Ions in Biological Systems. Marcel Dekker Inc., New York, 1995, 31: 595–616

58. Beguinot F, Kahn CR, Moses AC, Smith RJ: Distinct biologically active receptors for insulin, insulin-like growth factor 1, and insulin-like growth factor II in cultured skeletal muscle cells. J Biol Chem 260: 15892–15898, 1985

59. Giorgino F, Smith RJ: Dexamethasone enhances insulin-like growth factor-1 effects on skeletal muscle cell proliferation. Role of specific intracellular signaling pathways. J Clin Invest 96: 1473–1483, 1995

60. Kadota S, Fantus IG, Deragon G, Guyda HJ, Hersh B, Posner BI: Peroxide(s) of vanadium: A novel and potent insulin-mimetic agent which activates the insulin receptor kinase. Biochem Biophys Res Commun 147: 259–266, 1987

61. Skolnik EY, Baker A, Li N, Lee CH, Lowenstein E, Mohammadi M, Margolis B, Schlessinger J: The function of GRB2 in binding the insulin receptor to ras signaling pathways. Science 260: 1953–1955, 1993

62. Milarski KL, Saltiel AR: Expression of catalytically inactive syp phosphatase in 3T3 cells blocks stimulation of mitogen-activated protein kinase by insulin. J Biol Chem 269: 21239–21243, 1994

63. Xiao S, Rose DW, Sasaoka T, Maegawa H, Burke RT Jr, Roller PP, Shoelson SE, Olefsky JM: Syp (SH-PTP2) is a positive mediator of growth factor-stimulated mitogenic signal transduction. J Biol Chem 269: 21244–21248, 1994

64. Noguchi T, Matozaki T, Horita K, Fujioka Y, Kasuga M. Role of SH-PTP2, a protein tyrosine phosphatase with Src homology 2 domains, in insulin stimulated Ras activation. Mol Cell Biol 14: 6674–6682, 1994

65. Morgan DO: Principles of CDK regulation. Science 374: 131–134, 1995

66. Sebastian B, Kakizuka A, Hunter T: cdc25M2 activation of cyclin-dependent kinases by dephosphorylation of threonine-14 and tyrosine-15. Proc Natl Acad Sci USA 90: 30521–3524, 1993

67. Krek W, DeCaprio JA: Cell synchronization. Meth Enzymol 254: 114–124, 1995

68. Hamaguchi T, Sudo T, Osada H: RK-682, a potent inhibitor of tyrosine phosphatase arrested the mammalian cell cycle progression at G1 phase. FEBS Lett 372: 54–58, 1995

69. D'Onofrio F, Le MQ, Chiasson J-L, Srivastava AK: Activation of mitogen-activated protein (MAP) kinases by vanadate is independent of insulin receptor autophosphorylation. FEBS Lett 340: 269–275, 1994

70. Pandey SK, Chiasson JL, Srivastava AK: Vanadium salts stimulate mitogen-activated protein (MAP) kinases and ribosomal S6 kinases. Mol Cell Biochem 153: 69–78, 1995

71. Band CJH, Posner BI: Phosphatidylinositol 3'-kinase and p70[S6K] are required for insulin but not bisperoxovanadium 1, 10 phenanthroline [bpV(phen)] inhibition of insulin-like growth factor binding protein gene expression. Evidence for MEK-independent activation of mitogen-activated protein kinase by bpV(phen). J Biol Chem 272: 138–145, 1997

72. Sun H, Tonks NK, Bar-Sagi D: Inhibition of Ras-induced DNA synthesis by expression of the phosphatase MKP-1. Science 266: 285–288, 1994

73. Groom LA, Sneddon AA, Alessi DR, Dowd S, Keyse SM: Differential regulation of the MAP, SAP and RK/p38 kinases by Pystl, a novel cytosolic dual-specificity phosphatase. EMBO J 15: 3621–3632, 1996

74. Davis RJ: The mitogen-activated protein kinase signal transduction pathway. J Biol Chem 268: 14553–14556, 1993

75. Marshall CJ: Specificity of receptor tyrosine kinase signaling: Transient versus sustained extracellular signal-regulated kinase activation. Cell 80: 179–186, 1995

76. Lamphere L, Lienhard GE: Components of signaling pathways for insulin and insulin like growth factor-I in muscle myoblasts and myotubes. Endocrinology 131: 2196–2202, 1992

77. deVries-Smits AMM, Burgering BMT, Leevers SJ, Marshall CJ, Bos JL: Involvement of p21ras in activation of extracellular signal-regulated kinase 2. Nature 357: 602–604, 1992

78. Shisheva A, Shechter Y: Role of cytosolic tyrosine kinase in mediating insulin-like actions of vanadate in rat adipocytes. J Biol Chem 268: 6463–6469, 1993

79. Shisheva A, Shechter Y: Mechanism of pervanadate stimulation and potentiation of insulin-activated glucose transport in rat adipocytes: Dissociation from vanadate effect. Endocrinology 133: 1562–1568, 1993

80. Huyer G, Liu S, Kelly J, Moffat J, Pryette P, Kennedy B, Tsaprailis G, Gesser MJ, Ramachandran C: Mechanism of inhibition of protein-tyrosine phosphatase by vanadate and pervanadate. J Biol Chem 272: 843–851, 1997

81. Tsiani E, Sorisky A, Fantus IG: Vanadate and pervanadate increase glucose uptake in L6 skeletal muscle cells by a mechanism independent of PI3K. 10th International Congress of Endocrinology. Program and Abstracts, Vol. 11: OR66-6, 1997

82. Ida M, Imai K, Hashimoto S, Kawashima H: Pervanadate stimulation of wortmannin-sensitive and -resistant 2-deoxyglucose transport in adipocytes. Biochem Pharmacol 51: 1061–1067, 1996

83. Folli F, Saad MJA, Backer JM, Kahn CR: Regulation of phosphatidyl-inositol 3-kinase activity in liver and muscle of animal models of insulin-resistant and insulin-deficient diabetes mellitus. J Clin Invest 92: 1787–1794, 1993

84. Goodyear U, Giorgino F, Sherman LA, Carey J, Smith RJ, Dohm GL: Insulin receptor phosphorylation, insulin receptor substrate-1 phosphorylation and phosphatidylinositol 3-kinase activity are decreased in intact skeletal muscle strips from obese subjects. J Clin Invest 95: 2195–2204, 1995

85. Bjornholm M, Kawano Y, Lehtihet M, Zierath JR: Insulin receptor substrate-1 phosphorylation and phosphatidylinositol 3-kinase activity in skeletal muscle from NIDDM subjects after *in vivo* insulin stimulation. Diabetes 46: 524–527, 1997

86. Goldfine AB, Simonson DC, Folli F, Patti M-E, Kahn CR: *In vivo* and *in vitro* studies of vanadate in human and rodent diabetes mellitus. Mol Cell Biochem 153: 217–231, 1995

87. Cohen N, Halberstam M, Shlimovich, Chang CJ, Shamoon H, Rossetti L: Oral vanadyl sulfate improves hepatic and peripheral insulin sensitivity in patients with non-insulin-dependent diabetes mellitus. J Clin Invest 95: 2501–2509, 1995

88. DeFronzo RA, Ferrannini E: Insulin resistance: A multifaceted syndrome responsible for NIDDM, obesity, hypertension, dyslipidemia and atherosclerotic cardiovascular disease. Diabetes Care 14: 173–194, 1991

89. Tlsty T, Briot A, Poulose B: Analysis of cell cycle checkpoint status in mammalian cells. Meth Enzymol 254: 125–133, 1995

Molecular and Cellular Biochemistry **182**: 121–133, 1998.

Regulation of the Na+/K+-ATPase by insulin: Why and how?

Gary Sweeney and Amira Klip

Division of Cell Biology, The Hospital for Sick Children, Toronto, Ontario, Canada

Abstract

The sodium-potassium ATPase (Na+/K+-ATPase or Na+/K+-pump) is an enzyme present at the surface of all eukaryotic cells, which actively extrudes Na+ from cells in exchange for K+ at a ratio of 3:2, respectively. Its activity also provides the driving force for secondary active transport of solutes such as amino acids, phosphate, vitamins and, in epithelial cells, glucose. The enzyme consists of two subunits (α and β) each expressed in several isoforms. Many hormones regulate Na+/K+ -ATPase activity and in this review we will focus on the effects of insulin. The possible mechanisms whereby insulin controls Na+/K+-ATPase activity are discussed. These are tissue- and isoform-specific, and include reversible covalent modification of catalytic subunits, activation by a rise in intracellular Na+ concentration, altered Na+ sensitivity and changes in subunit gene or protein expression. Given the recent escalation in knowledge of insulin-stimulated signal transduction systems, it is pertinent to ask which intracellular signalling pathways are utilized by insulin in controlling Na+/K+-ATPase activity. Evidence for and against a role for the phosphatidylinositol-3-kinase and mitogen activated protein kinase arms of the insulin-stimulated intracellular signalling networks is suggested. Finally, the clinical relevance of Na+/K+-ATPase control by insulin in diabetes and related disorders is addressed. (Mol Cell Biochem **182**: 121–133, 1998)

Key words: Na+/K+-ATPase, insulin

Introduction

The Na+/K+-ATPase (Na+/K+-pump) maintains the electrochemical gradients of Na+ and K+ and thus the membrane potential generated by the differential membrane permeabilities to these ions. The pump catalyzes the transfer of three intracellular Na+ ions in exchange for two extracellular K+ ions per molecule of ATP hydrolysed [1]. This disproportionate movement of cations across the cell membrane (the electrogenic Na+/K+-pump current) results in a further increased negativity of several millivolts in the cell membrane potential. The Na+/K+-pump is ubiquitously distributed throughout cells of higher organisms and its importance is highlighted by the fact that it represents the only mechanism for Na+ extrusion in mammalian cells. Na+/K+-ATPase activity contributes to the control of cellular pH, osmotic balance and therefore cell volume, as well as to the Na+-coupled transport of nutrients such as amino acids and vitamins in all cells. In addition, it provides the driving force for the uptake of glucose into intestinal and renal epithelial cells [2]. It has been estimated that the number of Na+/K+-ATPase molecules per cell can vary from 200 to several million and that, at rest, the function of the Na+/K+-ATPase accounts for between 5–40% of cellular ATP consumption [3, 4]. The K_m of the Na+/K+-ATPase for ATP is 0.5–0.8 mM, thus the pump is likely to be saturated by ATP under physiological conditions. However, the intracellular concentration of Na+ ($[Na^+]_i$) under prevailing basal conditions is not saturating, therefore increases in $[Na^+]_i$ alone suffice to increase pump activity by mass action. The activity of the Na+/K+-ATPase is not constant but instead it adapts to challenges to ion distribution. Activity can be modulated in a short (within seconds) or long-term (within hours) fashion. This regulation has been shown to involve mechanisms such as changes in intrinsic activity, subcellular distribution or subunit expression [5]. These changes can be elicited by several hormones such as aldosterone, thyroid hormone, endothelin, acetylcholine and, of significant relevance to this review, insulin in muscle, fat and kidney cells [5]. Even though the stimulation elicited by hormones

Address for offprints: A. Klip, Division of Cell Biology, The Hospital for Sick Children, Toronto, Ontario, M5G 1X8, Canada

is only approximately 5% of the increase in Na$^+$/K$^+$-ATPase activity caused by maximal electrical stimulation of muscle [6], hormonal regulation remains a key mechanism for the control of Na$^+$ and K$^+$ gradients. Inhibition of the Na$^+$/K$^+$-ATPase represents the mechanism of action of cardiac glycosides such as ouabain, commonly used therapeutically as positive inotropic agents, and there is evidence suggesting that a cardiac glycoside-like factor exists endogenously and controls Na$^+$/K$^+$-ATPase activity [7].

In addition to being a major anabolic hormone regulating carbohydrate metabolism, insulin plays a pivotal role in Na$^+$ and K$^+$ homeostasis. Indeed, a decrease in extracellular K$^+$ concentration is one of the first reported actions of insulin [8] and there is evidence for regulation of Na$^+$, K$^+$ and Ca^{2+} currents by insulin [9]. Insulin-stimulated active K$^+$ uptake into muscle is accomplished largely through stimulation of the Na$^+$/K$^+$-pump. The importance of insulin (and catecholamine) stimulated Na$^+$/K$^+$-ATPase activity in restoring contractility to K$^+$-paralysed rat muscle has been documented in detail [10]. Indeed, muscle exposed to high K$^+$ concentrations (10–12.5 mM) exhibited a reduction in isometric twitch and tetanic force which could be restored, at least partially, by insulin. Force recovery upon insulin treatment was suppressed by pre-exposure of the tissue to ouabain, suggesting that insulin's action is due to stimulation of the Na$^+$/K$^+$-ATPase [10]. This phenomenon is of physiological relevance in conditions such as exercise-induced hyperkalemia when muscle releases K$^+$ upon exercise and it is the rapid stimulation of the Na$^+$/K$^+$-pump which returns plasma K$^+$ levels to resting values, thereby preventing K$^+$ from reaching levels that could cause depolarization [11]. In addition, the stimulation of this enzyme prevents changes in [Na$^+$]$_i$ that would otherwise occur due to insulin-driven increases in metabolic activity via activation of Na$^+$/H$^+$ exchange and increased Na$^+$-coupled amino acid uptake. Insulin also plays a major role in controlling renal sodium reabsorption and haemodynamics [12]. Control of ionic homeostasis is therefore a prominent yet often underestimated feature of insulin action.

This review will focus on how insulin produces changes in Na$^+$/K$^+$-ATPase activity and what intracellular signalling molecules are involved in insulin action. In addition, the level of control of the Na$^+$/K$^+$-ATPase by insulin might play a role in the complications of diabetes and in the etiology of associated disorders such as hypertension and obesity [13, 14]. The importance of and evidence for hypo- or hyper-insulinemic induced changes in Na$^+$/K$^+$-ATPase activity in the pathogenesis of such disorders will be reviewed.

Structure of the Na$^+$/K$^+$-ATPase

To understand the regulation of the Na$^+$/K$^+$-ATPase by insulin it is necessary to consider in detail the structure

of the pump. The enzyme is composed of a 112 kDa subunit and a 35 kDa subunit (which is a heavily glycosylated protein that migrates on SDS-PAGE at 42–55 kDa). The α (112 kDa) subunit is thought to span the plasma membrane eight or ten times and both its NH$_2$- and COOH-termini face the intracellular space [15]. The β subunit spans the plasma membrane once and has a cytosolic NH$_2$-terminus [15]. Typically, the Na$^+$/K$^+$-ATPase is thought to reside in the plasma membrane as an α-β heterodimer, however intracellular pools of Na$^+$/K$^+$-ATPase molecules and excess β over α subunits have also been reported [15a]. Both subunits exist in multiple isoforms which are regulated developmentally and in a tissue-specific manner [16]. Cloning of the various subunit cDNAs in the last decade has made it possible to study the tissue-specific distribution, regulation and function of these subunits by developmental and hormonal cues. Through point mutations, deletions and other modifications it has been possible to elucidate the function of distinct domains of the enzyme [17].

The α subunit contains binding sites for Na$^+$, K$^+$, Mg^{2+}, ATP and cardiac glycosides and is thus responsible for the catalytic activity of the enzyme. Four isoforms of the α subunit have been identified to date [18], all products of different genes rather than resulting from alternative splicing. Amino acid sequence identity amongst α subunits is approximately 82%, with the major differences occurring in the N-terminal region. Expression of α subunits varies between tissues (see Table 1). For example, α1 isoform is ubiquitously distributed, α2 is the predominant isoform in adult muscle, whilst adult rat brain expresses all isoforms. Developmental regulation is also apparent: only α1 subunits are expressed in 3T3-L1 fibroblasts, yet upon differentiation

Table 1. Expression of α subunits of the Na$^+$/K$^+$-ATPase in mammalian tissues

	α1	α2	α3
Tissue			
Adipocyte (r)	[55]	[55]	–
Brain (h)	[103]	[103]	[103]
Eye (h)	[104]	[104]	[104]
Heart (h)	[105]	[105]	[105]
Heart (r)	[20]	[20]	[106]
Liver (r)	[107]	–	–
Kidney (h)	[103]	[20]	–
Kidney (r)	[108]	[108]	[108]
Lung (h)	[103]	[20]	–
Skeletal muscle (h)	[103]	[103]	–
Skeletal muscle (r)	[31]	[31]	[20]
Vascular smooth muscle (r)	[109]	[109]	[109]
Intestinal epithelium (r)	[110]	–	–
Thyroid (h)	[103]	[103]	
Uterus (h)	[103]	–	–

References cited describe expression of protein or mRNA in rat (r) and human (h) adult tissues.

into adipocytes α1 levels are decreased and α2 expression is induced [19] and, in the rat, α3 expression drops and α2 appears in the transition from fetal to adult heart [20]. Heterogeneity amongst α subunits allows for the possibility of distinct cell Na$^+$/K$^+$-ATPase activity, since depending on the complement of isoforms expressed, the enzyme exhibits different affinity for ions and cardiac glycosides and also different responses to hormones or other effector molecules. The isoforms differ in their affinity for intracellular Na$^+$ (α1 = α2 > α3) and for extracellular K$^+$ (α3 > α1 = α2) [20]. In rodents, an established functional difference among α subunits lies in their sensitivity to inhibition by the cardiac glycoside ouabain. The rodent α1 isoform exhibits low affinity (K_i = 1–5 × 10^{-5} M) for ouabain while the α2 and α3 isoforms have higher ouabain affinity (K_i = 1–5 × 10^{-7} M) [21]. Amino acid residues in the first extracellular region of the α subunit determine the differential ouabain sensitivity exhibited by rodent α subunit isoforms [22]. In contrast, no marked difference in ouabain sensitivity is observed among the α subunits of primates. Insulin has the capacity to regulate Na$^+$/K$^+$-ATPase activity in an isoform-selective fashion. In adipocytes where both α1 and α2 subunits are expressed, the α1 isoform is responsible for the majority of Na$^+$/K$^+$-ATPase activity under resting conditions [23, 24] and insulin preferentially activates the component of high ouabain sensitivity (α2) [25, 26], although the α1 subunit is also stimulated but to a lesser extent [27]. Preferential activation by insulin of the α2 subunit has also been shown in brain [28] and skeletal muscle [23] and selective induction of α1 in response to insulin in rat astrocytes has also been shown [29].

At least three isoforms of the β subunit have been identified in mammalian cells [18, 30], amongst which there is a fairly low amino acid sequence identity (β1 and β2 are 40% homologous and β3 is 38 and 48% homologous to β1 and β2, respectively). As was the case for the α subunit, expression of β subunit isoforms varies among tissues (see Table 2). Interestingly, β1 subunit expression in skeletal muscle correlates with oxidative red muscle phenotype, whilst β2 is selectively expressed in glycolytic fast-twitch muscles [31]. The precise role of the β subunit is uncertain, although its association with the α subunit may be required for efficient assembly of functional pumps [32], to regulate delivery of α subunits to the plasma membrane [33], determine its polarized distribution [34] and may also contribute to K$^+$ binding [35] and K$^+$ transport [36].

The existence of multiple α and β subunit isoforms suggests that preferential association of distinct subunits [37] might regulate the functional properties of the Na$^+$/K$^+$-ATPase [38, 39]. Any α subunit can associate with any β subunit and domains required for α-β and even for α-α

Table 2. Expression of β subunits of the Na$^+$/K$^+$-ATPase in mammalian tissues

	β1	β2	β3
Tissue			
Brain (r)	[111]	[111]	[30]
Eye (r)	[112]	[112]	–
Heart (r)	[113]	–	–
Liver (r)	[107]	–	[30]
Kidney (r)	[114]	[114]	[30]
Adipocyte (r)	–	–	–
Lung (r)	–	–	[30]
Skeletal muscle (h)	[115]	–	–
Skeletal muscle (r)	[31]	[31]	–
Vascular smooth muscle (r)	[116]	–	–
Intestine (r)	[117]	–	[30]
Thyroid (r)	–	–	–
Uterus (r)	–	–	–
Neurons (r)	[118]	–	–
Testes (r)	–	–	[30]

References cited describe expression of protein or MRNA in rat (r) and human (h) adult tissues.

association have been characterized [40, 41]. Although there is evidence that the targeting of newly synthesized pump subunits to the plasma membrane requires prior subunit assembly [42], excess β subunits are found in the cell membrane [15a]. In addition, despite the common assumption that an α-β heterodimer is required for enzymatic activity, recent evidence suggests that α subunits expressed alone in insect cells display a Mg^{2+}-dependent ATPase activity [43].

Na$^+$/K$^+$-ATPase regulation

The mechanisms of regulation of pump availability and activity have principally been studied for the α1 subunit, and in response to hormones other than insulin. Short term regulation of α1 Na$^+$/K$^+$-ATPase activity can be elicited by mechanisms such as increased [Na$^+$]$_i$ secondary to increased Na$^+$ influx, phosphorylation/dephosphorylation, translocation of subunits to the plasma membrane or increased Na$^+$ affinity [5]. It has been shown that long term regulation of this subunit can involve changes in gene transcription, translation and degradation of the protein [5, 44]. The mechanisms of hormonally-induced changes in Na$^+$/K$^+$-ATPase activity are discussed in detail below. These changes are achieved via complex networks of intracellular signals, often involving activation or inhibition of protein kinases and phosphatases. The intracellular signalling pathways that may be implicated in Na$^+$/K$^+$-ATPase regulation by insulin are also discussed.

*Mechanisms leading to stimulation of the Na⁺/
K⁺-ATPase by insulin*

Translocation of pump subunits
Translocation of pump subunits from an intracellular storage site to the plasma membrane represents the main regulatory mechanism for Na⁺/K⁺-ATPase activity in response to insulin in skeletal muscle. This mechanism has been supported by measurements of a higher Na⁺/K⁺-ATPase hydrolytic activity and increased ouabain binding to membranes prepared from insulin-treated amphibian muscle, with concomitant decreases in both parameters in purified intracellular membrane fractions [45, 46]. We have shown that in rat muscle, the $\alpha 2$ subunit is preferentially translocated by insulin [47]. Thus, upon 30 min of insulin action *in vivo* followed by isolation of skeletal muscle preparations, there was an increase in $\alpha 2$ and $\beta 1$ isoform abundance at the plasma membrane and a decrease in these isoforms in intracellular membranes, with no change in $\alpha 1$ or $\beta 2$ distribution [39, 47]. The specific redistribution of $\alpha 2$ and $\beta 1$ subunits was further verified by immunogold detection under the electron microscope, using ultracryosections of muscles treated with insulin or saline solution *in vivo* [47]. The recruitment of $\alpha 2$ and $\beta 1$ subunits to the muscle surface was observed in the slow-twitch oxidative soleus muscle, but not in white gastrocnemius (a preferentially fast-twitch glycolytic muscle) which expresses the $\beta 2$ but not $\beta 1$ isoform. This demonstrates that, *in vivo*, insulin can regulate the Na⁺/K⁺-ATPase subunit distribution within muscle cells. The fact that the response is specific for the $\alpha 2$ isoform and only when the β isoform present is $\beta 1$ suggests the possibility that the $\beta 1$ subunit may be preferentially associated with $\alpha 2$ subunit and confers some degree of targeting specificity upon it. The recently discovered $\beta 3$ isoform has only been studied at the level of its mRNA expression [30], and future studies should reveal whether its subcellular distribution can be regulated by insulin. The selective regulation exhibited by insulin seems to differ from that of exercise, since a one hour exercise bout increased primarily the level of expression of all the subunits of the pump, albeit to different degrees [48].

There are interesting correlates between insulin-regulated Na⁺/K⁺-ATPase activity and insulin-regulated glucose transport, beginning with the observation that insulin elicits translocation of the subunit/transporter from an intracellular storage compartment to the plasma membrane. These systems mediate the uptake into cells of K⁺ and glucose, respectively. Both of these compounds are major constituents of the food which are absorbed during a meal, both are insulin secretagogues, and both are rapidly stored in the muscle in response to insulin. Like GLUT1, the $\alpha 1$ subunit is ubiquitous and of a house-keeping nature. Like GLUT4, the $\alpha 2$ subunit is tissue specific and typical of muscle, heart and fat tissues (see Table 3). In addition, it is

recognized that there may be insulin resistance of the pump in diabetes, just as there is insulin resistance of glucose uptake [49, 50].

We have shown that both GLUT4 and the $\alpha 2$ subunit of the pump (accompanied by the $\beta 1$ subunit) translocate from intracellular membranes to the cell surface in response to insulin [47, 51]. However by electron microscopy we found that the intramuscular location of GLUT4 and the $\alpha 2$ subunit differs. Moreover, immunopurified GLUT4-containing vesicles were found to be largely devoid of Na⁺/K⁺-pump subunits [52, 53]. This suggests that despite the many similarities highlighted above, the signalling pathway mediating the translocation process might be different in each case.

In contrast to the recruitment of pump units from an intracellular storage site to the plasma membrane seen in oxidative muscle, insulin does not change the subcellular distribution of pump subunits in fat cells or fat cells in culture [54, 55], as determined by subcellular fractionation. This implies that additional and cell type-specific mechanisms operate in the regulation of the Na⁺,K⁺-ATPase by insulin, as outlined next.

Increased intracellular Na⁺ concentration
As mentioned previously, an increase in $[Na^+]_i$ alone is sufficient to increase Na⁺/K⁺-ATPase activity. Insulin can indeed increase Na⁺ influx by stimulating the Na⁺/K⁺/2Cl⁻ cotransporter, the Na⁺/H⁺ exchanger or a Na⁺ channel as reviewed below.

The Na⁺/K⁺/2Cl⁻ cotransporter has been identified, cloned and characterized in various tissues [56]. This ubiquitous transporter moves the ions Na⁺, K⁺ and Cl⁻ in an electrically neutral fashion, and acts in concert with the Na⁺/K⁺-ATPase to regulate $[Na^+]_i$ and $[K^+]_i$. This is particularly important in kidney tubule cells where the Na⁺/K⁺/2Cl⁻ cotransporter is predominantly expressed on the apical surface (and whose inhibition represents the mechanism of action of diuretics such as bumetanide) and the Na⁺/K⁺-ATPase on the basolateral surface. The two entities mediate in tandem the reabsorption of Na⁺ from the lumen to the blood [57]. Similarly, in vascular endothelial cells the Na⁺/K⁺/2Cl⁻ cotransporter is responsible for the major portion of K⁺ influx [58] and therefore, together with the Na⁺/K⁺-ATPase,

Table 3. Correlation of tissue expression of glucose transporters (GLUTS) and α subunits of the Na⁺/K⁺-ATPase

	GLUT 1	GLUT 3	GLUT 4	GLUT 5
$\alpha 1$	Ubiquitous			
$\alpha 2$			Muscle, Fat, Heart	
$\alpha 3$		Brain		
$\alpha 4$				Testis

Information gathered from data in refs. [20, 119].

it plays a central role in the maintenance of intracellular volume [59]. We recently showed that insulin stimulates the $Na^+/K^+/2Cl^-$ cotransporter in 3T3-L1 fibroblasts and adipocytes, and that this stimulation is required at least in part for the subsequent stimulation of Na^+/K^+-ATPase activity. This was documented from increases in both bumetanide- and ouabain-sensitive $^{86}Rb^+$ uptake in these cells [55]. It is possible that activation of the cotransporter causes a transient increase in $[Na^+]_i$. The coupled change in activity of the cotransporter followed by pump activation would result in an efficient uptake of K^+ from the extracellular millieu, and would also lead to changes in $[KCl]_i$, in Cl^- content and in cell volume. The latter may have important metabolic consequences [60], whereas Cl^- has been proposed as a regulator of other ion channels within the cell [61], and K^+ has been shown to stimulate transcription [62]. Thus, the concerted stimulation of the cotransporter and the pump may have far reaching implications.

A second route for insulin-dependent Na^+ influx is the Na^+/H^+ antiporter. Na^+/H^+ exchange via the antiporter is activated in response to insulin [63] presumably for the purpose of extruding metabolically produced H^+. An influx of Na^+ is the energetic cost of this reaction. Indeed, an insulin-dependent alkalinization has been observed in several cell types [64], which is prevented by amiloride, an inhibitor of the Na^+/H^+ antiporter. Blocking Na^+/H^+ exchange with amiloride was found to inhibit insulin stimulation of the Na^+/K^+-ATPase in rat hepatocytes [65] and BC_3H1 myocytes [66], suggesting that the increase in Na^+ caused by the antiporter may be required for the stimulation of the pump in these cells. It has been shown in cultured skeletal muscle cells that insulin-stimulated Na^+/K^+-ATPase activity was insensitive to amiloride or tetrodotoxin [67], suggesting that in this tissue changes in $[Na^+]_i$ do not underlie the mechanism of activation. Similarly, activity of Na^+/K^+-ATPase in skeletal muscle does not increase during K^+ depletion, despite elevated $[Na^+]_i$ [68] and indeed insulin has been shown to lower $[Na^+]_i$ in rat soleus muscle [69]. An increased $[Na^+]_i$ may, however, at least partly regulate Na^+/K^+-ATPase activity in muscle cells in culture, where translocation of subunits represents the commonly accepted means of activation, as treatment of skeletal myotubes with tetrodotoxin (which blocks voltage-dependent Na^+ channels and thus lowers Na^+) reduced the number of pump units detected by ouabain binding and also decreased $^{86}Rb^+$ uptake [70]. With regard to this, it is known that insulin can stimulate a μ-conotoxin-sensitive Na^+ channel which is primarily responsible for the increased Na^+ entry that sustains activity of the Na^+/K^+-ATPase in skeletal muscle [71]. It has also been suggested that insulin elevates $[Na^+]_i$ by directly activating a Na^+ channel in 3T3-F422A adipocytes [72]. Thus, stimulation of the Na^+/K^+-ATPase by insulin subsequent to increased $[Na^+]_i$ is apparent in many cell types, although the contribution of this mechanism in skeletal muscle cells remains controversial.

Phosphorylation
The Na^+/K^+-ATPase can become phosphorylated upon exposure to kinases such as protein kinase A (PKA) and protein kinase C (PKC), or to agonists that stimulate their activity *in vitro*. These kinases can also modify the catalytic activity of the Na^+/K^+-ATPase [73, 74]. Although extremely exciting, these observations are mostly correlative and there is considerable discussion on whether a simplified model can be derived relating degree of phosphorylation with degree of activity. The effects of Na^+/K^+-ATPase phosphorylation on activity vary and can result in either activation [75] or more commonly inhibition, depending on the system studied [44]. In the case of PKC, this may represent the fact that different combinations of PKC and Na^+/K^+-ATPase isoforms are expressed in various tissues. This recent literature will not be reviewed here as it has not been implicated yet in insulin action. Moreover, all studies to date have concerned the α1 and not the more insulin-sensitive α2 subunit. However, as a general corollary, it is important to consider that changes in phosphorylation may potentially cause conformational changes which could potentially alter the affinity of the Na^+/K^+-ATPase for one or more of its substrates. Alternatively, changes in phosphorylation states may alter the catalytic turnover of the α subunit, and can even be envisaged to control the rate of subunit internalization and subsequent degradation. It is hoped that future studies should address the potential regulation of the pump by insulin via phosphorylation of the α or the β subunit isoforms.

Na^+ sensitivity
The antinatriuretic effect of insulin is mediated via regulation of the Na^+/K^+ATPase in the kidney [76]. A set of related studies conducted by Feraille *et al.* using either rat proximal convoluted tubule or cortical collecting duct have shown that the stimulation of the pump by insulin in this tissue is not mediated via changes in either $[Na^+]_i$, translocation of subunits or Na^+/K^+-ATPase hydrolytic activity [77, 78]. Instead, insulin appears to increase the sensitivity of the Na^+/K^+-ATPase to Na^+, as evidenced by a decrease in the dissociation constant of the Na^+ binding site [77]. This reflects the exquisite tissue-specificity of the mechanism of action of insulin on the Na^+/K^+-pump.

Biosynthesis
Upon long-term exposure of cells to insulin, an additional level of control of Na^+/K^+-ATPase emerges, which includes changes at the levels of gene expression. Long-term treatment of 3T3-L1 fibroblasts with insulin induced α2 mRNA and reduced levels of β1 mRNA [79]. Similarly, in vascular smooth muscle cells, α2 mRNA was selectively upregulated by high insulin levels, without change in the mRNA of the

126

more abundantly expressed α1 isoform [80]. Changes at the protein level have also been detected, which may arise from either increased transcription, mRNA stabilization and/or protein translation and stabilization. This is documented by studies in cultured rat astrocytes, in which insulin specifically increases the levels of the α1 protein [29], an effect apparently mediated via IGF-1 receptors. In 3T3-L1 fibroblasts we and others [79] have observed that prolonged insulin treatment selectively increased α2 isoform expression with no change in α1 expression (Fig. 1). This increase may potentially be mediated by a rise in $[Na^+]_i$, since increasing $[Na^+]_i$ artificially using veratridine can increase the number of α subunits expressed in the plasma membrane, an effect blocked by inhibitors of protein synthesis [70].

A pictorial representation of the various mechanisms discussed in this section leading to stimulation of the Na^+/K^+-ATPase by insulin is presented in Fig. 2.

Intracellular signalling pathways involved in the control of Na^+/K^+-ATPase activity by insulin

Whereas a substantial amount of knowledge has been gained on the molecular mechanisms of regulation of glucose uptake and metabolism, relatively little is known about the events that underlie stimulation of K^+ uptake and Na^+ extrusion through the Na^+/K^+-pump by insulin. Given the significant recent advances in our understanding of the insulin signal transduction system, it is pertinent to ask which pathways participate in the control of Na^+/K^+-ATPase

activity by insulin. Emerging evidence is discussed next and summarized in Fig. 3.

Activation of phosphatidylinositol 3-kinase (PI3K) is required for the insulin-dependent stimulation of glucose transport in muscle and fat cells [81, 82]. PI3K is comprised of an 85 kDa regulatory subunit and a 110 kDa catalytic subunit. The regulatory subunit contains SH2 domains which can associate with phosphorylated tyrosine residues on IRS-1, resulting in activation of the catalytic subunit [83]. We have preliminary evidence that in 3T3-L1 fibroblasts the stimulation of the Na^+/K^+-ATPase by insulin is inhibited by preincubation of the cells with the PI3K inhibitors wortmannin and LY294002 (Sweeney and Klip, submitted 1998), suggesting an involvement of PI3K. In some cells, activation of PI3K can lead to the subsequent stimulation of p70 S6 kinase [82]. However, the stimulation of the Na^+/K^+-ATPase by insulin in 3T3-L1 fibroblasts could not be blocked by preincubation with the p70 S6 kinase inhibitor rapamycin (Sweeney and Klip, submitted 1998). In search for other activatory signals situated downstream from PI3K one must consider that certain isoforms of PKC, in particular PKC-ζ, is stimulated by insulin via activation of PI3K [84]. Given the ability of PKC to phosphorylate the α subunit of the Na^+/K^+-ATPase stated earlier, it is worth examining the possible role of PKC-ζ in the activation of the Na^+/K^+-ATPase by insulin. More complex mechanisms involving PKC might be envisaged as PI3K can also stimulate phospholipase D which results in hydrolysis of phosphatidylcholine and production of phosphatidic acid and diacylglycerol, which might in turn activate PKC [84].

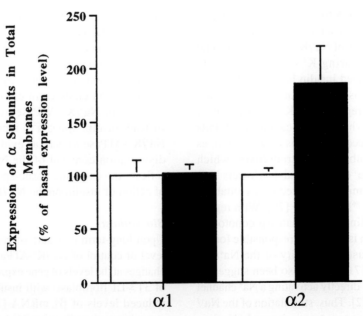

Fig. 1. Expression of Na^+/K^+-ATPase α subunits in total membranes from cultured 3T3-L1 fibroblasts (open bars) and after treatment of cells with insulin (filled bars).

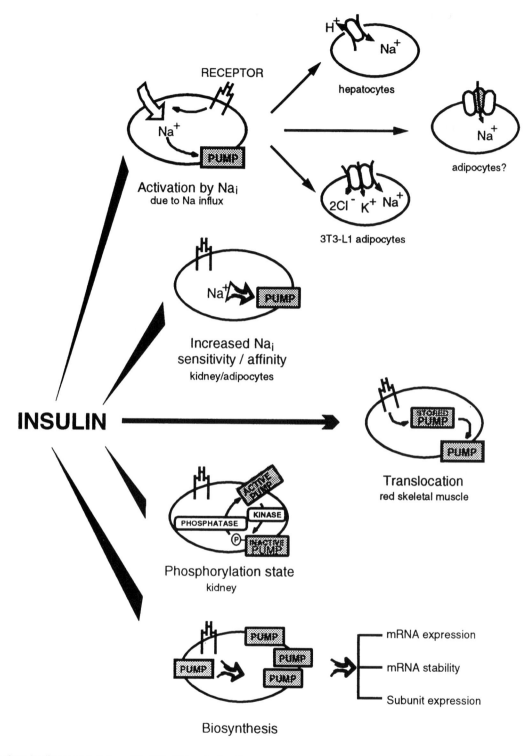

Fig. 2. Mechanisms leading to stimulation of Na⁺/K⁺-ATPase by insulin.

Intriguingly, we have recently observed that SB203580, an inhibitor of the MAP kinase homologue p38MAP kinase (also termed reactivating kinase), can inhibit insulin-stimulated Na⁺/K⁺-ATPase activity (Sweeney and Klip, unpublished observation). If not regulating Na⁺/K⁺-ATPase activity by direct phosphorylation, it is conceivable that p38MAP kinase may phosphorylate and activate phospho–lipase A₂, a response that has been shown elsewhere [85].

128

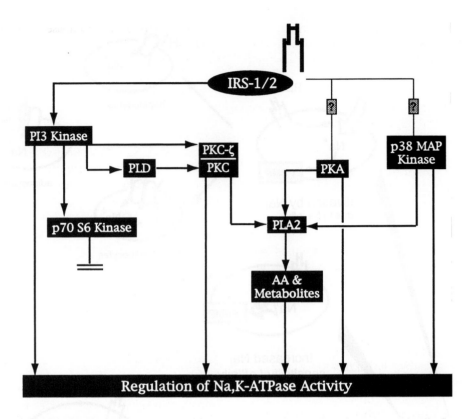

Fig. 3. Possible intracellular signalling pathways involved in the regulation of Na$^+$/K$^+$-ATPase activity in response to insulin. See text for details. Note that none of these have been unequivocally demonstrated.

This would lead to the generation of arachidonic acid and its metabolites which are known to regulate the Na$^+$/K$^+$-ATPase [86]. Since phospholipase A$_2$ can also be activated by PKA and PKC [86, 87], it would be important to analyze whether both PI3K and p38 MAP kinase activate phospholipase A$_2$ in the signalling pathway that regulates Na$^+$/K$^+$-ATPase activity in response to insulin.

Na$^+$/K$^+$-ATPase in diabetes, insulin resistance and hypertension

A decrease in Na$^+$/K$^+$-ATPase activity, and consequently Na$^+$ retention, are common features observed in diabetes [49]. In the streptozotocin-induced diabetic rat model, changes in Na$^+$/K$^+$-ATPase activity and subunit expression have been extensively studied. Table 4 presents a summary of these studies and highlights the fact that changes in activity and subunit expression are highly tissue-specific and although insulin therapy restores blood glucose in these animals, the changes in Na$^+$/K$^+$-ATPase are not always reversed [88].

Alterations in Na$^+$/K$^+$-ATPase activity and expression have been claimed to be associated with hypertension, as well as with other diabetic complications such as neuropathy, retinopathy and nephropathy. In each case, it is possible that

defects in the regulation of the pump by insulin (i.e. insulin resistance of the pump) could contribute to the abnormality. We review below some examples of this possibility.

Hypertension

The importance of abnormalities in Na$^+$ metabolism in the development of hypertension is well documented [89]. For

Table 4. Changes in Na$^+$/K$^+$-ATPase subunit expression and activity in the streptozotocin rat model of diabetes

Tissue	Change in subunit expression	Change in activity	Ref
Skeletal muscle	Increased α1 and α2	No change	[120]
Skeletal muscle	No change	Decrease	[121]
Kidney	Increased α1 and β1	Increase	[120]
Cardiac muscle	Decreased α2 and β1 No change α1	Decrease	[120]
Cerebral microvessels	–	No change	[122]
Cerebral cortex	Decreased α1 α3 β1 β2 No change α2	Decrease	[88]
Nerve	Decreased α1 No change α2 and α3	Decrease	[96]
Nerve	–	Decrease	[123]
Retina	–	Decrease	[124]
Intestine	–	Increase	[125]

example, increased proximal tubule Na^+ reabsorption [90] and reduced Na^+/K^+-ATPase activity of circulating cells [13] are associated with hypertension. Moreover, platelets from type 2 diabetic patients exhibit higher $[Na^+]_i$ levels which correlate with increased systolic and diastolic blood pressure [91], and impaired Na^+/K^+-ATPase activity may underlie hypoxia-induced pulmonary hypertension [92]. Insulin-like growth factor-1 (IGF-1), produced in the vasculature [93], stimulates Na^+/K^+-ATPase activity [29, 94]. Therefore, one could envisage that insulin/IGF-1 resistance of the pump could manifest itself as an inability of vascular cells to antagonize vasoconstriction. This is because vasoconstricting agents act in part by elevating cytosolic Ca^{2+} levels, which would be normally excluded via the Na^+/Ca^{2+} exchanger which is driven by the electrochemical Na^+ gradient. A weakened Na^+ gradient or a gain in $[Na^+]_i$ would prevent the effective removal of cytosolic Ca^{2+} and would promote vasoconstriction.

Neuropathy and retinopathy
Several studies have identified a defect in Na^+/K^+-ATPase function in diabetic neuropathy. The Na^+/K^+-ATPase activity measured in red blood cell membranes of diabetic patients was found to be lower in the subgroup that was afflicted by neuropathy [95]. Streptozotocin diabetic rats experience a decrease in motor nerve conduction velocity, accompanied by a reduction in Na^+/K^+-ATPase activity and $\alpha 1$ isoform expression [96]. Greene *et al.* [97] have advanced the hypothesis that a decrease in Na^+/K^+-ATPase activity in peripheral nerve may be a direct consequence of hyperglycemia. This hypothesis proposes that high glucose entering peripheral nerves would be converted into sorbitol via the aldose reductase. An increase in sorbitol would lead to a decrease in myoinositol and consequently to a reduction in PKC activity by an undefined mechanism. In peripheral nerve, PKC is thought to have a stimulatory action on the Na^+/K^+-pump, which would be dampened in diabetes. More recently, Greene has proposed that the reduction in pump activity is due to alterations in nitric oxide production [98], as a result of the increase in sorbitol and decrease in myoinositol. We submit that, if insulin or IGF-1 are natural activators of the pump in peripheral nerve, then it is possible that failure of insulin to stimulate the pump could contribute to the generation of the neuropathy syndrome. Thus, reduction in pump activity associated with insulin resistance could occur in addition to the defects associated with hyperglycemia via the polyol pathway.

An alternative route involving PKC and the Na^+/K^+-ATPase in the generation of diabetic complications, especially related to retinopathy and perhaps vascular complications, has been proposed by King [99]. The hypothesis states that hyperglycemia results in elevations in diacylglycerol which in turn activate a specific isoform of PKC, PKC β-II. This enzyme appears to activate phospholipase A_2 to generate arachidonic acid which may ultimately regulate Na^+/K^+-ATPase activity. A strategy to relieve the complications of retinopathy has been proposed if sufficiently effective inhibitors of PKC β-II could be generated [100].

Possible insulin resistance of the pump in nephropathy and obesity
A decrease in Na^+/K^+-ATPase activity has been claimed to be associated with the development of diabetic nephropathy [101] and it is conceivable that this arises from an inability of insulin to adequately stimulate the pump. Similarly, because of the high degree of consumption of ATP by the pump, it has been hypothesized that a reduction in pump activity could accompany certain types of obesity [14, 102]. The reverse can also occur. Thus, sudden supraphysiological elevations in insulin levels, which occur in thyrotoxicosis, can lead to thyrotoxic periodic paralysis, due to extreme reduction in circulating K^+ levels.

Conclusions

The regulation of the Na^+/K^+-ATPase by insulin is especially important in kidney, muscle and fat. Given the significance of this enzyme there is a need for the identification of mechanisms regulating its activity and of the signalling molecules that mediate its control by insulin. It is likely that both of these will be tissue-specific. In the kidney the predominant α subunit is the $\alpha 1$ isoform and this may be the form regulated by insulin. In contrast, in muscle and fat the $\alpha 2$ subunit is regulated by the hormone. Whereas in muscle it is translocated to the cell surface, in fat cells it is activated without change in location. Whether insulin causes post-translational modification of the $\alpha 2$ subunit is not known and should be addressed. Preliminary studies in 3T3-L1 fibroblasts suggest a complex mechanism involving PI3K and p38 MAP Kinase in mediating insulin-stimulated Na^+/K^+-ATPase activity, partly via the $Na^+/K^+/2Cl^-$ cotransporter. Control of Na^+/K^+-ATPase by insulin is of particular clinical relevance in conditions associated with insulin resistance such as diabetes, hypertension and obesity. It is hoped that by delineating the mechanisms involved in insulin resistance of the pump, strategies could be conceived to alleviate it. This could contribute to the treatment of hypertension and of diabetic neuropathy, in which a reduction in pump activity has been implicated as an etiological factor.

Acknowledgements

We thank Dr. Philip J. Bilan and Mr Romel Somwar for careful reading of this manuscript, and Mr. Rob Sargeant for

130

participation in the results presented in Fig. 1. We thank all the members of our group who contributed to the studies quoted: Dr. H.S. Ewart, Dr. H. Hundal, Dr. L. Lavoie, Dr. A. Marette, Dr. Y. Mitsumoto, Mr. T. Ramlal, Dr. T. Tsakiridis, Miss P. Wong. Work in the laboratory of A.K. on this subject is supported by, a grant (MT-12601)from the Medical Research Council of Canada.

References

1. Lingrel JB, Kuntzweiler T: Na^+,K^+-ATPase. J Biol Chem 269: 19659–19662, 1994
2. Jorgensen PL: Structure, function and regulation of Na,K-ATPase in the kidney. Kidney Int 29: 10–20, 1986
3. Kostic MM, Zivkovic RV: Energy metabolism of reticulocytes: Two different sources of energy for Na^+K^+-ATPase activity. Cell Biochem Func 12: 107–112, 1994
4. Gruwel ML, Alves C, Schrader J: Na^+K^+-ATPase in endothelial cell energetics: ^{23}Na nuclear magnetic resonance and calorimetry study. Am J Physiol 268: H351–H358, 1995
5. Ewart HS, Klip A: Hormonal regulation of the Na^+-K^+-ATPase: Mechanisms underlying rapid and sustained changes in pump activity. Am J Physiol 269: C295–C311, 1995
6. Clausen T: The Na^+, K^+ pump in skeletal muscle: Quantification, regulation and functional significance. Acta Physiol Scand 156: 227–235, 1996
7. Rose AM, Valdes R Jr: Understanding the sodium pump and its relevance to disease. Clin Chem 40: 1674–1685, 1994
8. Briggs AP, Koenig I, Doisy EA, Weber CJ: Some changes in the composition of blood due to the injection of insulin. J Biol Chem 58: 721–730, 1924
9. Zierler K, Wu FS: Insulin acts on Na, K, and Ca currents. Trans Assoc Am Phys 101: 320–325, 1988
10. Clausen T, Andersen SL, Flatman JA: Na^+-K^+ pump stimulation elicits recovery of contractility in K^+-paralysed rat muscle. J Physiol 472: 521–536, 1993
11. Somjen GG: Extracellular potassium in the mammalian central nervous system. Annu Rev Physiol 41: 159–177, 1979
12. Stenvinkel P, Bolinder J, Alvestrand A: Effects of insulin on renal haemodynamics and the proximal and distal tubular sodium handling in healthy subjects. Diabetologia 35: 1042–1048, 1992
13. Weder AB: Sodium metabolism, hypertension, and diabetes. Am J Med Sci 307: S53–S59, 1994
14. Martinez FJ, Sancho-Rof JM: Epidemiology of high blood pressure and obesity. Drugs 46: 160–164, 1993
15. Mercer RW: Structure of the Na,K-ATPase. Int Rev Cytol 137C: 139–168, 1993
15a. Lavoie L, Levenson R, Martin-Vasallo P, Klip A: The molar ratios of α and β subunits of the Na^+-K^+-ATPase differ in distinct subcellular membranes from rat skeletal muscle. Biochemistry 36: 7726-7732, 1997
16. Herrera VL, Cava T, Sassoon D, Ruiz-Opazo N: Developmental cell-specific regulation of Na^+K^+-ATPase alpha 1-, alpha 2-, and alpha 3-isoform gene expression. Am J Physiol 266: C1301–C1312, 1994
17. Andersen JP, Vilsen B: Structure-function relationships of cation translocation by Ca^{2+}- and Na^+, K^+-ATPases studied by site-directed mutagenesis. FEBS Lett 359: 101–106, 1995
18. Levenson R: Isoforms of the Na,K-ATPase: Family members in search of function. Rev Physiol Biochem Pharma 123: 1–45, 1994
19. Russo JJ, Manuli MA, Ismail-Beigi F, Sweadner K, Edelman IS: Na^+-K^+-ATPase in adipocyte differentiation in culture. Am J Physiol 259: C968–C977, 1990
20. Lingrel JB: Na,K-ATPase: Isoform structure, function, and expression. J Bioener Biomem 24: 263–270, 1992
21. Repke KR, Megges R, Weiland J, Schon R: Location and properties of the digitalis receptor site in Na^+/K^+-ATPase. FEBS Lett 359: 107–109, 1995
22. Canessa CM, Horisberger JD, Louvard D, Rossier BC: Mutation of a cysteine in the first transmembrane segment of Na,K-ATPase alpha subunit confers ouabain resistance. EMBO J 11: 1681–1687, 1992
23. Lytton J, Lin JC, Guidotti G: Identification of two molecular forms of (Na^+,K^+)ATPase in rat adipocytes. Relation to insulin stimulation of the enzyme. J Biol Chem 260: 1177–1184, 1985
24. McGill DL, Guidotti G: Insulin stimulates both the α1 and the α2 isoforms of the rat adipocyte (Na^+,K^+) ATPase. J Biol Chem 266: 15824–15831, 1991
25. Resh MD, Nemenoff RA, Guidotti G: Insulin stimulation of (Na^+,K^+)-adenosine triphosphatase-dependent $^{86}Rb^+$ uptake in rat adipocytes. J Biol Chem 255: 10938–10945, 1980
26. Lytton J: Insulin affects the sodium affinity of the rat adipocyte (Na^+,K^+)-ATPase. J Biol Chem 260: 10075–10080, 1985
27. McGill DL, Guidotti G: Insulin stimulates both the alpha 1 and the alpha 2 isoforms of the rat adipocyte (Na^+,K^+) ATPase. Two mechanisms of stimulation. J Biol Chem 266: 15824–15831, 1991
28. Brodsky JL: Insulin activation of brain Na^+-K^+-ATPase is mediated by alpha 2 form of enzyme. Am J Physiol 258: C812–C817, 1990
29. Matsuda T, Murata Y, Kawamura N, Hayashi M, Tamada K, Takuma K, Maeda S, Baba A:. Selective induction of alpha 1 isoform of $(Na^+ + K^+)$-ATPase by insulin/insulin-like growth factor-I in cultured rat astrocytes. Arch Biochem Biophys 307: 175–182, 1993
30. Malik N, Canfield VA, Beckers M, Gros P, Levenson R: Identification of the mammalian Na,K-ATPase β3 subunit. J Biol Chem 271: 22754–22758, 1996
31. Hundal HS, Marette A, Ramlal T, Liu Z, Klip A: Expression of beta subunit isoforms of the Na^+, K^+-ATPase is muscle type-specific. FEBS Lett 328: 253–258, 1993
32. Jaunin P, Jaisser F, Beggah AT, Takeyasu K, Mangeat P, Rossier BC, Horisberger JD, Geering K: Role of the transmembrane and extra-cytoplasmic domain of beta subunits in subunit assembly, intracellular transport, and functional expression of Na,K-pumps. J Cell Biol 123: 1751–1759, 1993
33. McDonough AA, Geering K, Farley RA: The sodium pump needs its beta subunit. FASEB J 4: 1598–1605, 1990
34. Schmalzing G, Gloor S, Omay H, Kroner S, Appelhans H, Schwarz W: Up-regulation of sodium pump activity in Xenopus laevis oocytes by expression of heterologous beta 1 subunits of the sodium pump. Biochem J 279: 329–336, 1991
35. Eakle KA, Kim KS, Kabalin MA, Farley RA: High-affinity ouabain binding by yeast cells expressing Na^+, K^+-ATPase alpha subunits and the gastric H^+, K^+-ATPase beta subunit. Proc Natl Acad Sci USA 89: 2834–2838, 1992
36. Geering K, Beggah A, Good P, Girardet S, Roy S, Schaer D, Jaunin P: Oligomerization and maturation of Na,K-ATPase: Functional interaction of the cytoplasmic NH_2 terminus of the β subunit with the α subunit. J Cell Biol 133: 1193–1204, 1996
37. Fambrough DM, Lemas MV, Hamrick M, Emerick M, Renaud KJ, Inman EM, Hwang B, Takeyasu K: Analysis of subunit assembly of the Na-K-ATPase. Am J Physiol 266: C579–C589, 1994
38. Eakle KA, Lyu RM, Farley RA: The influence of beta subunit structure on the interaction of Na^+/K^+-ATPase complexes with Na^+. A chimeric beta subunit reduces the Na^+ dependence of phosphoenzyme formation from ATP. J Biol Chem 270: 13937–13947, 1995
39. Hundal HS, Marette A, Mitsumoto Y, Ramlal T, Blostein R, Klip A: Insulin induces translocation of the alpha 2 and beta 1 subunits of the Na^+/K^+-ATPase from intracellular compartments to the plasma membrane in mammalian skeletal muscle. J Biol Chem 267: 5040–5043, 1992

40. Lemas MV, Hamrick M, Takeyasu K, Fambrough DM: 26 amino acids of an extracellular domain of the Na,K-ATPase alpha-subunit are sufficient for assembly with the Na,K-ATPase beta-subunit. J Biol Chem 269: 8255–8259, 1994

41. Koster JC, Blanco G, Mercer RW: A cytoplasmic region of the Na,K-ATPase alpha-subunit is necessary for specific alpha1alpha association. J Biol Chem 270: 14332–14339, 1995

42. Horowitz B, Eakle KA, Scheiner~Bobis G, Randolph GR, Chen CY, Hitzeman RA, Farley RA: Synthesis and assembly of functional mammalian Na,K-ATPase in yeast. J Biol Chem 265: 4189–4192, 1990

43. Blanco G, De Tomaso AW, Koster J, Xie ZJ, Mercer RW: The alpha subunit of the Na,K-ATPase has catalytic activity independent of the beta-subunit. J Biol Chem 269: 23420–23425, 1994

44. McDonough AA, Farley RA: Regulation of Na,K-ATPase activity. Curr Opin Nephrol Hyper 2: 725–734, 1993

45. Kanbe M, Kitasato H: Stimulation of Na,K-ATPase activity of frog skeletal muscle by insulin. Biochem Biophys Res Commun 134: 609–616, 1986

46. Omatsu-Kanbe M, Kitasato H: Insulin stimulates the translocation of Na$^+$/K$^+$-dependent ATPase molecules from intracellular stores to the plasma membrane in frog skeletal muscle. Biochem J 272: 727–733, 1990

47. Marette A, Krischer J, Lavoic L, Ackerley C, Carpentier JL, Klip A: Insulin increases the Na$^+$-K$^+$-ATPase alpha 2-subunit in the surface of rat skeletal muscle: Morphological evidence. Am J Physiol 265: C1716–C1722, 1993

48. Tsakiridis T, Wong PC, Liu Z, Rodgers CD, Vranic M, Klip A: Exercise increases the plasma membrane content of the Na$^+$/K$^+$ pump and its mRNA in skeletal muscles. J App Physiol 80: 699–705, 1996

49. Weidmann P, Ferrari P: Central role of sodium in hypertension in diabetic subjects. Diabetes Care 14: 220–32, 1991

50. Mueckler M: The molecular biology of glucose transport: Relevance to insulin resistance and non-insulin-dependent diabetes mellitus. J Diabetes Comp 7: 130–141, 1993

51. Hundal HS, Klip A: Regulation of glucose transporters and the Na/K-ATPase by insulin in skeletal muscle. Adv Exp Med Biol 334: 6378, 1993

52. Aledo JC, Hundal HS: Sedimentation and immunological analyses of GLUT4 and a2-Na,K-ATPase subunit-containing vesicles from rat skeletal muscle: Evidence for segregation. FEBS Lett 376: 211–215, 1995

53. Lavoie L, He L, Ramlal T, Ackerley C, Marette A, Klip A: The GLUT4 glucose transporter and the alpha 2 subunit of the Na$^+$,K$^+$-ATPase do not localize to the same intracellular vesicles in rat skeletal muscle. FEBS Lett 366: 109–114, 1995

54. Sargeant R, Mitsumoto Y, Hundal H, Marette A, Liu Z, Klip A: Na$^+$/K$^+$-ATPase subunit expression and localization in cells in culture: Regulation by insulin. Biophys J 64: A331, 1993

55. Sargeant RJ, Liu Z, Klip A: Action of insulin on Na$^+$-K$^+$-ATPase and the Na$^+$-K$^+$-2Cl$^-$ cotransporter in 3T3-L1 adipocytes. Am J Physiol 269: C217–C225, 1995

56. Payne JA, Forbush Br: Molecular characterization of the epithelial Na-K-Cl cotransporter isoforms. Curr Opin Cell Biol 7: 493–503, 1995

57. Bertorello AM, Katz AI: Short-term regulation of renal Na-K-ATPase activity: Physiological relevance and cellular mechanisms. Am J Physiol 265: F743–F755, 1993

58. O'Donnell ME: Role of Na-K-Cl cotransport in vascular endothelial cell volume regulation. Am J Physiol 264: C1316–C1326, 1993

59. O'Donnell ME, Martinez A, Sun D: Endothelial Na-K-Cl cotransport regulation by tonicity and hormones: phosphorylation of cotransport protein. Am J Physiol 269: C1513–C1523, 1995

60. Haussinger D: The role of cellular hydration in the regulation of cell function. Biochem J 313: 697–710, 1996

61. Tohda H, Foskett JK, O'Brodovich H, Marunaka Y: Cl$^-$ regulation of a Ca^{2+}activated nonselective cation channel in beta- agonist- treated fetal distal lung epithelium. Am J Physiol 266: C104–C109, 1994

62. Bird IM, Word RA, Clyne C, Mason Jl, Rainey WE: Potassium negatively regulates angiotensin II type I receptor expression in human adrenocortical H295R cells. Hypertension 25: 1129–1134, 1995

63. Klip A, Ramlal T, Cragoe EJ Jr: Insulin-induced cytoplasmic alkalinization and glucose transport in muscle cells. Am J Physiol 250: C720–C728, 1986

64. Klip A: Action of insulin on Na$^+$/H$^+$ exchange. In: S Grinstein (ed). CRC 'Na$^+$/H$^+$ Exchange'. 1988, pp 285–303

65. Lynch CJ, Mader AC, McCall KM, Ng YC, Hazen SA: Okadaic acid stimulates ouabain-sensitive ^{86}Rb$^+$-uptake and phosphorylation of the Na$^+$/K$^+$-ATPase alpha-subunit in rat hepatocytes. FEBS Lett 355: 157–162, 1994

66. Rosic NK, Standaert ML, Pollet RJ: The mechanism of insulin stimulation of (Na$^+$,K$^+$)-ATPase transport activity in muscle. J Biol Chem 260: 6206–6212, 1985

67. Sampson SR, Brodie C, Alboim SV: Role of protein kinase C in insulin activation of the Na-K pump in cultured skeletal muscle. Am J Physiol 266: C751–C758, 1994

68. McDonough AA, Thompson CB: Role of skeletal muscle sodium pumps in the adaptation to potassium deprivation. Acta Physiol Scand 156: 295–304, 1996

69. Weil E, Sasson S, Gutman Y: Mechanism of insulin-induced activation of Na$^+$-K$^+$-ATPase in isolated rat soleus muscle. Am J Physiol 261: C224–C230, 1991

70. Brodie C, Sampson SR: Regulation of the sodium-potassium pump in cultured rat skeletal myotubes by intracerular sodium ions. J Cell Physiol 140: 131–137, 1989

71. McGeoch JE, Morielli AD: An insulin-sensitive cation channel controls [Na$^+$]$_i$ via [Ca^{2+}]-regulated Na$^+$ and Ca^{2+} entry. Mol Biol Cell 5: 485–496, 1994

72. Brodsky JL: Characterization of the (Na$^+$ (+) K$^+$)-ATPase from 3T3-F442A fibroblasts and adipocytes. Isozymes and insulin sensitivity. J Biol Chem 265: 10458–10465, 1990

73. Comelius F, Logvinenko N: Functional regulation of reconstituted Na,K-ATPase by protein kinase A phosphorylation. FEBS Lett 380: 277–280, 1996

74. Garg LC, Saha PK, Mohuczy-Dominiak D: Cholinergic inhibition of Na-K-ATPase via activation of protein kinase C in Madin-Darby canine kidney cells. J Am Soc Nephrol 4: 195–205, 1993

75. Aperia A, Holtback U, Syren ML, Svensson LB, Fryckstedt J, Greengard P: Activation/deactivation of renal Na$^+$,K$^+$-ATPase: A final common pathway for regulation of natriuresis. FASEB J 8: 436–439, 1994

76. Feraille E, Marsy S, Cheval L, Barlet-Bas C, Khadouri C, Favre H, Doucet A: Sites of antinatriuretic action of insulin along rat nephron. Am J Physiol 263: F175–F179, 1992

77. Feraille E, Carranza ML, Rousselot M, Favre H: Insulin enhances sodium sensitivity of Na-K-ATPase in isolated rat proximal convoluted tubule. Am J Physiol 267: F55–F62, 1994

78. Feraille E, Rousselot M, Rajerison R, Favre H: Effect of insulin on Na$^+$,K$^+$-ATPase in rat collecting duct. J Physiol 488: 171–180, 1995

79. Russo JJ, Sweadner KJ: Na$^+$-K$^+$-ATPase subunit isoform pattern modification by mitogenic insulin concentration in 3T3-L1 preadipocytes. Am J Physiol 264: C311–C316, 1993

80. Tirupattur PR, Ram JL, Standley PR, Sowers JR: Regulation of Na$^+$,K$^+$-ATPase gene expression by insulin in vascular smooth muscle cells. Am J Hyper 6: 626–629, 1993

81. Tsakiridis T, McDowell HE, Walker T, Downes CP, Hundal HS, Vranic M, Klip A: Multiple roles of phosphatidylinositol 3-kinase in regulation of glucose transport, amino acid transport, and glucose transporters in L6 skeletal muscle cells. Endocrinology 136: 4315–4322, 1995

82. Cheatham B, Vlahos CJ, Cheatham L, Wang L, Blenis J, Kahn CR: Phosphatidylinositol 3-kinase activation is required for insulin stimulation of pp70 S6 kinase, DNA synthesis, and glucose transporter translocation. Mol Cell Biol 14: 4902–4911, 1994

132

83. Fry MJ: Structure, regulation and function of phosphoinositide 3-kinases. Biochim Biophys Acta 1226: 237–268, 1994

84. Standaert ML, Avignon A, Yamada K, Bandyopadhyay G, Farese RV: The phosphatidylinositol 3-kinase inhibitor, wortmannin, inhibits insulin-induced activation of phosphatidylcholine hydrolysis and associated protein kinase C translocation in rat adipocytes. Biochem J 313: 1039–1046, 1996

85. Xing M, Insel PA: Protein kinase C-dependent activation of cytosolic phospholipase A_2 and mitogen-activated protein kinase by alpha 1-adrenergic receptors in Madin-Darby canine kidney cells. J Clin Invest 97: 1302–1310, 1996

86. Xia P, Kramer RM, King GL: Identification of the mechanism for the inhibition of Na^+,K^+-adenosine triphosphatase by hyperglycemia involving activation of protein kinase C and cytosolic phospholipase A_2. J Clin Invest 96: 733–740, 1995

87. Satoh T, Cohen HT, Katz AI: Intracellular signaling in the regulation of renal Na-K-ATPase. I. Role of cyclic AMP and phospholipase A_2. J Clin Invest 89: 1496–1500, 1992

88. Ver A, Csermely P, Banyasz T, Kovacs T, Somogyi J: Alterations in the properties and isoform ratios of brain Na^+/K^+-ATPase in streptozotocin diabetic rats. Biochim Biophys Acta 1237: 143–150, 1995

89. Lijnen P: Alterations in sodium metabolism as an etiological model for hypertension. Cardiovasc Drugs Ther 9: 377–399, 1995

90. Leese GP, Vora JP: The management of hypertension in diabetes: With special reference to diabetic kidney disease. Diabet Med 13: 401–410, 1996

91. Tepel M, Bauer S, Husseini S, Raffelsiefer A, Zidek W: Increased cytosolic free sodium concentrations in platelets from type 2 (non-insulin-dependent) diabetic patients is associated with hypertension. J Endocrinol 138: 565–572, 1993

92. Tamaoki J, Tagaya E, Yamawaki I, Konno K: Hypoxia impairs nitro-vasodilator induced pulmonary vasodilation: role of Na-K-ATPase activity. Am J Physiol 271: L172–L177, 1996

93. Sowers JR: Effects of insulin and IGF-1 on vascular smooth muscle glucose and cation metabolism. Diabetes 45: S47–S51, 1996

94. Dorup 1, Clausen T: Insulin-like growth factor I stimulates active Na^+-K^+ transport in rat soleus muscle. Am J Physiol 268: E849–E857, 1995

95. Raccah D, Gallice P, Pouget J, Vague P: Hypothesis: Low Na/K-ATPase activity in the red cell membrane, a potential marker of the predisposition to diabetic neuropathy. Diab Metabol 18: 236–241, 1992

96. Fink DJ, Datta S, Mata M: Isoform specific reductions in Na^+,K^+-ATPase catalytic (alpha) subunits in the nerve of rats with streptozotocin-induced diabetes. J Neurochem 63: 1782–1786, 1994

97. Greene DA, Lattimer SA, Sima AA: Sorbitol, phosphoinositides, and sodium potassium-ATPase in the pathogenesis of diabetic complications. New Eng J Med 316: 599–606, 1987

98. Stevens MJ, Dananberg J, Feldman EL, Lattimer SA, Kamijo M, Thomas TP, Shindo H, Sima AA, Greene DA: The linked roles of nitric oxide, aldose reductase and, (Na^+,K^+)-ATPase in the slowing of nerve conduction in the streptozotocin diabetic rat J Clin Invest 94: 853–859, 1994

99. King GL, Kunisaki M, Nishio Y, Inoguchi T, Shiba T, Xia P: Biochemical and molecular mechanisms in the development of diabetic vascular complications. Diabetes 45: S105–S108, 1996

100. Jirousek MR, Gillig JR, Gonzalez CM, Heath WF, McDonald H Jr, Neel DA, Rito CJ, Singh U, Stramm LE, Melikian-Badalian A, Baevsky M, Ballas LM, Hall SE, Winneroski LL, Faul MM: (S)-13-[(dimethyl-amino)methyll-10,11,14,15-tetrahydro-4,9:16, 21-dimetheno-1H, 13H-dibenzo[e,k]pyrrolo[3,4h][1,4,13]oxadiazacyclohexadecene-1,3(2H)-dione (LY333531) and related analogues: Isozyme selective inhibitors of protein kinase C beta. J Med Chem 39: 2664–2671, 1996

101. Mimura M, Makino H, Kanatsuka A, Asai T, Yoshida S: Reduction of erythrocyte $(Na^+$-$K^+)$ATPase activity in type 2 (non-insulin-dependent) diabetic patients with microalbuminuria. Horm Metab Res 26: 33–38, 1994

102. Clerico A, Giampietro O: Is the endogenous digitalis-like factor the link between hypertension and metabolic disorders as diabetes mellitus, obesity and acromegaly? Clin Physiol Biochem 8: 153–168, 1990

103. Akopyanz NS, Broude NE, Bekman EP, Marzen EO, Sverdlov ED: Tissue-specific expression of Na,K-ATPase beta-subunit. Does beta 2 expression correlate with tumorigenesis? FEBS Lett 289: 8–10, 1991

104. Martin-Vasallo P, Ghosh S, Coca-Prados M: Expression of Na,K-ATPase alpha subunit isoforms in the human ciliary body and cultured ciliary epithelial cells. J Cell Physiol 141: 243–252, 1989

105. Shamraj OI, Melvin D, Lingrel JB: Expression of Na,K-ATPase isoforms in human heart. Biochem Biophys Res Comm 179: 1434–1440, 1991

106. Zahler R, Sun W, Ardito T, Zhang ZT, Kocsis JD, Kashgarian M: The alpha3 isoform protein of the Na^+, K^+-ATPase is associated with the sites of cardiac and neuromuscular impulse transmission. Circ Res 78: 870–879, 1996

107. Gick GG, Hatala MA, Chon D, Ismail-Beigi F: Na,K-ATPase in several tissues of the rat: tissue-specific expression of subunit mRNAs and enzyme activity. J Mem Biol 131: 229–236, 1993

108. Clapp WL, Bowman P, Shaw GS, Patel P, Kone BC: Segmental localization of mRNAs encoding Na^+-K^+-ATPase alpha- and beta-subunit isoforms in rat kidney using RT-PCR. Kidney Int 46: 627–638, 1994

109. Songu-Mize E, Liu X, Stones JE, Hymel LJ: Regulation of Na^+,K^+-ATPase alpha-subunit expression by mechanical strain in aortic smooth muscle cells. Hypertension 27: 827–832, 1996

110. Giannella RA, Orlowski J, Jump ML, Lingrel JB: Na^+-K^+-ATPase gene expression in rat intestine and Caco-2 cells: Response to thyroid hormone. Am J Physiol 265: G775–G782, 1993

111. Martin-Vasallo P, Dackowski W, Emanuel JR, Levenson R: Identification of a putative isoform of the Na,K-ATPase beta subunit. Primary structure and tissue-specific expression. J Biol Chem 264: 4613–4618, 1989

112. Coca-Prados M, Femandez-Cabezudo MJ, Sanchez-Torres J, Crabb J-W, Ghosh S: Cell-specific expression of the human Na^+,K^+-ATPase beta 2 subunit isoform in the nonpigmented ciliary epithelium. Invest Ophth Vis Sci 36: 2717–2728, 1995

113. Book CB, Moore RL, Semanchik A, Ng YC: Cardiac hypertrophy alters expression of Na^+, K^+-ATPase subunit isoforms at mRNA and protein levels in rat myocardium. J Mol Cell Cardiol 26: 591–600, 1994

114. Ahn KY, Madsen KM, Tisher CC, Kone BC: Differential expression and cellular distribution of mRNAs encoding alpha- and beta-isoforms of Na^+-K^+-ATPase in rat kidney. Am J Physiol 265: F792–F801, 1993

115. Hundal HS, Maxwell DL, Ahmed A, Darakhshan F, Mitsumoto Y, Klip A: Subcellular distribution and immunocytochemical localization of Na,K-ATPase subunit isoforms in human skeletal muscle. Mol Mem Biol 11: 255–262, 1994

116. Ikeda U, Takahashi M, Okada K, Saito T, Shimada K: Regulation of Na-KATPase gene expression by angiotensin II in vascular smooth muscle cells. Am J Physiol 267: H1295–H1302, 1994

117. Khan I, Collins SM: Altered expression of sodium pump isoforms in the inflamed intestine of *Trichinella spiralis*-infected rats. Am J Physiol 264: G1160–G1168, 1993

118. Fink DJ, Fang D, Li T, Mata M: Na,K-ATPase beta subunit isoform expression in the peripheral nervous system of the rat. Neurosci Lett 183: 206–209, 1995

119. Gould GW, Holman GD: The glucose transporter family: Structure, function and tissue-specific expression. Biochem J 295: 329–341, 1993

120. Ng YC, Tolerico PH, Book CB: Alterations in levels of Na^+-K^+-ATPase isoforms in heart, skeletal muscle, and kidney of diabetic rats. Am J Physiol 265: E243–E251, 1993

121. Nishida K, Ohara T, Johnson J, Wallner JS, Wilk J, Sherman N, Kawakami K, Sussman KE, Draznin B: Na^+/K^+-ATPase activity and its alpha II subunit gene expression in rat skeletal muscle: Influence of diabetes, fasting, and refeeding. Metab Clin Exper 41: 56–63, 1992

122. Mooradian AD, Grabau G, Bastani B: Adenosine triphosphatases of rat cerebral microvessels. Effect of age and diabetes mellitus. Life Sci 55: 1261–1265, 1994

123. Nowak TV, Castelaz C, Ramaswamy K, Weisbruch JP: Impaired rodent vagal nerve sodium-potassium-ATPase activity in streptozotocin diabetes. J Lab Clin Med 125: 182–186, 1995

124. Ottlecz A, Bensaoula T: Captopril ameliorates the decreased Na^+,K^+-ATPase activity in the retina of streptozotocin-induced diabetic rats. Invest Ophth Vis Sci 37: 1633–1641, 1996

125. Madsen KL, Ariano D, Fedorak RN: Vanadate treatment rapidly improves glucose transport and activates 6-phosphofructo-l-kinase in diabetic rat intestine. Diabetologia 38: 403–412, 1995

Molecular and Cellular Biochemistry **182**: 135–141, 1998.

Potential mechanism(s) involved in the regulation of glycogen synthesis by insulin

Ashok K. Srivastava and Sanjay K. Pandey
Research Center, CHUM, Hôtel-Dieu Campus, Department of Medicine, University of Montréal, Montréal, Québec, Canada

Abstract

Stimulation of glycogen synthesis is one of the major physiological responses modulated by insulin. Although, details of the precise mechanism by which insulin action on glycogen synthesis is mediated remains uncertain, significant advances have been made to understand several steps in this process. Most importantly, recent studies have focussed on the possible role of *glycogen synthase kinase*-3 (GSK-3) and *glycogen bound protein phosphatase*-1 (PP-1G) in the activation of *glycogen synthase* (GS) – a key enzyme of glycogen metabolism. Evidence is also accumulating to establish a link between insulin receptor induced signaling pathway(s) and glycogen synthesis. This article summarizes the potential contribution of various elements of insulin signaling pathway such as *mitogen activated protein kinase* (MAPK), *protein kinase B* (PKB), and *phosphatidyl inositol 3-kinase* (PI3-K) in the activation of GS and glycogen synthesis. (Mol Cell Biochem **182**: 135–141, 1998)

Key words: glycogen synthesis, glycogen synthase, glycogen synthase kinase-3, protein kinase B, protein phosphatase-1, phosphatidyl inositol 3-kinase

Introduction

In mammalian tissues, carbohydrate is stored mainly in the form of glycogen and the major sites of glycogen storage are skeletal muscle and liver. The concentration of glycogen is higher in the liver than in the muscle but, due to its much greater mass, muscle stores more glycogen. Insulin regulates the synthesis of glycogen at two steps, first by controlling the uptake and transport of glucose – the building block for the synthesis of glycogen molecule, and secondly, by regulating the phosphorylation and activation states of enzymes involved in the synthesis and degradation of glycogen [1]. Insulin-induced glucose uptake is mediated by insulin sensitive glucose transporter proteins called GLUT4 [2]. For the biosynthesis of glycogen, the intracellular glucose undergoes several modifications to generate 'active glucose,' uridine diphosphate glucose (UDP-G). This reaction takes place in three sequential steps. First glucose is phosphorylated by hexokinase to generate glucose-6-phosphate (G-6-P), which in the second step is converted to glucose-1-phosphate (G-1-P) by phosphoglucomutase, and finally the active glucose or UDP-G is synthesized from G-1-P and uridine triphosphate (UTP) in a reaction catalyzed by UDP-glucose pyrophosphorylase. UDP-G thus formed, serves as glucosyl group donor for glycogen synthesis catalyzed by glycogen synthase (Fig. 1). Glycogen synthase can add glucosyl residue only on preexisting glycogen containing at least 4 glucosyl residues. In addition to glycogen synthase, a specialized initiator protein glycogenin and branching enzymes also contribute to overall glycogen synthesis (Fig. 1).

In recent years, significant advances have been made to understand the mechanisms of insulin-induced activation of glycogen synthesis however, the precise steps involved in this process remain uncertain. In this article we will present a brief overview on the current status on insulin-induced activation of glycogen synthase and glycogen synthesis.

Glycogen synthase

Glycogen synthase is a multimeric protein consisting of 4 subunits of 85 kD whose activity is regulated by phos-

Address for offprints: A.K. Srivastava, Research Center, CHUM, Hôtel-Dieu Campus, 3840 St. Urbain Street, Montréal, Québec, H2W 1T8, Canada

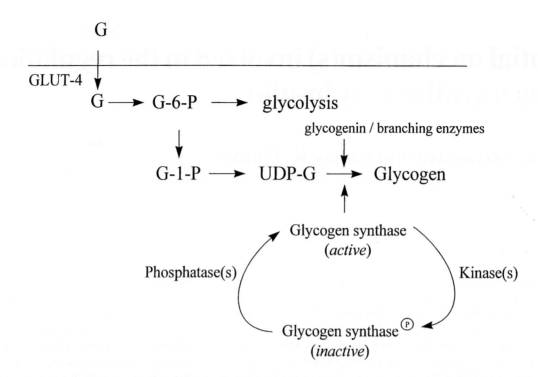

Fig. 1. Schematic model showing sequential steps of glycogen synthesis. Glucose (G) enters the cells of insulin-sensitive tissues by a facilitative transport process mediated by insulin-sensitive glucose transporter protein, GLUT 4. GLUT 4 is translocated from cytosol to the plasma membrane in response to insulin. G is then converted to glucose-6-phosphate (G-6-P) by hexokinase. which then either enters into glycolysis and/or is converted to glucose-1-phosphate (G-1-P) by phosphoglucomutase. G-1-P, in the presence of UTP and UDP-glucose pyrophosphorylase is converted to UDP-G, an active glucose donating molecule, which donates glucosyl residue to glycogen molecule. Glycogen synthase catalyzes the transfer of glucosyl units from UDP-glucose to the nonreducing ends of glycogen molecule. Glycogenin/branching enzyme help add glucose to growing chain of glycogen molecule. Glycogen synthase itself cycles between an active and inactive state. Phosphorylation by protein kinase(s) leads to decrease in glycogen synthase activity (i.e. inactive form) whereas dephosphorylation catalyzed by protein phosphatase(s) causes activation of the enzyme (i.e. active state).

phorylation and dephosphorylation mechanism [3, 4]. The phosphorylation of glycogen synthase, catalyzed by several protein kinases which phosphorylate multiple serine residues (at least 9), in both amino as well as carboxy terminus of glycogen synthase subunit, generally leads to a decrease in the catalytic activity of glycogen synthase. However, the site of phosphorylation determines the degree of inactivation of the catalytic activity [3, 4]. For example, in the case of rabbit skeletal muscle glycogen synthase, *in vitro* experiments have shown that phosphorylation of sites 2 (Ser^7) and 2a (Ser^{10}) [5, 6], 3a (Ser^{640}) and 3b (Ser^{644}) [7] caused significant decrease in the catalytic activity whereas, phosphorylation at site 3c (Ser^{648}) [7], site 4 (Ser^{1332}) [8], site 5 (Ser^{656}) [9, 10], or sites 1a (Ser^{697}) or site 1b (Ser^{710}) [11, 12] did not significantly alter the catalytic activity of glycogen synthase. Glucose-6-P serves as an allosteric modulator and can stimulate the catalytic activity of even highly phosphorylated form of glycogen synthase. The dephosphorylated form of glycogen synthase is insensitive to allosteric stimulation by G-6-P [1]. Phosphorylation of various sites on glycogen synthase is catalyzed by many protein kinases which include protein kinase A, Casein

kinase, protein kinase C and GSK-3 [3, 4]. In most cases these phosphorylations are 'ordered' and follow 'hierarchal' pattern [3, 4].

Protein phosphatase-1

Insulin causes a decrease in the serine phosphorylation of sites 2, 2a, 3a, 3b, and 3c of glycogen synthase leading to its activation. [1, 4, 13]. This decrease could be accomplished either by activating the protein phosphatases which dephosphorylate the glycogen synthase and/or by inhibiting the protein kinases which catalyzes the phosphorylation of glycogen synthase. In fact, insulin activates type 1 serine/threonine protein phosphatase-1 (PP-1) activity in skeletal muscle and isolated cells [14–18]. Since PP-1 can dephosphorylate all the sites in glycogen synthase, its activation by insulin has been attributed to a decrease in the phosphorylation of glycogen synthase [19, 20]. Muscle PP-1 comprises of a 37 kDa catalytic subunit (PP1-C) and a 160 kDa glycogen targeting subunit (PP-1G) [19]. Free PP-1C is 5–10 times less active as a glycogen synthase phosphatase than the holoenzyme complex of PP-1 C/PP-1G bound to glycogen and it is believed that insulin induces an increase in the association of PP-1C with PP-1G to form a holoenzyme [19].

Insulin-induced serine phosphorylation of PP-1G at site 1 has been attributed to an increase in the rate of association of PP-1G to PP-1C and binding to glycogen [19, 21]. Initially, an involvement of MAPK/90 kDa ribosomal S6 kinase (p90rsk) in insulin-induced PP-1G phosphorylation as well as glycogen synthase dephosphorylation and activation was suggested [21] however, recent studies using specific inhibitors of MAPK pathway [22] and cells expressing a dominant negative Ras mutant with an attenuated MAPK signaling [23, 24] have raised doubts about this model. Thus, alternate mechanisms for insulin-induced activation of PP-1 must exist. Recently, a novel glycogen targeting subunit of PP-1 termed PTG, for protein targeting to glycogen, has been described [25]. PTG, in addition to binding to PP-1C and glycogen is also able to complex with glycogen synthase, glycogen phosphorylase and glycogen phosphorylase kinase and its overexpression resulted in an increase in both basal as well as insulin-stimulated glycogen synthesis [25]. Overexpression of PP-1G in L6 rat skeletal muscle has also been shown to result in an increase in insulin-stimulated glycogen synthesis and glycogen synthase functional activity without affecting the basal activity [26]. Moreover, depletion of endogenous PP-1G by using an antisense RNA caused a reduction in basal PP-1C activity and blocked PP-1 activation [26]. Interestingly, similar to the effects on glycogen synthesis, overexpression and depletion of PP-1G resulted in an increased and attenuated uptake of 2-deoxy-glucose in response to insulin respectively [26]. Based on these observations a critical role of PTG/PP-1G in glycogen metabolism has been proposed however, the precise mechanism by which these targeting subunits regulate PP-1 activity and stimulate glycogen synthase remains to be clarified.

Glycogen synthase kinase-3

The protein kinase whose activity is inhibited by insulin is GSK-3. GSK-3 which phosphorylates a series of serine residues (sites 3a, b, c, and 4) at the carboxy terminus of muscle glycogen synthase and inactivates it [12], is expressed as two related isoforms, α and β, having molecular sizes of 51 and 47 kDa respectively [27]. Both isoforms are constitutively active in the basal state and are transiently and partly inhibited by serine phosphorylation near the amino-terminus in response to insulin and other growth factors in various cell types [28–30]. Several insulin-activated protein kinases including p90rsk, 70 kDa ribosomal S6 kinase (P70s6k) and PKB (also known as Akt or RAC) have been shown to catalyze the serine phosphorylation and inactivation of GSK-3 *in vitro* [31–34]. However, recent studies using

specific inhibitors of p70s6k and MAPK/p90rsk have demonstrated that these protein kinases may not catalyze the *in vivo* phosphorylation and inhibition of GSK-3 activity [30]. Rapamycin, an inhibitor of p70s6k, failed to inhibit insulin-induced inactivation of GSK-3 in CHO cells overexpressing insulin receptor [31], L6 myotubes [30] rat adipocytes [35], rat skeletal muscle [36] and human myoblasts [37]. Rapamycin was also without effect on insulin-stimulated glycogen synthesis in rat adipocytes [35], mouse skeletal muscle [36], CHO cells overexpressing insulin receptor [31]. However, some studies in which rapamycin treatment was shown to inhibit insulin-induced glycogen synthesis [37–40] have suggested the existence of additional rapamycin-sensitive mechanism, perhaps independent of an effect on GSK-3, to regulate the activation of glycogen synthase and glycogen synthesis. Park Davis inhibitor, PD098059, which inhibits MAPK kinase activity and thus MAPK and p90rsk [41], did not prevent insulin-induced inhibition of GSK-3 in L6 myotubes [33] and human myotubes [37]. This compound was also without effect on insulin-stimulated glycogen synthase activation, glycogen synthesis and PP-1 activation in rat adipocytes and skeletal muscle [22, 40]. Moreover, additional support for the lack of a role of MAPK pathway in the regulation of glycogen synthesis has come from experiments using cells transfected with dominant negative mutants which despite an attenuated activation of MAPK in response to insulin exhibit a normal activation of glycogen synthesis/glycogen synthase [23]. Recently PKB has emerged as the most potential protein kinase catalyzing the phosphorylation and inactivation of GSK-3 [33].

PKB and PI3-kinase

PKB'S, are cellular homolog of the retroviral v-Akt [42–44]. These kinases have sequence homology to both protein kinase A and protein kinase C but in addition have an amino-terminal pleckstrin homology (PH) domain [45]. At least three isoforms of PKB termed PKB α [46] PKB β [47, 48] and PKB-γ [49] have been identified. PKB is rapidly activated by insulin in L6 myotubes [33], skeletal muscle [33], and adipose tissue [36, 50]. Insulin-induced activation of PKB in isolated cells is associated with increased phosphorylation of Thr 308 in the catalytic domain and ser 473 in the carboxy terminal tail and appears to be critical to generate a high level of PKB activity [46]. The activation as well as the phosphorylation of these residues in PKB is blocked by wortmannin, an inhibitor of PI3-kinase [46]. PI3-kinase is primarily a lipid kinase which phosphorylates D3 position of phosphatidylinositol (PI), PI(4) P, PI(4, 5) P_2 to generate PI(3)P, PI(3, 4) P_2 and PI(3, 4, 5,) P_3 respectively (reviewed in [51, 52]). PI3-kinase is stimulated by insulin [53] as well as many other stimuli [51, 52] and it exists in

138

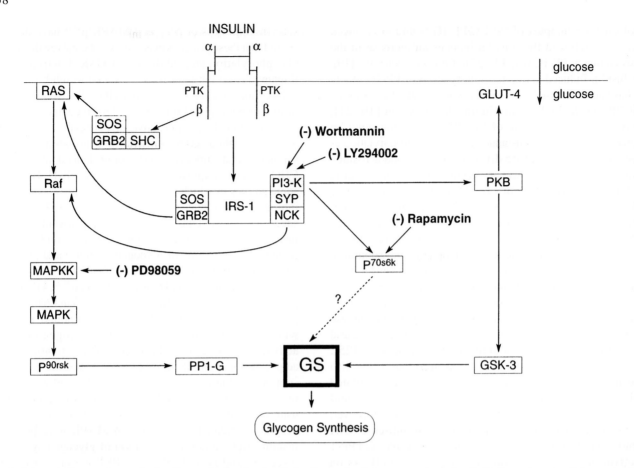

Fig. 2. Schematic diagram showng the signaling mechanisms involved in insulinstimulated glycogen synthesis. Insulin binding to its extracellular α-subunits causes activation of its intracellular β-subunit tyrosine kinase activity. The activated β-subunit in turn phosphorylates several substrate proteins including IRS-1 and Shc. Tyrosine phosphorylated IRS-1 recruits several proteins with Src homology (SH2)-containing domains. These include SOS, a guanine nucleotide exchange factor, adapter proteins Grb-2 and Nck, Syp-a phosphatase and phosphatidyl-3-inositol kinase (PI3-K). These IRS-1 associated proteins form a complex, which either stimulates Ras-RafMAPKK-MAPK-p90rsk pathway or P13-k/PKB-GSK-3 cascade. Activated Ras recruits Raf, a serine/threonine kinase to plasma membrane and activates it. Raf in turn activate MAPKK (or MEK), a dual specificity kinase. This MAPKK activates MAPK by phosphorylating on tyrosine and threonine residues. It was proposed that MAPK activates p[90rsk] which further phosphorylates and activates glycogen bound form of protein phosphatase 1 (PP1-G). PP1-G dephosphorylates glycogen synthase (GS) which results in its activation leading to glycogen synthesis. However, this pathway of activation of PP1-G by insulin has shortcomings and has been excluded in mediating this response. Thus, signaling mechanisms responsible for the activation PP1-G remains elusive at this moment. On the other hand evidence is accumulating to suggest a central role of PI3-k and serine-threonine protein kinase, PKB in GSK-3 inhibition and insulin-induced activation of GS and glycogen synthesis. A role of P[70s6k] in the regulation of glycogen synthesis has also been suggested which appears to be cell-specific based on the use of rapamycin-an inhibitor of p[70s6k] activation. Both activation of PP1-G and inhibition of GSK-3 might contribute to insulin-stimulated GS. Specific inhibitors such as PI-3k inhibitors, wortmannin and LY294002; MAPKK inhibitor PD98059 and P[70s6k] inhibitor rapamycin, have been indicated in bold letters.

several isozymic forms [51, 52]. The most widely studied isozymic form is a heterodimer consisting of 110-kDa (p110) catalytic subunit and 85-kDa (p85) regulatory subunit, which has Src homology 2 (SH2) domains. The SH2 domain of p85 interacts with tyrosine phosphorylated insulin receptor substrate-1 (IRS-1). This interaction results in the stimulation of the catalytic activity of p110 subunit [54]. The lipid products of PI3-kinase reaction have been shown to directly activate PKB activity *in vitro* [55–57]. It was observed that PI(3, 4) P_2 and not PI(3, 4, 5) P_3 exerted an stimulatory effect on PKB activity [57]. In addition, a protein

kinase which phosphorylates Thr 308 in PKB-α has recently been purified [58]. This protein kinase is activated by both PI(3, 4) P_2 and PI(3, 4, 5) P_3 and has been named *phos*-phatidylinositol (3, 4) P2/PI (3, 4, 5) P_3 *d*ependent *k*inase-1 (PDK-1) [58] however, the precise role of PDK-1 in insulin -induced activation of PKB and inactivation of GSK-3 needs to be defined.

Nearly all the studies which suggested a role of PI3-k in insulin-stimulated glycogen synthesis have utilized two cell permeable inhibitors of PI3-k, wortmannin, a fungal metabolite and Elliy Lilly compound LY294002. Wortmannin inhibits PI3-k by binding tightly and irreversibly to the p110 subunit of PI3-k [59] and has been shown to block insulin-induced activation of glycogen synthase in 3T3-adipocytes [38], PC12 cells [23], rat adipocytes [60], CHO cells overexpressing insulin receptor [61] and human myoblasts [37]. Wortmannin also blocked insulin-stimulated glycogen synthesis in 3T3-L1 adipocytes [38]. LY294002, which is structurally unrelated to wortmannin and inhibits PI3-k activity by acting at the ATP binding site of p110 subunit of PI3-kinase [62] has been also shown to block insulin-induced activation of glycogen synthase in PC12 cells [23] and CHO cells [63].

Role of PI3-k in insulin-induced activation of glycogen synthase has also been examined by using cells over-expressing dominant negative mutant of p85 subunit [61] or by co-expressing inter SH2 region of p85 subunit and p110 α subunit of PI3-k [64]. Overexpression of a mutant form of p85 subunit which is unable to bind to p110 subunit in CHO cells resulted in marked attenuation of insulin-stimulated PI3-kinase activity while glycogen synthase activation remained unaffected [61]. Similarly, constitutive activation of PI3-k activity by co-expressing p85 and p110 α subunit did not alter the basal glycogen synthase activity in 3T3-L1 adipocytes [64]. Clearly these results are at variance with the observations made by using pharmacological inhibitors of PI3-kinase and suggest the involvement of additional unidentified wortmannin and LY294002 inhibitable targets for glycogen synthesis.

Several isoforms of p110 and p85 subunits have been described [51, 52, 65] however, the specificity of these forms in mediating the effect on glycogen synthase remains unexplored. Moreover, specific targeting and generation of phosphorylated lipid products of PI3-kinase reaction in the appropriate cellular compartments might also play a critical role in eliciting the effect of insulin on glycogen synthase [65].

Conclusion

During the recent years major advances have been made to understand the cellular mechanism involved in insulin-induced activation of glycogen synthase and glycogen synthesis. Considerable attention has focussed on PP-1G and GSK-3 since insulin-induced activation and inhibition respectively of these two molecules can contribute to an overall decrease in the phosphorylation state of glycogen synthase resulting in its activation. However, the mechanistic details linking insulin receptor activation with PP-1G stimulation and GSK-3 inhibition appear to be complex and need elaboration. PP-1G has been suggested to be activated by a phosphorylation reaction but the nature of this putative protein kinase remains elusive at this point. Primarily, based

on the use of pharmacological inhibitors of PI3-k a role of this lipid kinase in insulin-induced inhibition of GSK-3 via PKB activation has been suggested but the results using dominant negative mutants of p85 subunit or overexpression of p110 subunit of PI3-k have raised doubts about role of this enzyme. It should also be noted that activation of PKB and inhibition of GSK-3 induced by β-adrenergic agonist does not involve wortmannin sensitive mechanism and is not associated with the activation of glycogen synthase. However, the contribution of other isozymic form of PI3-k and/or the wortmannin and LY294002 inhibitable new signaling intermediates in the glycogen synthase activation can not be ruled out. A simplified schematic model indicating potential mechanisms to link insulin binding with glycogen synthase activation and glycogen synthesis is depicted in Fig. 2. Clearly, several steps in this model need to be further elaborated and require additional investigation.

Acknowledgements

The work in the authors laboratory is supported by a grant from Medical Research Council of Canada (A.K.S) and a studentship from Fonds pour la Formation de Chercheur et l'Aide a la Recherche (FCAR), Quebec, (S.K.P). We thank Mrs. Susanne Bordeleau-Chenier for excellent secretarial assistance.

References

1. Cohen P: In Control of Enzyme Activity, 2nd Edition. Chapman and Hall, New York, 1983, pp 42–71
2. Mueckler M: Facilitative glucose transporters Eur J Biochem 219: 713–725, 1994
3. Roach PJ: Control of glycogen synthase by hierarchal protein phosphorylation. FASEB J: 2961–2968, 1990
4. Roach PJ: Multisite and hierarchal protein phosphorylation. J Biol Chem 266: 14139–14142, 1991
5. Flotow H, Roach PJ: Synergistic phosphorylation of rabbit muscle glycogen synthase by cyclic AMP-dependent protein kinase and casein kinase I. Implication for hormonal regulation of glycogen synthase. J Biol Chem 264: 9126–9128, 1989
6. Nakielny S, Campbell DG, Cohen P: The molecular mechanism by which adrenalin inhibits glycogen synthesis. Eur J Biochem 199: 713–722, 1991
7. Wang Y, Roach P: Inactivation of rabbit muscle glycogen synthase by glycogen synthase kinase-3. Dominant role of ser-640 (site-3a). J Biol Chem 268: 23876–23980, 1993
8. Poulter L, Ang SG, Gibson BW, Williams DH, Holmer GFB, Caudwell FB, Pitcher J, Cohen P: Analysis of the *in vivo* phosphorylation state of rabbit skeletal muscle glycogen synthase by fast-atom-bombardment mass spectrometry. Eur J Biochem 175: 497–510, 1988
9. De Paoli-Roch AA, Ahmad Z, Roach PJ: Characterization of a rabbit skeletal muscle protein kinase (PCO.7) able to phosphorylate muscle glycogen synthase and phosvitin. J Biol Chem 256: 8955–8962, 1981

10. Cohen P, Yelawfees D, Aitken A, Donelia-Dean A, Hemmings BA, Parker PJ: Separation and characterization of glycogen synthase kinase-3, glycogen synthase kinase-4, and glycogen synthase kinase-5 from rabbit skeletal muscle. Eur J Biochem 124: 21–35, 1982

11. Embi N, Parker PJ, Cohen P: A reinvestigation of the phosphorylation of rabbit skeletal muscle glycogen synthase by cyclic AMP-dependent protein kinase. Identification of the third site of phosphorylation as serine-7. Eur J Biochem 115: 405–413, 1981

12. Parker PJ, Embi N, Caudwell FB, Cohen P: Glycogen synthase from rabbit skeletal muscle. State of phosphorylation of seven phosphoserine residues in vivo in the presence and absence of adrenaline. Eur J Biochem 124: 47–55, 1982

13. Lawrence JC Jr: Signal transduction and protein phosphorylation in the regulation of cellular metabolism by insulin. Ann Rev Physiol 54: 177–193, 1994

14. Toth B, Boilen M, Stalmans W: Acute regulation of hepatic protein phosphatases by glucagon, insulin and glucose. J Biol Chem 263: 723–728, 1991

15. Olivier AR, Ballou LM, Thomas G: Differential regulation of S6 phosphorylation by insulin and EGF in Swiss mouse 3T3 cells. Insulin activation of type 1 phosphatase. Proc Natl Acad Sci USA 85: 4720–4724, 1988

16. Olivier AR, Thomas G: Three forms of phosphatase type-1 in Swiss 3T3 fibroblasts. J Biol Chem 265: 22460–22466, 1990

17. Srinivasan M, Begum N: Regulation of protein phosphatase 1 and 2A activities by insulin during myogenesis in rat skeletal muscle cells in culture. J Biol Chem 269: 12514–12520, 1994

18. Begum N: Stimulation of protein phosphatase-1 activity by insulin in rat adipocytes. Evaluation of the role of mitogen activated protein kinase pathway. J Biol Chem 270: 709–714, 1995

19. Hubbard MJ, Cohen P: On target with a new mechanism for the regulation of protein phosphatase. Trends Biochem Sci 18: 172–177, 1993

20. Begum N: Role of protein serine/threonine phosphatase 1 and 2A in insulin action. Adv Prot Phos 2: 263–281, 1995

21. Dent P, Lavoinne A, Nakienly S, Caudwell FB, Watt P, Cohen P: The molecular mechanism by which insulin stimulates glycogen synthesis in mammalian skeletal muscle. Nature 348: 302–308, 1990

22. Lazar DF, Weise RJ, Brady MJ, Mastik CC, Watyers SB, Yamuchi K, Pessin JE, Cuatrecasas P, Saltiel AR: Mitogen-activated protein kinase kinase inhibition does not block the stimulation of glucose utilization by insulin. J Biol Chem 270: 20801–20807, 1995

23. Yamamoto-Honda R, Tobe K, Kaburagi Y, Ueki K, Asai S, Yachi M, Shirouzu M, Ydoi J, Akanuma Y, Yokoyama S, Yazaki Y, Kadowaki T: Upstream mechanisms of glycogen synthase activation by insulin and insulin-like growth factor-I. J Biol Chem 270: 2729–2734, 1995

24. Dorrestijn J, Ouwens Dm, Vadenheede JR, Bos JJ, Massen JA: Expression of a dominant negative Ras mutant does not affect stimulation of glucose uptake and glycogen synthesis by insulin. Diabetologia 39: 558–563, 1996

25. Printon JA, Brady MJ, Saltiel AR: PTG, a protein phosphatase-1 binding protein with a role in glycogen metabolism. Science 275: 1475–1478, 1997

26. Ragolia L, Begum N: The effect of modulating the glycogen-associated regulatory subunit of protein phosphatase-1 on insulin action in rat skeletal muscle cells. Endocrinology 138: 2398–2404, 1997

27. Woodgett JR: Molecular cloning and expression of glycogen synthase kinase 3/Factor A. EMBO J 9: 2431–2438, 1990

28. Hughes K, Ramakrishna S, Benjamin WB, Woodgett JR: Identification of multifunctional ATP-citrate lyase kinase as the α isoform of glycogen synthase kinase-3. Biochem J 288: 309–314, 1992

29. Welsh GI, Proud CG: Glycogen synthase kinase-3 is rapidly inactivated in response to insulin and phosphorylates eukaryotic initiation factor eIF-2B. Biochem J 294: 625–629, 1993

30. Cross DAE, Alessi DR, Vadenheede JR, McDowell HE, Hundai HS, Cohen P: The inhibition of glycogen synthase kinase-3 by insulin or insulin-like growth factor 1 in the rat skeletal muscle cell line is blocked by wortmannin but not by rapamycin. Evidence that wortmannin blocks activation of the mitogen-activated protein kinase in L6 cells between Ras and Raf. Biochem J 303: 21–26, 1994

31. Sutherland C, Leighton I, Cohen P: Inactivation of glycogen synthase kinase 3b by MAP kinase-activated protein kinase-1 (RSK-2) and p70s6 kinase, new kinase connections in insulin and growth factor signaling Biochem J 296: 15–19, 1993

32. Sutherland C, Cohen P: The α isoform of glycogen synthase kinase-3 from rabbit skeletal muscle is inactivated by p70 s6 kinase or MAP kinase-activated protein kinase-1 in vitro. FEBS Lett 338: 37–42, 1994

33. Cross DAE, Alessi DR, Cohen P, Andjelkovich M, Hemmings B: Inhibition of glycogen synthase kinase-3 by insulin mediated by protein kinase B. Nature 378: 785–789, 1995

34. Welsh GI, Foulstone EJ, Young SW, Tavare JM, Proud CG: Wortmannin inhibits the effects of insulin and serum on the activities of glycogen synthase kinase-3 and mitogen activated protein kinase Biochem J 303: 15–20, 1994

35. Moule SK, Edgell NJ, Welsh GI, Diggle TA, Foulstone EJ, Heesom KJ, Proud CG, Denton RM: Multiple signaling pathways involved in the stimulation of fatty acid and glycogen synthesis by insulin in rat epidydymal fat cells. Biochem J 311: 595–601, 1995

36. Cross DAE, Watt PW, Shaw M, Kaay JVD, Downes CP, Holder JC, Cohen P: Insulin activates protein kinase B, inhibits glycogen synthase kinase-3 and activates glycogen synthase by rapamycin-insensitive pathways in skeletal muscle and adipose tissue. FEBS Lett 406: 211–215, 1997

37. Hurel SJ, Rochford JJ, Borthwick AC, Wells AM, Vandenheede JR, Turnbull DM, Yeaman SJ: Insulin action in cultured human myoblasts: Contribution of different signalling pathways to regulation of glycogen synthesis. Biochem J 320: 871–877, 1996

38. Shepard PR, Navé BT, Siddle K: Insulin stimulation of glycogen synthesis and glycogen synthase activity is blocked by wortmannin and rapamycin in 3T3-L1 adipocytes: Evidence for the involvement of phosphoinositide 3-kinase and ribosomal protein-s6 kinase. Biochem J 305: 25–28, 1995

39. Moxham CM, Tabrizchi A, Davis RJ, Malbon C: C-jun N-terminal kinase mediates activation of skeletal muscle glycogen synthase by insulin in vivo. J Biol Chem 271: 30765–30773, 1996

40. Azpiazu I, Saltiel AR, Depaol-Roach AA, Lawrence JC Jr: Regulation of both glycogen synthase and PHAS-1 by insulin in rat skeletal muscle involves mitogen-activated protein kinase-independent and rapamycin-sensitive pathways. J Biol Chem 271: 5033–5039, 1996

41. Dudley P, Pang L, Decker S, Bridges A, Saltiel A: A synthetic inhibitor of the mitogen-activated protein kinase cascade. Proc Natl Acad Sci USA 92: 7686–7689, 1995

42. Jones PF, Jakubowicz T, Pitossi FJ, Maurer F, Hemmings BA: Molecular cloning and identification of a serine/threonine protein kinase of the second-messenger subfamily. Proc Natl Acad Sci USA 88: 4171–4175, 1991

43. Coffer PJ, Woodgett JR: Molecular cloning and characterization of a novel putative protein-serine kinase related to cAMP-dependent and protein kinase C families. Eur J Biochem 201: 475–481, 1991

44. Bellacosa A, Testa JR, Staal SP, Tsichlis PN: A retroviral oncogene, akt, encoding a serine-threonine kinase containing an SH2-like region. Science 254: 274–277, 1991

45. Cohen GB, Ren R, Baltimore D: Modular binding domains in signal transduction proteins. Cell 80: 237–248, 1995

46. Alessi DR, Andjelkovic M, Caudwell B, Corn P, Morrice N, Cohen P, Hemming BA: Mechanism of activation of protein kinase B by insulin and IGF-I. EMBO J 15: 6541–6551, 1996

47. Jones PF, Jakubowicz T, Hemming BA: Molecular cloning of a second form of protein kinase. Cell Reg 2: 1001–1004, 1991

48. Cheng JQ, Godwin AK, Bellacosa A, Taguchi T, Franke TF, Hamilton TC, Tsichlis PN, Testa JR: Akt 2, a putative oncogene encoding a member of a subfamily of protein serine-threonine kinase, is amplified in human ovarian carcinoma. Proc Natl Acad Sci USA 89: 9267–9271, 1992

49. Konishi H, Kuroda S, Tanaka M, Matsuzaki H, Ono Y, Kameyama K, Haga T, Kikkawa U: Molecular cloning and characterization of a new member of the RAC protein-kinase family-association of the pleckstrin homology domain of 3 types of RAC protein kinase with protein kinase C subspecies and beta-gamma subunits of G-proteins. Biochem Biophys Res Commun 216: 526–534, 1995

50. Moule SK, Welsh GI, Edgell NJ, Foulstone EJ, Proud CG, Denton RM: Regulation of protein kinase β and glycogen synthase kinase-3 by insulin and B-adrenergic agonists in rat epididymal fat cells. J Biol Chem 272: 7713–7719, 1997

51. Vanhaesebroek B, Leevere SJ, Panayotou G, Waterfield MD: Phospho-inositide 3-kinase: A conserved family of signal transducer. Trends Biochem Sci 22: 123–128, 1997

52. Toker A, Cantley LC: Signaling through the lipid products of phosphoinositide 3-OH kinase. Nature 387: 673–676, 1997

53. Ruderman N, Kapeller R, White MF, Cantley LC: Activation of phosphoinositol 3-kinase by insulin. Proc Natl Acad Sci USA 87: 1411–1415, 1990

54. Cheatham B, Kahn CR: Insulin action and insulin signaling network. Endocr Rev 16: 117–142, 1995

55. Franke TF, Kaplan DR, Cantley LC, Toker AA: Direct regulation of Akt protooncogene product by PI3,4P. Science 275: 665–668, 1997

56. Klippel A, Kavanaugh WM, Pot D, Williams LT: A specific product of PI3-k directly activates the protein kinase Akt through its pleckstrin homology domain. Mol Cell Biol 17: 338–344, 1997

57. Frech M, Andjelkovic M, Ingley E, Reddy KK, Falck JR, Hemming BA: High affinity binding of inositol phosphate and phosphoinositides to the pleckstrin homology domain of RAC/protein kinase B and their influence on kinase activity. J Biol Chem 272: 8474–8481, 1997

58. Alessi DR, James SR, Downes CP, Holmes AB, Gaffney PR, Rees CB, Cohen P: Characterization of a 3- phosphoinositide-dependent protein kinase which phosphorylates and activates PKB-α. Curr Biol 7: 261–269, 1997

59. Wymann MP, Bulgarelli-leva G, Zvelebil MJ, Pirola L, Vanhaesebroek B, Waterfield MD, Panayotou G: Wortmannin inactivates phosphoinositide 3-kinase by covalent modification of Lys-803, a residue involved in the phosphate transfer reaction. Mol Cell Biol 16: 1722–1733, 1996

60. Standaert ML, Bandyopadhyay G, Farese RV: Studies with wortmannin suggest a role for phosphatidylinositol 3-kinase in the activation of glycogen synthase and mitogen-activated protein kinase by insulin in rat adipocytes: Comparison of insulin and protein kinase C modulators Biochem Biophys Res Commun 209: 1082–1088, 1995

61. Sakaue H, Hara K, Noguchi T, Matozaki T, Kotani K, Ogawa W, Ynezawa K, Waterfield MD, Kasuga M: Ras-independent and wortmannin-sensitive activation of glycogen synthase by insulin in Chinese hamster ovary cells. J Biol Chem 270: 11304–11309, 1995

62 Vlahos CJ, Matter WF, Hui KY, Brown RF: A specific inhibitor of phosphatidylinositol 3-kinase, 2-(4-morpholinyl)-8-phenyl-4H-1 benzopyran-4-one (LY294002). J Biol Chem 269: 5241–5248, 1994

63. Pandey SK, Anand-Srivastava MB, Srivastava AK: Implication of phosphatidylinositol 3-kinase (PI3-k) in vanadyl sulfate(VS)-stimulated glycogen synthesis. Canadian J Diab Care 21: p39, 1997

64. Frevert EU, Kahn BB: Differential effect of constitutively active PI3-k on glucose transport, glycogen synthase activity and DNA synthesis in 3T3-L1 adipocytes. Mol Cell Biol 17: 190–198, 1997

65. Inuki K, Funaki M, Ohigara T, Katagiri H, Kanda A, Anai M, Fukushima Y, Hosaka T, Suzuki M, Shin B, Takata K, Yazaki Y, Kikuchi M, Oka Y, Asano T: p85 α gene generates three isoforms of regulatory subunit for phosphatidylinositol 3-kinase(PI3-kinase), p50α, p55α, and p85α with different PI3-kinase activity elevating responses to insulin. J Biol Chem 272: 7873–7882, 1997

Molecular and Cellular Biochemistry **182**: 143–152, 1998.
© 1998 *Kluwer Academic Publishers. Printed in the Netherlands.*

Metabolic and therapeutic lessons from genetic manipulation of GLUT4

Maureen J. Charron and Ellen B. Katz
Department of Biochemistry, Albert Einstein College of Medicine, Bronx, New York, USA

Abstract

This review focuses on the effects of varying levels of GLUT4, the insulin-sensitive glucose transporter, on insulin sensitivity and whole body glucose homeostasis. Three mouse models are discussed including MLC-GLUT4 mice which overexpress GLUT4 specifically in skeletal muscle, GLUT4 null mice which express no GLUT4, and the MLC-GLUT4 null mice which express GLUT4 only in skeletal muscle. Overexpressing GLUT4 specifically in the skeletal muscle results in increased insulin sensitivity in the MLC-GLUT4 mice. In contrast, the GLUT4 null mice exhibit insulin intolerance accompanied by abnormalities in glucose and lipid metabolism. Restoring GLUT4 expression in skeletal muscle in the MLC-GLUT4 null mice results in normal glucose metabolism but continued abnormal lipid metabolism. The results of experiments using these mouse models demonstrates that modifying the expression of GLUT4 profoundly affects whole body insulin action and consequently glucose and lipid metabolism. (Mol Cell Biochem **182**: 143–152, 1998)

Key words: GLUT4, transgenic mice, glucose transporter, insulin resistance, diabetes mellitus, obesity

Introduction

Normal blood glucose levels are maintained by the regulation of absorption of glucose in the small intestine, the production of glucose by the liver and the uptake of glucose in peripheral tissues. These processes are controlled through the action of receptors that specifically bind to the hormones insulin and glucagon and by other membrane proteins that transport glucose across the plasma membrane. The membrane proteins which are thought to contribute greatly to the complex regulation of blood glucose belong to a family of structurally related glucose transport proteins which includes six identified members (GLUT1-GLUT5, and GLUT7) [1–4]. These proteins are the products of unique genes and are expressed in a tissue specific manner. GLUT1 is expressed in many tissues (i.e. blood-brain barrier, placenta, kidney) and transformed cell lines. GLUT2 is expressed primarily in cells in which a net release of glucose can occur (i.e. hepatocytes, kidney, small intestine and pancreatic β-cells). GLUT3, like GLUT1, is present in many tissues (i.e. brain, kidney, placenta). GLUT4 is unique in that it is expressed in tissues in which glucose uptake is stimulated by insulin (i.e. adipose cells, cardiac and skeletal muscle). GLUT5 is expressed primarily in absorptive epithelial cells (i.e. small intestine and kidney). GLUT7 is present in the membranes of the endoplasmic reticulum of hepatocytes and is believed to be responsible for transporting glucose out of this intracellular compartment.

Glucose homeostasis depends mainly on controlled changes in glucose transport in insulin-responsive tissues such as muscle and adipose cells. Both GLUT1 and GLUT4 are expressed in these tissues [5]. However, GLUT4 is only expressed in tissues where glucose transport is regulated by insulin (i.e. adipose cells, skeletal muscle, and heart) and is thought to be the major glucose transporter in skeletal muscle and adipose cells [6].

Insulin stimulates glucose transport in adipose cells by eliciting the translocation of an intracellular pool of GLUT1 and GLUT4 to the plasma membrane [7–10]. Approximately 90% of the glucose transporters expressed in adipose cells are GLUT4 [11–16]. The translocation of GLUT4 to the plasma membrane of adipose cells occurs in the presence of GTPγS, suggesting that a GTP-binding protein is required for GLUT4 translocation [17]. Immunoelectron microscopy

Address for offprints: M.J. Charron, Department of Biochemistry, Albert Einstein College of Mediciner, 1300 Morris Park Avenue, Bronx, New York 10641, USA

studies in rat adipose cells performed by one of us were the first to suggest that the carboxy-terminus of GLUT4 is 'masked' prior to insulin activation and insertion into the plasma membrane [18]. The molecular nature of the GLUT4 'mask' is still unknown, as is the role of insulin in the "unmasking" process. Insulin also stimulates GLUT4 translocation in skeletal muscle [19–24]. This process in skeletal muscle is also mediated by GTP-binding proteins [23]. GLUT1 is expressed in much lower levels in skeletal muscle membranes and does not translocate in response to insulin [25]. Insulin-like growth factors (IGF-I, IGF-II), insulin mimetics, and contraction/hypoxia have also been shown to stimulate translocation of GLUT4 [20, 23, 24, 26–33].

Insulin resistant metabolic states such as diabetes mellitus, fasting, and obesity are characterized by decreased *in vivo* glucose uptake in response to insulin as measured by euglycemic clamp [34–36]. It has been shown, in both skeletal muscle and adipose cells, that *in vivo* insulin resistance due to fasting and hyperresponsiveness after refeeding influence GLUT1 and GLUT4 expression in a glucose transporter-specific and tissue-specific manner [14, 37]. Studies on the effects of high fat diets and obesity on glucose homeostasis have been conducted in humans and rodents [14, 38–43]. In humans, high fat feeding and obesity have been shown to induce a state of insulin resistance and to decrease the number of GLUT4 transporters expressed in adipose cells [14, 38, 39]. GLUT4 gene expression is also decreased in adipose cells from both non-insulin dependent diabetic (NIDDM) and obese patients and in streptozotocin induced diabetic rats [14, 44, 45]. Studies using skeletal muscle from these individuals to measure the amount of GLUT4 present are variable, however, glucose transport rates have been shown to be decreased [14, 46–50]. In rats, a high fat diet results in a reduction in insulin stimulated glucose transport in muscle and fat and an increase in adipocyte size [37, 40, 41]. More recently it was demonstrated that high fat feeding induced insulin resistance in mice resulted in reduced PI3 kinase activity and reduced GLUT4 translocation to the plasma membrane of skeletal muscles [51].

Skeletal muscle is the most influential organ system involved in post-absorptive glucose uptake and utilization. Glucose transport into skeletal muscle can be rate-limiting for glucose utilization and is defective in insulin resistant states [24, 50]. Several studies have shown that GLUT4 translocation accounts for nearly all the increase in glucose uptake in peripheral tissue [14–16, 24]. With this knowledge we undertook studies in mice to determine the *in vivo* metabolic and cellular consequences of modulation of GLUT4 expression [52]. We have generated transgenic mice which overexpress GLUT4 in skeletal muscle [53]. These mice have been useful in exploring the therapeutic merit of stable overexpression of GLUT4 in muscle to enhance whole body glucose utilization and insulin action. Conversely, we utilized homologous recombination and embryonic stem cell technology to genetically ablate GLUT4 [54]. The resultant GLUT4 null mice have proven useful in understanding compensatory glucose transporter adaptations that may be novel targets in the treatment of diseases such as diabetes mellitus and obesity. By combining the genomes of GLUT4 null and GLUT4 transgenic mice to complement the skeletal muscle GLUT4 defect in the null mice, we have been able to understand the specific contribution of muscle GLUT4 to whole body glucose metabolism and adipose tissue mass [55]. Each of these mouse models is described below.

Transgenic overexpression of GLUT4 in muscle

Skeletal muscle is the major site for insulin stimulated glucose disposal and the most influential determinant of *in vivo* insulin sensitivity and insulin action [56, 57]. Alternations in skeletal muscle glucose transport mediated by GLUT4 are associated with insulin resistant states such as diabetes and obesity [5, 24, 58]. We generated transgenic mice that overexpress GLUT4 in muscle to test whether increased muscle glucose uptake mediated by GLUT4 would lead to enhanced whole body insulin action. Specifically, the myosin light chain 1f (MLC1f) promoter and enhancer elements were used to target GLUT4 overexpression to muscle [53]. The resultant transgenic mice, referred to as MLC-GLUT4, overexpress GLUT4 protein 3–4 fold in fast-twitch, glycolytic skeletal muscles (i.e. gastrocnemius and extensor digitorum longus, EDL) and 15–20% in heart. It is difficult to detect the MLC-GLUT4 transgene product in slow-twitch, oxidative skeletal muscles (i.e. soleus).

The increase in skeletal muscle GLUT4 resulted in increased glycemic control as evidenced by the accelerated clearance of a glucose load by the MLC-GLUT4 mice compared to controls. This was further supported using euglycemic/hyperinsulinemic clamps, the more rigorous test of whole body insulin action. A 2.5-fold increase in whole body insulin-stimulated glucose utilization (glucose turnover) was measured by this method (Table 1). Additionally, a very modest increase in basal glucose utilization was observed in MLC-GLUT4 mice. *In vivo* glucose uptake was assessed in hindlimb muscles using a tracer injection of 2-deoxyglucose (2-DOG). As expected, a 2.5 fold increase in both basal and insulin stimulated 2-DOG uptake was measured in gastrocnemius, composed of fast-twitch fibers which express the MLC-GLUT4 transgene, but not in soleus muscle of MLC-GLUT4 mice (Table 1). The increase in insulin stimulated glucose uptake resulted in a 2-fold increase in glycogen content in MLC-GLUT4 expressing muscles (Table 1).

Serum metabolites and insulin of MLC-GLUT4 mice

Young (2–3 months) MLC-GLUT4 mice clear an intra-peritoneal glucose load faster than control mice [53]. This is similar to the results of euglycemic/hyperinsulinemic clamp studies performed on 7–9 months old MLC-GLUT4 mice which demonstrated increased insulin sensitivity and glucose utilization by the transgenic mice. In accordance with this, young MLC-GLUT4 mice tend to require less insulin to maintain normal serum glucose levels in both the fasting and fed state. The overexpression of GLUT4 in skeletal muscle prevents decreased hexokinase II mRNA and enzyme activity in response to this mild insulinopenia [59]. With aging, MLC-GLUT4 mice exhibit significantly lower glucose levels without showing alterations in insulin under fasting conditions. The lower fasting serum glucose and higher fasting lactate are consistent with the increased glucose uptake and utilization by skeletal muscles (Table 1).

Lipid metabolism is also affected in MLC-GLUT4 mice as evidenced by elevations in serum free fatty acid levels (FFA) and β-hydroxybutyrate at 7–9 months of age [53]. These alterations in lipid metabolism are not associated with a change in body weight. However, body composition analysis remains to be performed to determine if there is any alteration in lean body mass. In summary, these results from experiments utilizing transgenic mice overexpressing GLUT4 exclusively in muscle demonstrate that skeletal muscle GLUT4 is a major determinant of whole body insulin action and glucose and lipid metabolism.

Targeted disruption of the GLUT4 locus

The murine GLUT4 locus was disrupted using homologous recombination in embryonic stem cells [54]. Through heterozygote knockout matings, mice deficient in GLUT4

mRNA and protein have been generated. The phenotype of GLUT4 null mice has been studied in three independent lines with consistent results. GLUT4 null mice exhibit numerous interesting features that provide insight into the integrated role of GLUT4 in normal glucose homeostasis, insulin action, and development. GLUT4 null mice are fertile but produce smaller litters than control mice and GLUT4 null mice have reduced longevity. GLUT4 null mice are approximately 20% smaller than age-matched controls and exhibit extreme reduction in fat mass (Table 2). In general, inguinal fat pads are approximately 10 times smaller than controls and the fat cells are at least 50% smaller in size. Another striking feature of the GLUT4 null mutation is the significant increase in heart mass (Fig. 1). The increase in heart mass is associated with a 50% increase in GLUT1 expression, the significance of which is under investigation.

Interestingly, GLUT4 null mice maintain normal plasma glucose levels and are able to clear a bolus of glucose as well as control mice [54]. This observation was initially surprising because it was predicted that GLUT4 null mice would exhibit alterations in glucose homeostasis and metabolism similar to a severely insulin resistant diabetic individual. GLUT4 null mice are in fact insulin resistant and this was first evident by the presence of hyperinsulinemia in the fed state. Further support of the insulin resistance in GLUT4 null mice comes from their impaired insulin tolerance (Fig. 2). As with the MLC-GLUT4 mice careful analysis of serum metabolites of GLUT4 null mice compared to controls reveal alterations in both glucose and lipid metabolism (Figs 3A and 3B). This finding demonstrates once again that a change in the expression of GLUT4 results in pleiotrophic metabolic alterations. Reduction in fed serum lactate levels may be indicative of reduced muscle glucose uptake and/or increased utilization by the liver as a gluconeogenic substrate. Similar reductions are noted in the concentration of fed serum FFA's which may be due to

Table 1. Characteristics of MLC-GLUT4 mice from euglycemic/hyperinsulinemic clamp study

		Control		MLC-GLUT4	
		Basal	Insulin	Basal	Insulin
Glucose turnover (mg/kg/min)		7.9 ± 0.7	13.8 ± 1.8	11.7 ± 2.6	34.4 ± 2.7*
Glycogen (mg/g)	Hindlimb	0.27 ± 0.08	1.49 ± 0.16	0.51 ± 0.18	3.21 ± 0.33*
	Gastrocnemius	0.64 ± 0.08	3.24 ± 0.26	1.53 ± 0.31*	5.75 ± 1.02*
	Soleus	0.69 ± 0.23	9.94 ± 3.15	0.37 ± 0.24	8.96 ± 2.33
2-DOG Uptake (ng/mg/min)	Hindlimb	2.60 ± 0.43	9.75 ± 2.73	10.14 ± 2.73*	24.74 ± 3.54*
	Gastrocnemius	3.04 ± 1.06	10.78 ± 2.83	9.82 ± 1.71*	31.83 ± 4.87*
	Soleus	2.15 ± 0.29	20.16 ± 7.58	4.17 ± 2.10	22.65 ± 3.28
Body Weight (g)		48.3 ± 1.6		47.2 ± 1.8	

Data are expressed as the mean ± S.E.M. 5–7 mice were analyzed in each group. Statistical significance between MLC-GLUT4 mice and age-matched controls under the same condition was determined using unpaired Student's *t*-test. *p ≤ 0.01.

146

Table 2. Body and ovarian fat pad weights and adipocyte cell size of GLUT4 null and MLC-GLUT4 null mice

	Body weight (g)	Ovarian pad weight (mg)	Adipocyte cell size (µg lipid/cell)
Control	28.9 ± 0.9	1340 ± 14	0.382 ± 0.10
GLUT4 null	$23.4 \pm 1.0*$	$126 \pm 49**$	$0.165 \pm 0.05*$
MLC-GLUT4 null	$24.3 \pm 0.9*$	$233 \pm 55**$	$0.119 \pm 0.05*$

Data are expressed as the mean ± S.E.M. and are derived from 8–15 mice per group. Statistical significance between control and age-matched GLUT4 null or MLC-GLUT4 null mice was determined using unpaired Student's t-test.*p < 0.06; **indicates p < 0.0001.

hyperinsulinemia and/or increased fatty acid oxidation in muscles. FFA levels of fasted GLUT4 null mice tend to be lower than that of control mice [54]. Additionally, 9-fold reductions in fasting ketones are noted which may be due to the reduced adipose tissue stores and FFA's [52, 54]. *In vivo* and *in vitro* analyses are underway to distinguish between these possibilities.

The upregulation of known facilitative glucose transporters in muscle and other organs could potentially compensate for GLUT4 ablation. The upregulation of GLUT1 in the heart noted above is unlikely to modify whole body glucose homeostasis since heart utilizes a small percentage of glucose under insulin action. Liver GLUT2 expression in GLUT4 null mice is elevated which may be indicative of altered glucose transport into or from the liver [54]. The brain is another site of altered glucose transporter expression in GLUT4 null mice. Preliminary studies have revealed measurable changes in the amount of GLUT1 and GLUT3 proteins (A.E. Stenbit, E.B. Katz, and M.J. Charron, un-

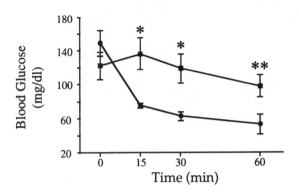

Male

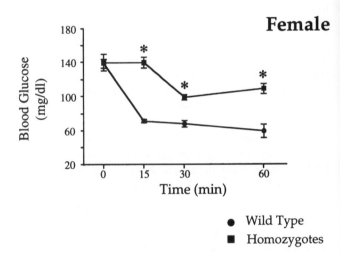

Female

● Wild Type
■ Homozygotes

Fig. 2. Insulin tolerance test (ITT) in GLUT4 null mice. After a 6 h fast, 10 week-old mice were given 0.75 U porcine insulin per kg body weight by peritoneal injection. Data was obtained from repeated measures test, expressed as mean ± S.E.M, and derived from 5 mice per group. *p < 0.05. Redrawn from [73].

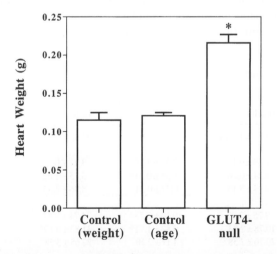

Fig. 1. Heart weight of female GLUT4 null mice compared to hearts of weight and age-matched controls. Data are expressed as mean ± S.E.M. and are derived from 8 mice per group. Statistical significance was determined using unpaired Student's t-test. *significance of p < 0.01.

published results). Though GLUT4 is expressed in the brain, it is of very low abundance [60]. The action of insulin on brain glucose uptake is controversial. The influence of altered GLUT1 and GLUT3 expression on basal and insulin stimulated conditions on brain function(s) and whole body glucose homeostasis remain to be elucidated. Glucose clamp studies are in progress to define alterations in glucose uptake and utilization in these and other tissues under ambient (basal) and hyperinsulinemic conditions.

Since skeletal muscle is the main site of insulin stimulated glucose disposal and GLUT4 is the most abundant glucose transporter, it was expected that the upregulation of another glucose transporter could partially account for the absence of overt hyperglycemia or diabetes in GLUT4 null mice. Northern and immunoblot analysis of GLUT4 null skeletal muscle has failed to detect increased GLUT1 expression or ectopic expression of any other known facilitative glucose transporter isoform [54, 61]. Present studies are focused on defining the role of GLUT1 in compensating for the lack of GLUT4 by examining its cellular distribution in skeletal muscle under basal and insulin stimulated conditions. Additionally, a screen for novel glucose transporter protein(s) is underway to identify the molecular mechanism(s) underlying the compensatory glucose transport activity in GLUT4 null muscle.

A.

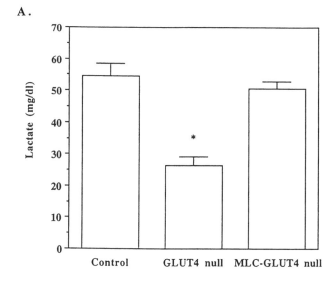

B.

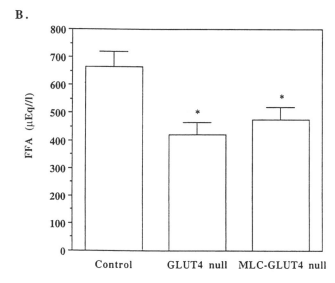

Fig. 3. Serum analysis of female GLUT4 null and MLC-GLUT4 null mice. The MLC-GLUT4 null mice result from crossing the GLUT4 null mice with the MLC-GLUT4 transgenic mice overexpressing GLUT4 only in skeletal muscle. The MLC-GLUT4 null mice express GLUT4 only in the fast-twitch skeletal muscle. Blood was drawn from mice in the fed state. Data are expressed as mean ± S.E.M. and are derived from 9 control, 8 GLUT4 null, and 25 MLC-GLUT4 null mice per group. Statistical significance was determined using unpaired Student's *t*-test. *significance of p < 0.05.

In vitro 2-DOG uptake studies were performed on two types of muscles from male and female GLUT4 null and control mice under basal and insulin stimulated (10–100 nM) conditions [61]. Basal glucose uptake was moderately reduced in EDL muscle of GLUT4 null mice. As may be expected, the EDL muscle did not exhibit an increase in glucose uptake in response to insulin. Interestingly, the glucose uptake characteristics in GLUT4 null soleus muscle were quite different than in EDL muscle and a sexual dimorphism was detected under basal conditions (Fig. 4). Specifically, basal glucose uptake in GLUT4 null females was unaltered where males exhibit a 2.3-fold increase compared to wild type soleus (Fig. 4A). Under insulin action the soleus muscle of female GLUT4 null mice is increased approximately 2-fold as compared to an approximate 3-fold increase in wild type females (Fig. 4B). The male GLUT4 null soleus does not exhibit a further increase in glucose

(A)

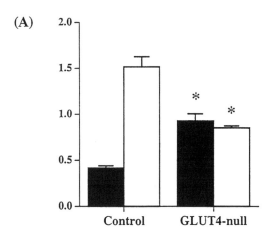

(B)

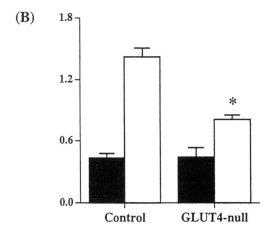

Fig. 4. In vitro 2 DOG uptake in GLUT4 null soleus. Data are expressed as mean ± S.E.M. and are derived from muscles from 6–8 mice per group. Panel A shows 2 DOG uptake in female GLUT4 null soleus; Panel B shows 2 DOG uptake in male GLUT4 null soleus. Black bars indicate basal uptake and white bars indicate insulin stimulated uptake. Statistical significance was determined using unpaired Student's *t*-test. *p < 0.05. Redrawn from [61].

uptake in response to insulin however, maximal glucose uptake is similar to that seen in insulin stimulated soleus from GLUT4 null females. Alterations in insulin stimulated glucose uptake in the muscles of GLUT4 null mice are independent of changes in autophosphorylation or tyrosine kinase activity of the insulin receptor [61]. Alterations in the expression, activity and phosphorylation levels of down-stream elements in the insulin receptor signalling cascade may occur as a response to the absence of GLUT4. Studies to test this hypothesis are underway. Insulin-stimulated glycogen synthesis was reduced in the soleus muscle and suppressed in the EDL muscle of GLUT4 null mice [61]. Furthermore, GLUT4 null muscles contain significantly less glycogen stores than wild type muscles most likely due to the reductions in glucose uptake described above. The functional and energetic consequences of reduced glucose uptake and glycogen synthesis and storage remain to be studied under various conditions including acute and prolonged stress (i.e. hypoxia, exercise).

The exciting observation of a glucose transport activity in the soleus muscle of GLUT4 null mice has raised the possibility that a novel glucose transport system may be activated to compensate for the loss of GLUT4 function. Since no alterations in the expression of other glucose transporters have been seen in GLUT4 null muscle, a previously unidentified glucose transporter or glucose-transporter-like gene product may be cloned. One could imagine that this transporter would have conserved the glucose binding sites utilized by other members of the facilitative glucose transporter family. By designing oligonucleotide primers from highly conserved regions of the GLUT family of cDNAs, including regions containing the putative glucose binding domain, novel transporters may be identified in GLUT4 null soleus muscle. Techniques such as RT-PCR analysis, differential display, and/or subtractive cloning will be useful in identifying candidates that compensate for the loss of GLUT4 function.

Developmental expression of GLUTs and the timing of compensations to the GLUT4 null mutation

The role of GLUT4 in embryonic development remains to be determined. Early mammalian embryogenesis may be regulated by both maternal and embryonic paracrine and autocrine factors. Expression of the insulin family of growth factors and receptors have been studied at the level of mRNA and protein in preimplantation mouse embryos [62]. Reverse transcriptase PCR analysis was used to detect insulin, IGF-I and IGF-II receptors, and IGF-II mRNAs as early as the 8-cell stage. This is the first point in development that an *in vitro* response to exogenously added insulin can be detected (i.e. increased DNA, RNA and protein synthesis)

[63]. Glucose becomes the preferred carbon source at the 8-cell stage of development [62]. Immunogold electron microscopic analysis revealed that both the trophoblastic and inner cell mass express the receptors for insulin, IGF-I and IGF-II and that insulin itself can be found as early as the morula stage of development [62]. The insulin present in preimplantation embryos is of maternal origin and may be directly altering gene expression since it is detected in the nuclei as well as the plasma membrane of the cells. The insulin growth factor family, along with their respective receptors, may play an important role in the metabolism, division and differentiation of early embryos [62]. The anabolic and proliferative effects of insulin on early mouse embryos is well established, however, the effects of insulin on glucose uptake are unclear. Gardner and Kaye [64] noted a 2-fold increase in glucose uptake in preimplantation mouse embryos in response to pharmacological concentrations of insulin (1 mg/ml). In adult fat cells, a 2-fold increase in plasma membrane bound GLUT1 is typically observed in response to insulin [1, 5, 14, 65]. As differentiation progresses, tissues become hormone responsive. Studies on preimplantation mouse embryos show that GLUT1 is expressed from the oocyte onward, GLUT2 mRNA is expressed from the 8-cell stage onward and GLUT4 is not expressed through the blastocyst stage [66, 67]. A Na^+-coupled glucose transporter has recently been detected in the apical membranes of mouse blastomeres [68]. It is not known exactly when during development the GLUT4 gene turns on and in which cells, however, our preliminary study using RT-PCR analysis identified GLUT4 expression in the ectoplacental cone and egg cylinder as early as day 7.5 *post coitum* (p.c.) (G. Schultz, E.B. Katz and M.J. Charron, unpublished result).

Immunoelectron microscopy studies on mouse blastocysts demonstrated that GLUT1 is widely distributed in the trophectoderm and inner cell mass cells and GLUT2 is expressed exclusively on trophectoderm membranes facing the blastocyst cavity [67]. Smith and Gridley studied GLUT1, GLUT2 and GLUT3 mRNA expression in early post-implantation (d7.5–d10.5 p.c.) mouse embryos using *in situ* hybridization analysis [69]. GLUT1 mRNA is expressed in the ectoplacental cone and both layers of the yolk sac at day 7.5 p.c.. By day 8.5 p.c. GLUT1 mRNA is widespread and on day 10.5 p.c. GLUT1 is down regulated with the strongest hybridization detected in the lens, pigmented retina and spinal cord. High level expression of GLUT2 mRNA in the visceral yolk sac endoderm is detected at days 7.5–8.5 p.c.. At day 10.5 p.c. GLUT2 mRNA is easily detected in the liver primordium, however, by day 12.5 p.c. GLUT2 mRNA is barely detectable. GLUT3 mRNA is expressed in yolk sac endoderm at day 7.5 p.c. and at day 10.5 p.c. low level expression of GLUT3 mRNA is largely confined to non-neural surface ectoderm and was not detected at day 14.5 p.c.. From this study it is evident that for all of the glucose transporters analyzed, expression levels begin to decrease near mid-pregnancy. In this study GLUT4 mRNA was not detected by Northern blot analysis of day 7.5 p.c. embryos; additional stages of development were not examined, nor were other modes of detection used.

GLUT4 expression has also been studied late in fetal development in rat [70]. GLUT4 mRNA was detected in heart at day 17 p.c. and GLUT4 protein at day 21 p.c.. Earlier time points were not examined. The pattern of expression of GLUT4 protein levels has been studied in neonates from day 1 through 34 [71]. Results of Western blot analysis show that GLUT4 levels increased early (day 7) and plateaued several weeks later in heart and diaphragm and fluctuated up and down in adipose cells before plateauing at day 34. Differences in the amount of GLUT4 protein in heart, skeletal muscle and adipose cells between males and females were also observed. At the time of weaning, rodents shift from a high fat diet of mother's milk to a high carbohydrate, low fat laboratory chow diet. Around this time, GLUT4 expression has reached stable, adult levels [71, 72]. One could imagine that any glucose transporter that is acting to compensate for the loss of GLUT4 may normally be expressed during these late embryonic, early neonatal periods of development. The genetic ablation of GLUT4 could promote the sustained expression or up-regulation of such a transporter to meet the metabolic demands of the developing mouse. Alternatively, the shift in dietary fat and/or carbohydrate at weaning could trigger expression of a novel transporter. Alterations in the expression of hormones such as insulin, cytokines such as TNFα, and/or growth factors, neuropeptides, and steroids in developing GLUT4 null mice may play a role in the activation of a compensatory glucose transport system. Finally, qualitative alterations in the targeting, trafficking and/or activity of GLUT1, without a measurable increase in total GLUT1 protein, may account for the glucose uptake characteristics seen in GLUT4 null soleus muscle. These and other possibilities are currently under investigation in our laboratory.

Transgenic replacement of GLUT4 into fast-twitch glycolytic muscles of GLUT4 null mice

GLUT4 null mice exhibit a plethora of metabolic and morphological alterations as a result of the loss of GLUT4 activity [54]. Severe adipose tissue reduction, reduced body weight, impaired insulin action and cardiac hypertrophy are among the most striking phenotypic alterations in mice lacking GLUT4. We have been able to add GLUT4 back to muscle of GLUT4 null mice to assess the relative role of GLUT4 in the etiology of the compound phenotype of the

null mice [55]. The MLC-GLUT4 transgene was mated into the background of the GLUT4 null mutation and mice expressing GLUT4 in fast-twitch glycolytic skeletal muscle, and to as lesser degree heart, have been generated. These mice are referred to as MLC-GLUT4 null. These mice express near normal amounts of GLUT4 protein in fast twitch glycolytic muscle (i.e. EDL) and almost no GLUT4 in slow-twitch oxidative muscle (i.e. soleus). As a result of restoring GLUT4 to fast twitch muscles, MLC-GLUT4 null mice display restored basal and insulin stimulated glucose uptake in EDL, but not soleus, muscle *in vitro* [55]. Like GLUT4 null mice, MLC-GLUT4 null mice weigh approximately 20% less than controls (Table 2). Further, MLC-GLUT4 null mice also have significantly reduced inguinal fat pads and small adipocytes compared to controls. This data demonstrates that complementation of muscle GLUT4 does not correct defects in adipose tissue due to GLUT4 ablation.

Transgenic complementation of GLUT4 in muscle of GLUT4 null mice corrected impaired glucose metabolism and insulin action seen in GLUT4 null mice [55]. Specifically, there is no difference in serum glucose or insulin levels between female MLC-GLUT4 null and controls. Male MLC-GLUT4 null mice exhibit modest reductions in fed glucose and insulin levels compared to controls suggesting enhanced whole body insulin action. Additionally, fed serum lactate levels of MLC-GLUT4 null mice are restored to normal levels (Fig. 3A). Insulin tolerance tests were performed to specifically test for improved whole body insulin action in MLC-GLUT4 null compared to GLUT4 null mice [55]. MLC-GLUT4 null mice clear glucose from their serum as efficiently as control mice. Results of this analysis demonstrated that the severe insulin intolerance seen in GLUT4 null mice is prevented by transgenic expression of GLUT4 in fast-twitch muscle. Combined, these data demonstrate that muscle GLUT4 is a major regulator of whole body glucose metabolism and that the defects in glucose metabolism and adipose tissue mass in GLUT4 null mice arise independently. Further evidence for the dissociation between defects in glucose and fat metabolism seen in GLUT4 null mice is seen in serum FFA levels. Despite the restoration of serum insulin to normal levels, fed FFA levels remained reduced in MLC-GLUT4 null mice compared to controls (Fig. 3B). Our initial studies using the MLC-GLUT4 transgene to restore GLUT4 to fast-twitch glycolytic muscle of GLUT4 null mice demonstrates the power of tissue specific transgenesis in analyzing compound phenotypes that result from genetic ablation. Using other tissue-specific GLUT4 transgenes we shall be able to dissect apart the individual contribution of adipose cells or cardiac muscle GLUT4 to the biology of each organ and importantly to whole body glucose and fat metabolism.

Conclusions

To assess the role of GLUT4 in whole body glucose homeostasis and insulin action, our laboratory has generated mice which express varying levels of GLUT4. MLC-GLUT4 mice, overexpressing GLUT4 specifically in skeletal muscle, possess increased insulin sensitivity. In contrast, GLUT4 null mice exhibit insulin intolerance and accompanying abnormalities in glucose and lipid metabolism. The alterations in glucose metabolism can be corrected by restoring GLUT4 expression in skeletal muscles only (MLC-GLUT4 null). The fat pad weight and fed FFA levels remain low in the MLC-GLUT4 null indicating that normal muscle expression of GLUT4 does not restore normal adipose tissue functions. The characterization of these mouse models has shown that modifying the level of expression of GLUT4 profoundly affects whole body insulin action and consequently glucose and lipid metabolism.

Acknowledgements

We wish to thank the following individuals for fruitful discussions and collaborations: Tsu-Shuen Tsao, Antine Stenbit, Jing Li, Rémy Burcelin, Lily Huang, Fredric Bone, Yannick Le Marchand-Brustel, Nadine Gautier, Juleen Zierath, and Karen Houseknecht. This work was supported by grants from the NIH (DK47425), American Diabetes Association, American Heart Association and the Pew Scholars Program in the Biomedical Sciences to M.J.C. and the Albert Einstein College of Medicine Cancer and Diabetes Centers. M.J.C. is a scholar of the Pew Charitable Trust.

References

1. Thorens B, Charron MJ, Lodish HF: Molecular physiology of glucose transporters. Diabetes Care 13: 209–218, 1990
2. Waddell ID, Zomerschoe AG, Voice MW, Burchell A: Cloning and expression of a hepatic microsomal glucose transport protein. Comparison with liver plasma-membrane glucose-transport protein GLUT 2. Biochem J 286: 173–177, 1992
3. Mueckler M: Family of glucose-transporter genes: implications for glucose homeostasis and diabetes. Diabetes 39: 6–11, 1990
4. Kayano T, Burant CF, Fukumoto H, Gould GW, Fan YS, Eddy RL, Byers MG, Shows TB, Seino S, Bell GI: Human facilitative glucose transporters. Isolation, functional characterization, and gene localization of cDNAs encoding an isoform (GLUT5) expressed in small intestine, kidney, muscle, and adipose tissue and an unusual glucose transporter pseudogene-like sequence (GLUT6). J Biol Chem 265: 13276–13282, 1990
5. Kahn BB: Glucose transport: Pivotal step in insulin action. Diabetes 45: 1644–1654, 1996
6. Charron MJ, Brosius FC III., Alper SL, Lodish HF: A glucose transport protein expressed predominantly in insulin-responsive tissues. Proc Natl Acad Sci USA 86: 2535–2539, 1989
7. Birnbaum MJ: Identification of a novel gene encoding an insulin-responsive glucose transporter protein. Cell 57: 305–315, 1989
8. James DE, Strube M, Mueckler M: Molecular cloning and characterization of an insulin-regulatable glucose transporter. Nature 338: 83–87, 1989
9. Kaestner KH, Christy RJ, McLenithan JC, Braiterman LT, Cornelius P, Pekala PH, Lane MD: Sequence, tissue distribution, and differential expression of mRNA for putative insulin-responsive glucose transporter in mouse 3T3-L1 adipocytes. Proc Natl Acad Sci USA 86: 3150–3154, 1989
10. Fukumoto H, Kayano T, Buse JB, Edwards Y, Pilch PF, Bell GI, Seino S: Cloning and characterization of the major insulin-responsive glucose transporter expressed in human skeletal muscle and other insulin-responsive tissues. J Biol Chem 264: 7776–7779, 1989
11. Cushman SW, Wardzala LJ: Potential mechanism of insulin action on glucose transport in the isolated rat adipose cell. Apparent translocation of intracellular transport systems to the plasma membrane. J Biol Chem 255: 4758–4762, 1980
12. Suzuki K, Kono T: Evidence that insulin causes translocation of glucose transport activity to the plasma membrane from an intracellular storage site. Proc Natl Acad Sci USA 77: 2542–2545, 1980
13. Satoh S, Nishimura H, Clark AE, Kozka IJ, Vannucci SJ, Simpson IA, Quon MJ, Cushman SW, Holman GD: Use of bismannose photolabel to elucidate insulin-regulated GLUT4 subcellular trafficking kinetics in rat adipose cells. Evidence that exocytosis is a critical site of hormone action. J Biol Chem 268: 17820–17829, 1993
14. Kahn BB: Facilitative glucose transporters: Regulatory mechanisms and dysregulation in diabetes. J Clin Invest 89: 1367–1374, 1992
15. Birnbaum MJ: The insulin-sensitive glucose transporter. Int Rev Cytol 137: 239–297, 1992
16. Kasanicki MA, Pilch PF: Regulation of glucose transporter function. Diabetes Care 13: 1990
17. Baldini G, Hohman R, Charron MJ, Lodish HF: Insulin and non-hydrolyzable GTP analogs induce translocation of GLUT4 to the plasma membrane in alpha-toxin-permeabilized rat adipose cells. J Biol Chem 266: 4037–4040, 1991
18. Smith RM, Charron MJ, Shah N, Lodish HF, Jarett L: Immunoelectron microscopic demonstration of insulin-stimulated translocation of glucose transporters to the plasma membrane of isolated rat adipocytes and masking of the carboxyl-terminal epitope of intracellular GLUT4. Proc Nat Acad Sci USA 88: 6893–6897, 1991
19. Klip A, Ramlal T, Young DA, Holloszy JO: Insulin-induced translocation of glucose transporters in rat hindlimb muscle. FEBS Lett 224: 224–230, 1987
20. Wallberg-Henriksson H: Glucose transport into skeletal muscle. Influence of contractile activity, insulin, catecholamines and diabetes mellitus. Acta Physiol Scand Suppl 564: 1–80, 1987
21. Hirshman MF, Goodyear LJ, Wardzala LJ, Horton ED, Horton ES: Identification of an intracellular pool of glucose transporters from basal and insulin-stimulated rat skeletal muscle. J Biol Chem 265: 987–991, 1990
22. Lund S, Holman GD, Schmitz O, Pedersen O: Glut4 content in the plasma membrane of rat skeletal muscle: Comparative studies of the subcellular fractionation method and the exofacial photolabelling technique using ATB-BMPA. FEBS Lett 330: 312–318, 1993
23. Etgen GJ Jr, Memon AR, Thompson GA Jr, Ivy JL: Insulin- and contraction-stimulated translocation of GTP-binding proteins and GLUT4 protein in skeletal muscle. J Biol Chem 268: 20164–20169, 1993
24. Zierath JR: In vitro studies of human skeletal muscle: Hormonal and metabolic regulation of glucose transport. Acta Physiol Scand 155: 1–96, 1995
25. Marette A, Burdett E, Douen A, Vranic M, Klip A: Insulin induces the translocation of GLUT4 from a unique intracellular organelle to transverse tubules in rat skeletal muscle. Diabetes 41: 1562–1569, 1992

26. Dohm GL, Eiton CW, Raju MS, Mooney ND, DiMarchi R, Pories WJ, Flickinger EG, Atkinson SM Jr, Caro JF: IGF-I-stimulated glucose transport in human skeletal muscle and IGF-I resistance in obesity and NIDDM. Diabetes 39: 1028–1032, 1990

27. Cartee GD, Douen AG, Ramlal T, Klip A, Holloszy JO: Stimulation of glucose transport in skeletal muscle by hypoxia. J Appl Physiol 70: 1593–1600, 1991

28. Zierath JR, Bang P, Galuska D, Hall K, Wallberg-Henriksson H: Insulin-like growth factor II stimulates glucose transport in human skeletal muscle. FEBS Lett 307: 379–382, 1992

29. Goodyear LJ, Hirshman MF, King PA, Horton ED, Thompson CM, Horton ES: Skeletal muscle plasma membrane glucose transport and glucose transporters after exercise. J Appl Physiol 68: 193–198, 1990

30. Ren JM, Semenkovich CF, Guive EA, Gao J, Holloszy JO: Exercise induces rapid increases in GLUT4 expression, glucose transport capacity, and insulin-stimulated glycogen storage in muscle. J Biol Chem 269: 14396–14401, 1994

31. Coderre L, Kandror KV, Vallega G, Pilch PF: Identification and characterization of an exercise-sensitive pool of glucose transporters in skeletal muscle. J Biol Chem 270: 27584–27588, 1995

32. Sherman LA, Hirshman MF, Cormont M, Le Marchand-Brustel Y, Goodyear LJ: Differential effects of insulin and exercise on Rab4 distribution in rat skeletal muscle. Endocrinol 137: 266–273, 1996

33. Cartee GD, Young DA, Sleeper MD, Zierath J, Wallberg-Henriksson H, Holloszy JO: Prolonged increase in insulin-stimulated glucose transport in muscle after exercise. Am J Physiol 256: E494–E499, 1989

34. Veroni MC, Proietto J, Larkins RG: Evolution of insulin resistance in New Zealand obese mice. Diabetes 40: 1480–1487, 1991

35. Greenfield MS, Doberne L, Kraemer F, Tobey T, Reaven G: Assessment of insulin resistance with the insulin suppression test and the euglycemic clamp. Diabetes 30: 387–392, 1981

36. DeFronzo RA, Tobin JD, Andres R: Glucose clamp technique: A method for quantifying insulin secretion and resistance. Am J Physiol 237: E214–E223, 1979

37. Charron MJ, Kahn BB: Divergent molecular mechanisms for insulin-resistant glucose transport in muscle and adipose cells in vivo. J Biol Chem 265: 7994–8000, 1990

38. Pedersen O, Kahn CR, Kahn BB: Divergent regulation of the GLUT1 and GLUT4 glucose transporters in isolated adipocytes from Zucker rats. J Clin Invest 89: 1964–1973, 1992

39. Ezaki O, Fukuda N, Itakura H: Role of two types of glucose transporters in enlarged adipocytes from aged obese rats. Diabetes 39: 1543–1549, 1990

40. Storlien LH, Jenkins AB, Chisholm DJ, Pascoe WS, Khouri S, Kraegen EW: Influence of dietary fat composition on development of insulin resistance in rats. Relationship to muscle triglyceride and omega-3 fatty acids in muscle phospholipid. Diabetes 40: 280–289, 1991

41. Storlien LH, James DE, Burleigh KM, Chisholm DJ, Kraegen EW: Fat feeding causes widespread in vivo insulin resistance, decreased energy expenditure, and obesity in rats. Am J Physiol 251: E576–E583, 1986

42. Wake SA, Sowden JA, Storlien LH, James DE, PW C, Shine J, Chisholm DJ: Effects of exercise training and dietary manipulation on insulin-regulatable glucose transporter mRNA in rat muscle. Diabetes 40: 275–279, 1991

43. Borkman M, Chisholm DJ, Furier SM, Storlien LH, Kraegen EW, Simons LA, Chesterman CN: Effects of fish oil supplementation on glucose and lipid metabolism in NIDDM. Diabetes 38: 1314–1319, 1989

44. Sinha MK, Raineri-Maidonado C, Buchanan C, Pories WJ, Carter-Su C, Pilch PF, Caro JO: Adipose tissue glucose transporters in NIDDM. Decreased levels of muscle/fat isoform. Diabetes 40: 472–477, 1991

45. Kahn BB, Charron MJ, Lodish HF, Cushman SW, Flier JS: Differential regulation of two glucose transporters in adipose cells from diabetic and insulin-treated diabetic rats. J Clin Invest 84: 404–411, 1989

46. Kahn BB, Rosen AS, Bak JF, Andersen PH, Damsbo P, Lund S, Pedersen O: Expression of GLUT1 and GLUT4 glucose transporters in skeletal muscle of humans with insulin-dependent diabetes mellitus: regulatory effects of metabolic factors. J Clin Endocrinol Metab 74: 1101–1109, 1992

47. Bonadonna RC, Saccomani MP, Seely L, Zych KS, Ferrannini E, Cobelli C, DeFronzo RA: Glucose transport in human skeletal muscle. The in vivo response to insulin. Diabetes 42: 191–198, 1993

48. Garvey WT, Maianu L, Hancock JA, Golichowski AM, Baron A: Gene expression of GLUT4 in skeletal muscle from insulin-resistant patients with obesity, IGT, GDM, and NIDDM. Diabetes 41: 465–475, 1992

49. Kahn BB, Rossetti L, Lodish HF, Charron MJ: Decreased in vivo glucose uptake but normal expression of GLUT1 and GLUT4 in skeletal muscle of diabetic rats. J Clin Invest 87: 2197–2206, 1991

50. Zierath JR, Houseknecht K, Kahn BB: Glucose transporters and diabetes. Sem Cell Dev Biol 7: 295–307, 1996

51. Zierath JR, Houseknecht KL, Gnudi L, Kahn BB: High-fat feeding impairs insulin-stimulated GLUT4 recruitment via an early insulin-signaling defect. Diabetes 46: 215–223, 1997

52. Katz EB, Burcelin R, Tsao TS, Stenbit AE, Charron MJ: The metabolic consequences of altered glucose transporter expression in transgenic mice. J Mol Med 74: 639–652, 1996

53. Tsao TS, Burcelin R, Katz EB, Huang L, Charron MJ: Enhanced insulin action due to targeted GLUT4 overexpression exclusively in muscle. Diabetes 45: 28–36, 1996

54. Katz EB, Stenbit AE, Hatton K, DePinho R, Charron MJ: Cardiac and adipose tissue abnormalities but not diabetes in mice deficient in GLUT4. Nature 377: 151–155, 1995

55. Tsao TS, Stenbit AE, Li J, Houseknecht K, Zierath JR, Katz EB, Charron MJ: Muscle-specific transgenic complementation of GLUT4-deficient mice normalizes glucose but not lipid metabolism. J Clin Invest 100: 671–677, 1997

56. DeFronzo RA: Lilly Lecture 1987: The triumvirate: β-cell, muscle, liver: A collusion responsible for NIDDM. Diabetes 37: 667–687, 1988

57. Shulman GI, Rothman DL, Jue T, Stein P, DeFronzo RA, Shulman RG: Quantitation of muscle glycogen synthesis in normal subjects and subjects with non-insulin-dependent diabetes by 13C nuclear magnetic resonance spectroscopy. N Eng J Med 322: 223–228, 1990

58. Zierath JR, He L, Guma A, Wahlstrom EO, Klip A, Wallberg-Henriksson H: Insulin action on glucose transport and plasma membrane GLUT4 content in skeletal muscle from patients with NIDDM. Diabetologia 39: 1180–1189, 1996

59. Tsao TS, Burcelin R, Charron MJ: Regulation of hexokinase II gene expression by glucose flux in skeletal muscle. J Biol Chem 271: 14959–14963, 1996

60. McCall AL, van Bueren AM, Huang L, Stenbit A, Celnik E, Charron MJ: Forebrain endothelium expresses GLUT4, the insulin- responsive glucose transporter. Brain Res 744: 318–326, 1997

61. Stenbit AE, Burcelin R, Katz EB, Tsao TS, Gautier N, Charron MJ, Le Marchand-Brustel Y: Diverse effects of Glut4 ablation on glucose uptake and glycogen synthesis in red and white skeletal muscle. J Clin Invest 98: 629–634, 1996

62. Heyner S, Smith RM, Schultz GA: Temporally regulated expression of insulin and insulin-like growth factors and their receptors in early mammalian development. BioEssays 11: 171–175, 1989

63. Gardner HG, Kaye PL: Characterization of glucose transport in preimplantation mouse embryos. Repro Fert Dev 7: 41–50, 1995

64. Gardner HG, Kaye PL: Insulin increases cell numbers and morphological development in mouse pre-implantation embryos in vitro. Repro Fert Dev 3: 79–91, 1991

65. Kahn BB: Facilitative glucose transporters: regulatory mechanisms and dysregulation in diabetes. J Clin Invest 89: 1367–1374, 1992

152

66. Hogan A, Heyner S, Charron MJ, Copeland NG, Gilbert DJ, Jenkins NA, Thorens B, Schultz GA: Glucose transporter gene expression in early mouse embryos. Development 113: 363–372, 1991

67. Aghayan M, Rao LV, Smith RM, Jarett L, Charron MJ, Thorens B, Heyner S: Developmental expression and cellular localization of glucose transporter molecules during mouse preimplantation development. Development 115: 305–312, 1992

68. Wiley LM, Lever JE, Pape C, Kidder GM: Antibodies to a renal Na+/giucose cotransport system localize to the apical plasma membrane domain of polar mouse embryo blastomeres. Dev Biol 143: 149–161, 1991

69. Smith DE, Gridley T: Differential screening of a PCR-generated mouse embryo cDNA library: Glucose transporters are differentially expressed in early postimplantation mouse embryos. Development 116: 555–561, 1992

70. Santalucia T, Camps M, Castello A, Munoz P, Nuel A, Testar X, Palacin M, Zorzano A: Developmental regulation of GLUT-1 (erythroid/Hep G2) and GLUT-4 (muscle/fat) glucose transporter expression in rat heart, skeletal muscle, and brown adipose tissue. Endocrinology 130: 837–846, 1992

71. Studelska DR, Campbell C, Pang S, Rodnick KJ, James DE: Developmental expression of insulin-regulatable glucose transporter GLUT-4. Am J Physiol 263: E102–E106, 1992

72. Girard J, Ferre P, Pegorier J-P, Duee P-H: Adaptations of glucose and fatty acid metabolism during perinatal period and suckling-weaning transition. Physiol Rev 72: 507–562, 1992

73. Katz EB, Stenbit AE, Hatton K, DePinho R, Charron MJ: Cardiac and adipose tissue abnormalities but not diabetes in mice deficient in GLUT4. Nature 377: 151–155, 1995

Molecular and Cellular Biochemistry **182**: 153–160, 1998.
© 1998 *Kluwer Academic Publishers. Printed in the Netherlands.*

Insulin action in skeletal muscle from patients with NIDDM

Juleen R. Zierath, Anna Krook and Harriet Wallberg-Henriksson
Department of Clinical Physiology, Karolinska Hospital, Karolinska Institute, Stockholm, Sweden

Abstract

Insulin resistance in peripheral tissues is a common feature of non insulin-dependent diabetes mellitus (NIDDM). The decrease in insulin-mediated peripheral glucose uptake in NIDDM patients can be localized to defects in insulin action on glucose transport in skeletal muscle. Following short term *in vitro* exposure to both submaximal and maximal concentrations of insulin, 3-O-methylglucose transport rates are 40–50% lower in isolated skeletal muscle strips from NIDDM patients when compared to muscle strips from nondiabetic subjects. In addition, we have shown that physiological levels of insulin induce a 1.6–2.0 fold increase in GLUT4 content in skeletal muscle plasma membranes from control subjects, whereas no significant increase was noted in NIDDM skeletal muscle. Impaired insulin-stimulated GLUT4 translocation and glucose transport in NIDDM skeletal muscle is associated with reduced insulin-stimulated IRS-1 tyrosine phosphorylation and PI3-kinase activity. The reduced IRS-1 phosphorylation cannot be attributed to decreased protein expression, since the IRS-1 protein content is similar between NIDDM subjects and controls. Altered glycemia may contribute to decreased insulin-mediated glucose transport in skeletal muscle from NIDDM patients. We have shown that insulin-stimulated glucose transport is normalized *in vitro* in the presence of euglycemia, but not in the presence of hyperglycemia. Thus, the circulating level of glucose may independently regulate insulin stimulated glucose transport in skeletal muscle from NIDDM patients via a down regulation of the insulin signaling cascade. (Mol Cell Biochem **182**: 153–160, 1998)

Key words: NIDDM, skeletal muscle, GLUT4, insulin, IRS-1, PI3-kinase

Pathogenesis of NIDDM

Although the primary defect in the pathogenesis of non-insulin-dependent diabetes mellitus (NIDDM) is unknown, a combination of genetic and environmental factors contribute to the manifestation this progressive metabolic disorder, which is usually not clinically apparent until later in life. Patients with NIDDM are characterized by fasting hyperglycemia, and by elevated, normal or low levels of insulin. Due to the progressive nature of the disease, elucidation of the 'primary defect' of this metabolic disorder has been difficult to establish in the newly diagnosed NIDDM patient [1]. Many of the affected individuals with fasting plasma glucose levels in the range between 6–12 mM are often asymptomatic and undiagnosed [2]. Once fasting plasma glucose levels reach 12 mM or higher, these individuals present various clinical symptoms and are diagnosed with NIDDM. Yet at this point, the disease has progressed

from impaired glucose tolerance to overt NIDDM, and has developed as a consequence of insulin secretory dysfunction [3] and/or insulin resistance [4, 5].

Defects in insulin secretion have been noted in lean or normal weight patients with NIDDM [4], whereas, peripheral insulin resistance has been reported in NIDDM patients regardless of weight [4–6]. However, not all Caucasian NIDDM patients demonstrate peripheral insulin resistance [7–9], suggesting that decreased insulin action in peripheral tissues might not be a primary defect in all patients with NIDDM.

A prospective study performed in Pima Indians has revealed that insulin resistance is a major risk factor for the development of NIDDM in this population [1]. Furthermore, neither increased glucose output from the liver during insulin infusion nor reduced insulin secretion from the β-cell in response to glucose, was shown to predict for NIDDM in Pima Indians [1]. In glucose tolerant first degree relatives to NIDDM patients, reduced peripheral insulin sensitivity has

Address for offprints: Harriet Wallberg-Henriksson, Department of Clinical Physiology, Karolinska Hospital, S-171 76, Stockholm, Sweden

been demonstrated [10–13]. However, differences in lean body mass or in the degree of physical fitness (VO_{2max}) contribute to reduced insulin sensitivity [14]. In future studies it will be important to match the first degree relatives to NIDDM patients and control subjects by degree of physical fitness (VO_{2max}) to determine whether the differences in insulin sensitivity would be less apparent. In many patients, NIDDM is likely to evolve from a combination of genetic and environmental factors [2], however in some cases, environmental factors such as inactivity, obesity, and stress are more important for the manifestation of the disease. When these factors are abolished, insulin sensitivity improves and the disease may regress [15].

Whole body insulin-resistance in NIDDM patients

In vivo studies reveal that in the postprandial state, insulin-mediated glucose uptake and utilization is markedly impaired in patients with NIDDM [16–18] and glucose tolerant first-degree relatives [13]. Skeletal muscle is the principal site of glucose uptake under insulin-stimulated conditions, accounting for approximately 75% of glucose disposal following glucose infusion [16, 19, 20]. Under euglycemic-hyperinsulinemic conditions, skeletal muscle has been identified as the most important site for insulin resistance in NIDDM patients [16]. From *in vivo* tracer studies, the defect in whole body glucose uptake observed in the NIDDM patients has been localized to the non-oxidative pathway for glucose metabolism [18, 19, 21]. This defect may result from a defect at the level of glucose transport, glucose phosphorylation or glycogen synthase.

Hyperglycemia *per se* has been reported to compensate for the reduced whole body insulin-mediated glucose uptake in NIDDM patients. When lean NIDDM patients and healthy individuals, matched for age, BMI, and VO_{2max}, are studied by the hyperinsulinemic clamp procedure under fasting ambient blood glucose concentrations (isoglycemic-hyperinsulinemic clamp), similar rates of insulin-mediated glucose utilization are observed [5] (Fig. 1). Under these conditions, hyperglycemia *per se* can fully compensate for defect(s) in fasting and insulin-stimulated rates of glucose oxidation, glucose storage and muscle glycogen synthase activation in NIDDM patients [22, 23]. In contrast, if the NIDDM patients and the controls are studied at the same glucose concentration, the NIDDM patients demonstrate marked whole body insulin resistance [24] (Fig. 1).

Glucose transport in skeletal muscle from NIDDM patients

We have utilized an *in vitro* muscle preparation to determine whether the reduced whole body insulin-mediated glucose

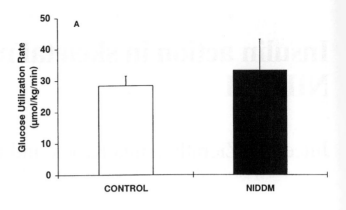

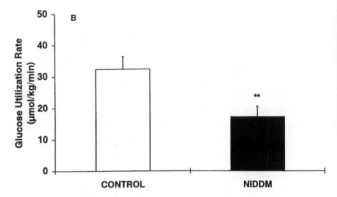

Fig. 1. Rates of whole body insulin-mediated glucose uptake in healthy individuals and NIDDM subjects assessed by the hyperinsulinemic clamp technique. (A) The subjects were clamped at their ambient glucose concentration; 5.6 ± 0.2 mM for the controls (n = 7) and 10.3 ± 0.9 mM for the NIDDM subjects (n = 7) (p < 0.001). Steady state insulin levels did not differ between the groups (588 ± 42 vs. 666 ± 72 pM for the controls and the NIDDM subjects, respectively, NS). (B) The subjects were clamped at the same blood glucose concentration; 5.5 ± 0.2 mM for the controls (n = 7) and 5.3 ± 0.9 mM for the NIDDM subjects (n = 9) (NS). Steady state insulin levels did not differ between the groups (401 ± 16 vs. 451 ± 44 pM for the controls and the NIDDM subjects, respectively, NS). Figure A is reprinted from Ref. [5] with permission from the journal.

uptake in NIDDM patients results from a defect at the level of glucose transport. Skeletal muscle biopsies were obtained from NIDDM patients and controls, and smaller muscle strips were incubated in the presence of increasing concentrations of insulin [24]. We found that the dose response curve for insulin-stimulated 3-O-methylglucose transport was markedly decreased in muscle specimens from the NIDDM patients compared to controls (Fig. 2). These results provided evidence to suggest that a post receptor defect may account for the reduced insulin-stimulated glucose transport in NIDDM muscle.

Recently *in vivo* rates of 3-O-methylglucose transport have been assessed in skeletal muscle from NIDDM patients and controls using the euglycemic clamp technique in conjunction with tracer techniques [26]. It is interesting to note that regardless of technique (*in vitro* vs. *in vivo*), insulin increased

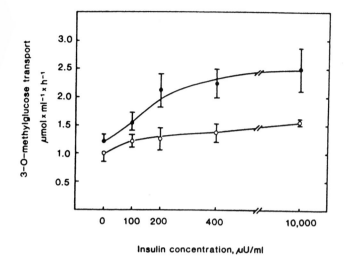

Fig. 2. Insulin-stimulated 3-O-methylglucose transport *in vitro* incubated muscle strips from controls and NIDDM patients. The rate of basal and insulin-stimulated 3-O-methylglucose transport was assessed in rectus abdominal skeletal muscle specimens obtained from 8 healthy controls (age 61 ± 6 years, BMI 24 ± 1) and 6 NIDDM patients (age 73 ± 3 years, BMI 26 ± 1). The NIDDM patients had significantly elevated fasting levels of plasma glucose (7.3 ± 0.6 vs. 4.8 ± 0.2 mM) with no significant alteration in the level of serum insulin (11.5 ± 3.6 vs. 8.6 ± 1.5 mU/l) compared to the controls. 3-O-methylglucose is expressed per ml of intracellular water. Reprinted from Ref. [25] with permission from the journal.

3-O-methylglucose 2–3 fold in the skeletal muscle from the controls, whereas, insulin had minimal effects on 3-O-methylglucose transport in skeletal muscle from the NIDDM patients. Furthermore, using either techniques, the rate of insulin-stimulated 3-O-methylglucose transport in skeletal muscle is reduced by ~50% in NIDDM patients [24, 25, 27]. Thus, the insulin signal transduction pathway to glucose transport appears to be impaired in NIDDM patients. Consequently, defects in the regulatory steps involved in glucose transport play a major role in determining peripheral insulin resistance in NIDDM. Whether this marked decrease in insulin-stimulated glucose transport in skeletal muscle is a permanent defect, or an acute down-regulation secondary to the diabetic state, remains to be elucidated.

Insulin signaling in NIDDM skeletal muscle

Intense interest has focused on whether the reduced insulin-mediated glucose transport in muscle from NIDDM patients results from alterations in the insulin signal transduction pathway or from alterations in the traffic and/or translocation of GLUT4 to the plasma membrane. Evidence from animal studies suggests that insulin signaling defects in muscle are associated with altered whole body glucose homeostasis [28–31]. In morbidly obese humans [32] and obese rodents [28–31], impaired insulin-stimulated glucose transport in skeletal

muscle is associated with decreased IR and IRS-1 protein content, decreased tyrosine phosphorylation of IR and IRS-1, and reduced PI3-kinase activity. TNF-α has been implicated to play a role in development of insulin resistance in obesity by increasing the serine phosphorylation of IRS-1, and thereby converting IRS-1 to an inhibitor of IR through increased serine phosphorylation of the IR [33]. In hyper-glycemic, hypoinsulinemic states, such as in the streptozo-tocin-induced diabetic rat, IR and IRS-1 phosphorylation, and PI3-kinase activity is enhanced in skeletal muscle [28, 29], despite reduced insulin-stimulated glucose transport [34]. Thus, changes in early insulin signaling elements may have a profound effect on whole body glucose metabolism.

The tyrosine kinase activity of the insulin receptor has been reported to be reduced in obese individuals, NIDDM subjects, as well as lean normoglycemic insulin resistant subjects [35–37]. We have recently assessed the effect of physiological hyperinsulinemia on IRS-1 tyrosine phosphorylation and PI3-kinase activity in skeletal muscle from six lean to moderately obese NIDDM and six healthy subjects [38]. We showed that a rise in serum insulin levels from ~60 to ~650 pmol/l was sufficient to increase IRS-1 tyrosine phosphory-lation 6 fold over basal levels in control muscle (p < 0.01), whereas no significant phosphorylation of IRS-1 was noted in NIDDM muscle (Fig. 3).

The reduced IRS-1 phosphorylation in NIDDM muscle did not appear to be related to changes in IRS-1 protein content, since IRS-1 protein expression was similar between control and NIDDM subjects (Fig. 4). Several amino acid poly-morphisms in the IRS-1 gene have been identified, and found to be present in both NIDDM subjects and controls. The phenotype of NIDDM subjects with IRS-1 variants does not seem to differ significantly from NIDDM patients in general [39–42]. Thus the overall significance of these variants, and possible contribution to the observed pathogenesis, remains to be determined.

Activation of the enzyme PI3-kinase appears to be a key step in insulin stimulation of glucose transport [43]. Thus, we investigated whether physiological hyperinsulinemia was sufficient to increase PI3-kinase activity in skeletal muscle *in vivo*. In muscle from control subjects, insulin induced a 2 fold increase in anti-phosphotyrosine immunoprecipable PI3-kinase activity (p < 0.01). Similar results were noted for IRS-1 associated PI3-kinase activity. In contrast, no signifi-cant increase in insulin-stimulated PI3-kinase activity was noted in the NIDDM muscle (Fig. 5). Furthermore, *in vitro* insulin-stimulated (600 pmol/l) 3-O-methylglucose transport was 40% lower in isolated muscle from these NIDDM subjects (p < 0.05). Thus, these findings couple both reduced insulin-stimulated IRS-1 tyrosine phosphorylation and PI3-kinase activity to the impaired insulin-stimulated glucose transport in skeletal muscle from lean to moderately obese NIDDM subjects.

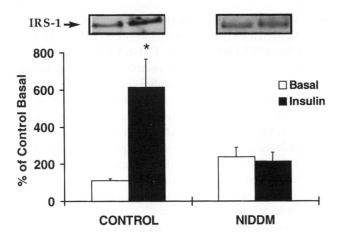

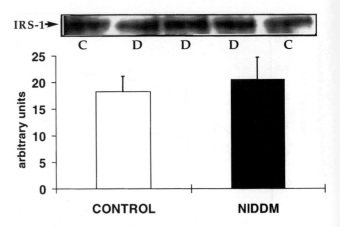

Fig. 3. *In vivo* stimulation of IRS-1 in skeletal muscle from control and NIDDM subjects. Muscle biopsies were obtained under fasting conditions (basal; open bar) and following a 40 min insulin infusion to raise plasma levels to 638 ± 44 or 729 ± 90 pmol/l (insulin-stimulated; closed bar) for control or NIDDM, respectively. Top: representative autoradiogram of tyrosine phosphorylated IRS-1 from basal or insulin-stimulated muscle. Bottom: tyrosine phosphorylated IRS-1 in control (basal n = 3 and insulin n = 6) and NIDDM (basal n = 4 and insulin n = 6) muscle. Results are expressed as mean ± S.E.M. Tyrosine phosphorylated IRS-1 was determined in IRS-1 immunoprecipitates. *p < 0.01 significantly different from control basal. Reprinted from Ref. [38] with permission from the journal.

Fig. 4. IRS-1 protein expression in skeletal muscle from controls and NIDDM subjects. Top: demonstrates a representative autoradiogram of immunoreactive IRS-1 in basal muscle (C = Control and D = NIDDM). Bottom: immunoreactive IRS-1 content in skeletal muscle from control (n = 11) and NIDDM (n = 11) subjects (age 54 ± 2 vs. 55 ± 2 years; BMI 27 ± 1 vs. 28 ± 1 kg /m^2 for the controls and the NIDDM patients, respectively, NS). The NIDDM patients demonstrated fasting hyperglycemia; 10.7 ± 0.7 mM vs. 5.6 ± 0.2 mM for the controls (p < 0.001). The fasting plasma insulin levels did not differ between the groups (45 ± 5 vs. 76 ± 14 pmol/l for the controls and the NIDDM subjects, respectively, NS).

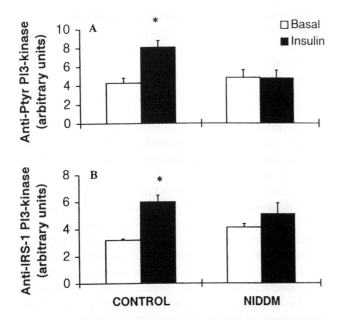

Fig. 5. Effect of physiological hyperinsulinemia on PI3-kinase in skeletal muscle from control and NIDDM subjects. Basal and insulin-stimulated muscle was obtained as described in Fig. 3 and PI3-kinase was measured in immunoprecipitates obtained with antibodies to phosphotyrosine (A) or to IRS-1 (B). Values are presented as phosphoimager units determined from quantification of the phosphoimage of the ^{32}P PI3 phosphate products. Results are expressed as mean ± S.E.M. for control and NIDDM subjects as indicated in Fig. 3. *p < 0.0.01 significantly different from control basal. Reprinted from Ref. [38] with permission from the journal.

GLUT4 in skeletal muscle

Since glucose transport is one of the first rate limiting steps in glucose metabolism which appears to be down-regulated in skeletal muscle from patients with NIDDM [24–27], GLUT4 has been an appropriate candidate gene for studies aimed to elucidate the molecular mechanism(s) involved in peripheral insulin resistance. The GLUT4 gene has been assessed for mutations in several populations and based upon these studies [44–46], genetic mutations in the GLUT4 gene are unlikely to contribute to the development of insulin-resistance observed in Caucasian NIDDM patients. In adipose tissue, GLUT4 mRNA and protein levels (total and plasma membrane) are decreased in patients with NIDDM [47]. However, in skeletal muscle, a tissue which is responsible for the majority of glucose uptake under insulin-mediated conditions [16], neither GLUT4 mRNA [48] nor total GLUT4 protein is altered in skeletal muscle biopsies from lean NIDDM patients [48–51] or from first degree relatives to NIDDM patients [52]. Thus, in skeletal muscle from NIDDM patients, the decrease in insulin-stimulated glucose transport cannot be accounted for by a decrease in the protein expression of GLUT4.

Recently, we have shown that an increase in fasting serum insulin levels from ~50 to ~600 pmol/l, induces a translocation of GLUT4 from an intracellular storage site to the plasma membrane in skeletal muscle from healthy individuals [5, 53, 54]. Conversely, a similar increase in insulin levels did not alter the plasma membrane GLUT4 content in muscle biopsies from NIDDM patients [5]. The latter observation provides evidence for a defect in insulin signaling and/or GLUT4 translocation in muscle from NIDDM patients.

Hexokinase II and glycogen synthesis in skeletal muscle from NIDDM patients

Once transported across the plasma membrane, glucose is phosphorylated by hexokinase to glucose-6-phosphate. Hexokinase II is the predominant hexokinase isoform expressed in insulin sensitive tissues such as skeletal muscle and fat [55]. Since glucose phosphorylation has been suggested to be one of the potential sites of skeletal muscle insulin resistance [56], hexokinase II has been considered to be a plausible candidate gene in the development of NIDDM and insulin resistance. With the cloning and characterization of the human hexokinase II gene [57], several groups have concluded that mutations in this gene do not contribute significantly to the development of insulin resistance of NIDDM [58–60].

The rate of skeletal muscle glycogen synthesis closely parallels the rate of non-oxidative glucose disposal when assessed under conditions of a hyperinsulinemic–euglycemic clamp [61] or conditions of a hyperinsulinemic–hyperglycemic clamp [18]. Glycogen synthase activity has been reported to be lower in muscle biopsies from NIDDM patients [18, 61–63] and young non-obese first-degree relatives of patients with NIDDM [11] compared to controls. Thus, a defect at the level of glycogen synthase has been suggested to contribute to the reduced non-oxidative glucose metabolism observed in skeletal muscle from NIDDM patients. The coding sequence of the glycogen synthase gene has been examined in Danish NIDDM subjects and no genetic defects reported [44]. Furthermore, no genetic abnormalities were identified in the genes encoding the upstream regulators glycogen synthase phosphatase/protein phosphatase 1 or the insulin stimulated protein kinase [64]. Despite the decreased insulin-stimulated glycogen synthase activity in skeletal muscle from patients with NIDDM, fasting levels of skeletal muscle glycogen are comparable between lean NIDDM patients and controls [11, 24]. Consequently, the decreased glycogen synthase activity observed in skeletal muscle from NIDDM patients under insulin-stimulated conditions may be a compensatory down-regulatory response to hyperglycemia, rather than a primary defect. The down-regulation of insulin-stimulated glycogen synthase in the NIDDM skeletal muscle may result from an accumulation of free intra-cellular glucose or may be secondary to a down regulation of the insulin-stimulated glucose transport.

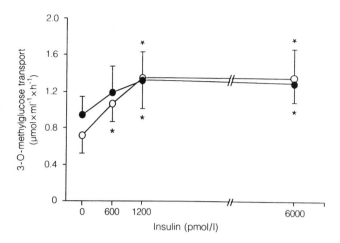

Fig. 6. Dose-response relationship for insulin-stimulated 3-O-methylglucose transport in skeletal muscle strips incubated at glucose concentrations near fasting plasma levels for the appropriate groups. Muscle from control subjects (open circles; n = 6) and NIDDM patients (closed circles n = 6) was incubated (2 h) in the presence of 5 mM (control subjects) or 8 mM (NIDDM) glucose and a further addition of insulin. Thereafter, an equimolar concentration of 3-O-methylglucose was substituted for glucose, and 3-O-methylglucose transport was assessed. Values are presented as mean ± S.E.M. *p < 0.05; †p < 0.01 significantly different from the corresponding basal. Reprinted from Ref. [24] with permission from the journal.

Hyperglycemia and skeletal muscle glucose transport in NIDDM patients

When isolated skeletal muscle strips from NIDDM patients and healthy controls were incubated at the same level of glycemia (Fig. 2), the insulin-dose response curve for 3-O-methylglucose transport was markedly impaired in the NIDDM muscle compared to the controls [25]. However, when the muscle strips from the NIDDM patients and healthy controls were incubated in the presence of near fasting ambient glucose concentrations, the non-insulin-mediated components of glucose transport compensated for the decrease in the insulin-stimulated components of glucose transport (Fig. 6). These results suggest that the decreased capacity for insulin-stimulated glucose transport in isolated skeletal muscle from patients with NIDDM may be fully compensated by the mass effect of glucose on glucose transport [24].

We have recently reported that experimentally induced euglycemia can restore insulin-stimulated glucose transport in isolated skeletal muscle from NIDDM patients [24]. When skeletal muscle strips from the NIDDM patients were incubated for 2 h in KHB buffer containing 4 mM glucose, insulin-stimulated glucose transport was restored to control levels (Fig. 7). Conversely, insulin resistance was maintained in skeletal muscle strips from NIDDM patients following a 2 h *in vitro* exposure to KHB containing 8 mM glucose. These results imply that chronic hyperglycemia

158

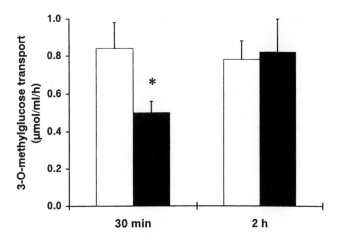

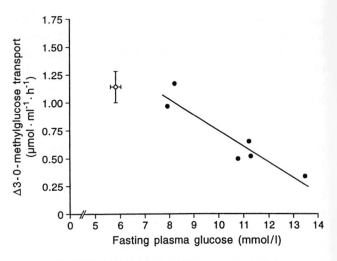

Fig. 7. Normalization of the decreased capacity for insulin-stimulated 3-O-methylglucose transport *in vitro*. Skeletal muscle strips from healthy controls (open bars n = 8) and NIDDM patients (closed bars n = 7) subjects were exposed (30 min or 2 h) to incubation media containing 100 mU/l insulin and 4 mM glucose. Thereafter, an equimolar concentration of 3-O-methylglucose was substituted for glucose, and 3-O-methylglucose transport was assessed as previously described [26]. Values are presented as mean ± S.E.M. *p < 0.05 significantly different from the control skeletal muscle. Reprinted from Ref. [24] with permission from the journal.

Fig. 8. Relationship between insulin-stimulated 3-O-methylglucose transport and the fasting plasma glucose. Skeletal muscle from patients with NIDDM (BMI 25.5 ± 0.5 kg/m², age 50 ± 2 years) and healthy control (BMI 26.1 ± 0.5 kg/m²; age 52 ± 2 years) were incubated in the presence of 5 mM glucose and insulin (100 mU/l) for 30 min. Thereafter, 3-O-methylglucose transport was assessed as described [24]. Fasting plasma glucose was determined by the glucose dehydrogenase method using a glucose reagent kit (Merck, Darmstadt, Germany). Individual values are presented for the NIDDM patients (closed circles) r = 0.93, p < 0.001 (equation $y = -0.14x + 2.15$). Mean values are presented for the control subjects (open circle). Reprinted from Ref. [24] with permission from the journal.

may directly contribute to the reduced insulin signaling and GLUT4 translocation in muscle from NIDDM patients.

Our *in vitro* studies reveal that in patients with NIDDM, the fasting plasma glucose level is negatively correlated with insulin-stimulated glucose transport (Fig. 8). This is consistent with an *in vivo* study demonstrating an inverse relationship between the insulin-stimulated metabolic clearance rate of glucose and fasting blood glucose in a group of 37 NIDDM patients [1]. Despite the relatively small group of NIDDM patients (Fig. 8), we noted that the insulin-stimulated increase in glucose transport demonstrated a highly significant, negative correlation with the fasting plasma glucose level. The patients with a near normal blood glucose level demonstrated an intact insulin response on cellular glucose transport. Conversely, the patients with a higher blood glucose level (>10 mM) demonstrated a markedly decreased cellular response of glucose transport to insulin. Taken together, these two studies suggest that the level of glycemia may play a role in determining insulin sensitivity.

Conclusion

Recently a 'compensation theory' has been proposed [65] which suggests that in the NIDDM patient, hyperglycemia and hyperinsulinemia are necessary to compensate for the cellular defect in insulin action in peripheral tissues. Thus, in order to overcome the insulin resistance at the level of

skeletal muscle, the non-insulin-mediated component of glucose transport may be 'up regulated' by the elevated level of extracellular glucose [65]. This hypothesis assumes that in NIDDM, a primary defect is located at the level of skeletal muscle and that hyperglycemia can compensate for this defect. An alternative theory considers the possibility that the non-insulin and insulin-mediated pathways for glucose transport work in concert to maintain an appropriate flux of glucose across the plasma membrane. With this hypothesis, the assumption is made that the primary defect in NIDDM is not located to skeletal muscle. Rather, the systems of glucose transport in skeletal muscle (non-insulin-mediated and insulin-stimulated), adapt to the prevailing glucose concentration. Thus, in the presence of elevated levels of glucose, the insulin-mediated pathway is down-regulated and the non-insulin-mediated pathway predominates. When euglycemia is restored, the insulin-mediated pathway is restored.

Acknowledgment

This study was supported by grants from the Swedish Medical Research Council (9517, and 12211, 11823), the Foundation for Strategic research (SSF), the Swedish Diabetes Association, NOVO Nordisk Insulin Foundation, Thuring's Foundation, and the Karolinska Institute Foundation.

References

1. Lillioja S, Mott DM, Spraul M, Ferraro R, Foley JE, Ravussin E, Knowler WC, Bennett PH, Bogardus C: Insulin resistance and insulin secretory dysfunction as precursors of non-insulin-dependent diabetes mellitus. N Engl J Med 329: 1988–1992, 1993

2. Turner RC, Hattersley AT, Shaw JTE, Levy JC: Type II diabetes: clinical aspects of molecular biological studies. Diabetes 44: 1–10, 1995

3. Porte Jr D: β-cells in Type II diabetes mellitus. Diabetes 40: 166–180, 1991

4. DeFronzo RA: The triumvirate: β-cell, muscle or liver. A collusion responsible for NIDDM. Diabetes 37: 667–687, 1988

5. Zierath JR, He L, Gumá A, Odegaard-Wahlström E, Klip A, Wallberg-Henriksson H: Insulin action on glucose transport and plasma membrane content in skeletal muscle from patients with NIDDM. Diabetologia 39:1180–1189, 1996

6. Kelley DE, Mokan M, Mandarino LJ: Metabolic pathways of glucose in skeletal muscle of lean NIDDM patients. Diabetes Care 16: 1158–1166, 1993

7. Efendic S, Östenson CG: Hormonal responses and future treatment of non-insulin-dependent diabetes mellitus (NIDDM). J Internal Med 234: 127–138, 1993

8. Efendic S, Kahn A, Östenson CG: Insulin release in Type 2 diabetes mellitus. Diabete et Metabolisme 20: 81–86, 1994

9. Galuska D, Nolte LA, Zierath JR, Wallberg-Henriksson H: Effect of metformin on insulin-stimulated glucose transport in isolated muscle obtained from patients with NIDDM. Diabetologia 37: 826–832, 1994

10. Laws A, Stefanick ML, Reaven, GM: Insulin resistance and hypertriglygeridemia in nondiabetic relatives of patients with noninsulin-dependent diabetes mellitus. J Clin Endocrinol Metab 69: 343–347, 1989

11. Vaag A, Henriksen JE, Beck-Nielsen H: Decreased insulin activation of glycogen synthase in skeletal muscles in young nonobese Caucasian first-degree relatives of patients with non-insulin-dependent diabetes mellitus. J Clin Invest 89: 782–788, 1992

12. Eriksson J, Koranyi L, Bourey R, Schalin C, Widén E, Mueckler M, Permutt AM, Groop LC: Insulin resistance in type 2 (non-insulin-dependent) diabetic patients and their relatives is not associated with a defect in the expression of the insulin-responsive glucose transporter (GLUT4) gene in human skeletal muscle. Diabetologia 35: 143–147, 1992

13. Martin BC, ·Warram JH, Krolewski AS, Bergman RN, Soeldner JS, Kahn CR: Role of glucose and insulin resistance in development of type 2 diabetes mellitus: results of a 25-year follow-up study. Lancet 340: 925–929, 1992

14. Seals DR, Hagberg JM, Allen WK, Hurley BF, Dalsky GP, Ehsani AA, Holloszy JO: Glucose tolerance in young and older athletes and sedentary men. J Appl Physiol 56: 1521–1525, 1984

15. Friedman JE, Dohm GL, Leggett-Frazier, N, Elton CW, Tapscott EB, Pories WP, Caro JF: Restoration of insulin responsiveness in skeletal muscle of morbidly obese patients after weight loss. J Clin Invest 89: 701–705, 1992

16. DeFronzo RA, Gunnarsson R, Björkman O, Olsson M., Wahren J: Effects of insulin on peripheral and splanchnic glucose metabolism in non-insulin-dependent (Type II) diabetes mellitus. J Clin Invest 76: 149–155, 1985

17. Eriksson E, Franssila-Kallunki A, Ekstrand A et al.: Early metabolic defects in persons at increased risk for non-insulin-dependent diabetes mellitus. N Eng J Med 321: 337–343, 1989

18. Shulman GI, Rothman DL, Jue T, Stein P, DeFronzo RA, Shulman RG: Quantitation of muscle glycogen synthase in normal subjects and subjects with non-insulin-dependent diabetes mellitus by ^{13}C nuclear magnetic resonance spectroscopy. N Engl J Med 322: 223–228, 1990

19. DeFronzo RA, Jacot E, Jequier E, Maeder E, Wahren J, Felber JP: The effect of insulin on the disposal of intravenous glucose. Results from indirect calorimetry and hepatic and femoral venous catheterization. Diabetes 30: 1000–1007, 1981

20. Nuutila P, Koivisto VA, Knuuti J, Ruotsalainen U, Teriis M, Haaparanta M, Bergman J, Solin O, Viopio-Pulkki L-M, Wegelius U, Yki-Jaryinen H: Glucose free fatty acid cycle operates in human heart and skeletal muscle in vivo. J Clin Invest 89: 1767–1774, 1992

21. Thiebaud D, Jacot E, DeFronzo RA, Maeder E, Jequier E, Felber JP: The effect of graded doses of insulin on total glucose uptake, glucose oxidation and glucose storage in man. Diabetes 31: 957–963, 1982

22. Kelley DE, Mandarino LJ: Hyperglycemia normalizes insulin-stimulated skeletal muscle glucose oxidation and storage in non-insulin dependent diabetes mellitus. J Clin Invest 86: 1999–2007, 1990

23. Vaag A, Damsbo P, Hother-Nielsen O, Beck-Nielsen H: Hyperglycaemia compensates for the defects in insulin-mediated glucose metabolism and in the activation of glycogen synthase in the skeletal muscle of patients with Type 2 (non-insulin-dependent) diabetes mellitus. Diabetologia 35: 80–88, 1992

24. Zierath JR, Galuska D, Nolte LA, Thörne A, Smedegaard Kristensen J, Wallberg-Henriksson H: Effects of glycemia on glucose transport in isolated skeletal muscle from patients with NIDDM: in vitro reversal of muscular insulin resistance. Diabetoliga 37: 270–277, 1994

25. Andréasson K, Galuska D, Thörne A, Sonnenfeld T, Wallberg-Henriksson H: Decreased insulin-stimulated 3-O-methylglucose transport in in vitro incubated muscle strips from type II diabetic subjects. Acta Physiol Scand 142: 255–260, 1991

26. Bonadonna RC, Del Prato S, Saccomani MP, Bonora E, Gulli G, Ferrannini E, Bier D, Cobelli C, DeFronzo RA: Transmembrane glucose transport in skeletal muscle of patients with non-insulin-dependent diabetes. J Clin Invest 92: 486–494, 1993

27. Zierath JR: In vitro studies of human skeletal muscle: Hormonal and metabolic regulation of glucose transport. Acta Physiol Scand 155 (Suppl. 626): 1–96, 1995

28. Saad MJA, Araki E, Miralpeix M, Rothenberg PL, White MF, Kahn CR: Regulation of insulin receptor substrate-1 in liver and muscle of animal models of insulin resistance. J Clin Invest 90: 1839–1849, 1992

29. Folli F, Saad MJA, Backer JM, Kahn CR: Regulation of phosphatidylinositol 3-kinase activity in liver and muscle of animal models of insulin-resistant and insulin deficient diabetes mellitus. J Clin Invest 92: 1787–1794, 1993

30. Saad MJA, Folli F, Kahn JA, Kahn CR: Modulation of insulin receptor, insulin receptor substrate-1, and phosphatidylinositol 3-kinase in liver and muscle of dexamethasone-treated rats. J Clin Invest 92: 2065–2072, 1993

31. Heydrick SJ, Jullien D, Gautier N, Tanti JF, Giorgetti S, Van Obberghen E, Marchand-Brustel Y: Defect in skeletal muscle phosphatidylinositol-3-kinase in obese insulin-resistant mice. J Clin Invest 91: 1358–1366, 1993

32. Goodyear LJ, Giorgino F, Sherman LA, Carey J, Smith RJ, Dohm GL: Insulin receptor phosphorylation, insulin receptor substrate-1 phosphorylation and phosphatidylinositol 3-kinase activity are decreased in intact skeletal muscle strips from obese subjects. J Clin Invest 95: 2195–2204, 1995

33. Hotamisligil GS, Peraldi P, Budavari A, Ellis R, White MF, Spegelman BM: IRS-1 mediated inhibition of insulin receptor tyrosine kinase activity in TNF-α-and Obesity-induced insulin resistance. Science: 271: 665–668, 1996

34. Wallberg-Henriksson H, Zetan N, Henriksson J: Reversibility of decreased insulin-stimulated glucose transport capacity in diabetic muscle with in vitro incubation: insulin is not required. J Biol Chem 262: 7665–7671, 1987

160

35. Nolan JJ, Freidenberg G, Henery R, Reichart D, Olefsky JM: Role of human skeletal muscle insulin receptor kinase in the *in vivo* insulin resistance of noninsulin-dependent diabetes mellitus and obesity. J Clin Endocrinol Metab 78: 471–477, 1994

36. Grasso G, Frittitta L, Anello M, Russo P, Sesti O, Trsichitta V: Insulin receptor tyrosine kinase activity is altered in both muscle and adipose tissue from non-obese normoglycaemic insulin-resistant subjects. Diabetologia 38: 55–61, 1995

37. Maegawa H, Shigeta Y, Egawa K, Kobayashi M: Impaired autophosphorylation of insulin receptors from abdominal skeletal muscles in nonobese subjects with NIDDM. Diabetes 40: 815–819, 1991

38. Björnholm M, Kawano Y, Lehtihet M, Zierath JR: Insulin receptor substrate-1 phosphorylation and phosphatidylinositol 3-kinase activity in skeletal muscle from NIDDM subjects after *in vivo* insulin stimulation. Diabetes 406: 524–527, 1997

39. Hitman GA, Hawrami K, McCarthy MI, Viswanathan M, Snehalatha C, Ramachandran A, Tuomilehto J, TuomilehtoWolf E, Nisinen A, Pedersen O: Insulin receptor substrate-1 gene mutations in NIDDM; implications for the study of polygenic disease. Diabetologia 38: 481–486, 1995

40. Imai Y, Fusco A, Suzuki Y, Lesniak MA, DAlfonso R, Sesti G, Bertoli A, Lauro R, Accili D, Taylor SI: Variant sequences of insulin receptor substrate-1 in patients with non-insulin-dependent diabetes mellitus. J Clin Endocrinol Metab 79: 1655–1658, 1994

41. Almind K, Bjørbæk C, Vestergaard H, Hansen T, Echwald S, Pedersen: Amino acid polymorphisms of insulin receptor substrate-1 in non-insulin-dependent diabetes mellitus. The Lancet 342: 828–832, 1993

42. Laakso M, Malkki M, Kekalainen P, Kuusisto J, Deeb SS: Insulin receptor substrate-1 variants in non-insulin-dependent diabetes. J Clin Invest 94:1141–1146, 1994

43. Shepherd PR, Nave BT, Rincon J, Nolte LA, Bevan AP, Siddle K, Zierath JR, Wallberg-Henriksson H: Differential regulation of phosphoinositide 3-kinase adapter subunit variants by insulin in human skeletal muscle. J Biol Chem 272: 19000–19007, 1997

44. Bjørbæk C, Echwald SM, Hubricht P, Vestergaard H, Hansen T, Zierath J, Pedersen O: Genetic variants in promoters and coding regions of the muscle glycogen synthase and the insulin-responsive GLUT4 genes in NIDDM. Diabetes 43: 976–983, 1994

45. Choi W-H, O'Rahilly S, Buse JB, Rees A, Morgan R, Flier JS, Moller DE: Molecular scanning of insulin-responsive glucose transporter (GLUT4) gene in NIDDM subjects. Diabetes 40: 1712–1718, 1991

46. O'Rahilly S, Krook A. Morgan R, Rees A, Flier JS, Moller DE: Insulin receptor and insulin-responsive glucose transporter (GLUT4) mutations and polymorphisms in a Welsh type 2 (non-insulin-dependent) diabetic population. Diabetologia 35: 486–489, 1992

47. Garvey WT, Maianu L, Hueeksteadt TP, Bimbaum MJ, Molina JM, Ciaraldi TP: Pretranslational suppression of a glucose transporter protein causes insulin resistance in adipocytes from patients with non-insulin-dependent diabetes mellitus and obesity. J Clin Invest 87: 1072–1081, 1991

48. Pedersen O, Bak JF, Andersen PH, Lund S, Moller DE, Flier JS, Kahn, BB: Evidence against altered expression of GLUT1 or GLUT4 in skeletal muscle of patients with obesity or NIDDM. Diabetes 39: 865–870, 1990

49. Handberg A, Vaag A, Damsbo P, Beck-Nielsen H, Vinten J: Expression of insulin regulatable glucose transporters in skeletal muscle from Type 2 (non-insulin dependent) diabetic patients. Diabetologia 33: 625–627, 1990

50. Garvey WT, Maianu L, Hancock JA, Golichowski AM, Baron A: Gene expression of GLUT4 in skeletal muscle from insulin-resistant patients with obesity, IGT, GDM, and NIDDM. Diabetes 41: 465–475, 1992

51. Andersen PH, Lund S, Vestergaard H, Junker S, Kahn BB, Pedersen O: Expression of the major insulin regulatable glucose transporter (GLUT4) in skeletal muscle of noninsulin-dependent diabetic patients and healthy subjects before and after insulin infusion. J Clin Endocrinol Metab 77: 27–32, 1993

52. Schalin-Jäntti C, Yki-Järvinen H, Koranyi L, Bourey R, Lindström J, Nikula-Ijäs P, Franssila-Kallunki A, Groop LC: Effect of insulin on GLUT4 mRNA and protein concentrations in skeletal muscle of patients with NIDDM and their first-degree relatives. Diabetologia 37: 401–407, 1994

53. Gumá A, Zierath JR, Wallberg-Henriksson H, Klip A: Insulin induces translocation of GLUT-4 glucose transporters in human skeletal muscle. Am J Physiol 268: E613–E622, 1995

54. Lund S, Holman GD, Zierath JR, Rincon J, Nolte LA, Clark AE, Pedersen O, Wallberg-Henriksson H: Effect of insulin on GLUT4 translocation and turnover rate in human skeletal muscle as measured by the exofacial bismannose photolabeling technique. Diabetes 46: 1965–1969, 1997

55. Postic C, Leturque A, Printz RL, Maulard P, Loizeau M, Granner DK, Girard J: Development and regulation of glucose transporter and hexokinase expression in rat. Am J Physiol 266: E548–E559, 1994

56. Rothman DL, Schulman RG, Schulman GI: ^{31}P Nuclear magnetic resonance measurements of muscle glucose-6-phosphate. Evidence for reduced insulin-dependent muscle glucose transport or phosphorylation activity in non-insulin-dependent diabetes mellitus. J Clin Invest 89: 1062–1075, 1992

57. Printz RL, Ardehali H, Koch S, Granner DK: Human hexokinase II mRNA and gene structure. Diabetes 44: 290–294, 1995

58. Echwald SM, Bjørbæk C, Hansen T, Clausen JO, Vestergaard H, Zierath JR, Printz RL, Granner DK, Pedersen O: Identification of four amino acid substitutions in hexokinase II and studies of relationships to NIDDM, glucose effectiveness, and insulin sensitivity. Diabetes 44: 347–353, 1995

59. Laakso M, Malkki M, Deeb SS: Amino acid substitutions in hexokinase II among patients with NIDDM. Diabetes 44: 330–334, 1995

60. Vidal-Puig A, Printz RL, Stratton IM, Granner DK, Moller DE: Analysis of the hexokinase II gene in subjects with insulin resistance and NIDDM and detection of a Gln^{142}→His substitution. Diabetes 44: 340–346, 1995

61. Bogardus C, Lillioja S, Stone K, Mott D: Correlation between muscle glycogen synthase activity and *in vivo* insulin action in man. J Clin Invest 73: 1185–1190, 1984

62. Roch-Noriand AE, Bergström J, Hultman E: Muscle glycogen and glycogen synthase in normal subjects and in patients with diabetes mellitus: effect of intravenous glucose and insulin administration. Scan J Clin Invest 30: 77–84, 1972

63. Vestergaard H, Bjørbæk C, Andersen PH, Bak JF, Pedersen O: Impaired expression of glycogen synthase mRNA in skeletal muscle of NIDDM patients. Diabetes 40: 1740–1745, 1991

64. Bjørbæk C, Vik TA, Echwald SM, Yang P-Y, Vestergaard H, Wang JP, Webb GC, Richmond K, Hansen T, Erikson RL, Miklos GLG, Cohen PTW, Pedersen O: Cloning of a human insulin-stimulated protein kinase (ISPK-1) gene and analysis of coding regions and mRNA levels of the ISPK-1 and the protein phosphatase-1 genes in muscle from NIDDM patients. Diabetes 44: 90–97, 1995

65. Beck-Nielsen H, Hother-Nielsen O, Vaag A, Alford F: Pathogenesis of type 2 (non-insulin-dependent) diabetes mellitus: the role of skeletal muscle glucose uptake and hepatic glucose production in the development of hyperglycemia. A critical comment. Diabetologia 37: 217–221, 1994

Molecular and Cellular Biochemistry **182**: 161–168, 1998.
© 1998 *Kluwer Academic Publishers. Printed in the Netherlands.*

Genetic manipulation of insulin action and β-cell function in mice

Betty Lamothe, Bertrand Duvillié, Nathalie Cordonnier, Anne Baudry, Susan Saint-Just, Danielle Bucchini, Jacques Jami and Rajiv L. Joshi
Institut Cochin de Génétique Moléculaire, INSERM U257, Paris, France

Abstract

Transgenic and gene targeting approaches have now been applied to a number of genes in order to investigate the metabolic disorders that would result by manipulating insulin action or pancreatic β-cell function in the mouse. The availability of such mutant mice will allow in the future to develop animal models in which the pathophysiologies resulting from polygenic defects might be reconstituted and studied in detail. Such animal models hopefully will lead to better understanding of complex polygenic diseases such as non-insulin-dependent diabetes mellitus (NIDDM). (Mol Cell Biochem **182**: 161–168, 1998)

Key words: transgenic mice, knockout mice, NIDDM, animal models

Introduction

Insulin action plays a vital role in glucose homeostasis by regulating carbohydrate, lipid and protein metabolism in the body. The three major sites for this action of insulin are muscle, liver and fat tissue. Besides normal insulin action in these tissues, proper pancreatic β-cell function is obviously mandatory for the normal control of glucose homeostasis. Recent studies focused on the search for candidate genes which might be involved in insulin resistance in non-insulin-dependent diabetes mellitus (NIDDM) have revealed that various kinds of mutations in several genes could lead to different levels of insulin resistance (reviews in [1, 2]). NIDDM is therefore considered today as a complex syndrome and the large spectrum of pathophysiologies that exist in this heterogeneous disease might be explained by cumulative molecular defects in different sets of diabeto-genes (reviews in [3, 4]). Since a limited number of animal models exist for NIDDM research (review in [5]), genetically engineered mice can represent interesting tools to investigate the metabolic disorders resulting from defined alterations in the structure or expression of a single gene or a set of genes and that would impair insulin action or lead to β-cell dysfunction. This review provides an overview on the development of such mouse models useful for NIDDM research generated using transgenic and gene targeting approaches. Our contribution in this area has consisted in the generation of insulin receptor-deficient mice [6].

Generation of transgenic and knockout mice

Figure 1 provides a brief description of methods used for generating transgenic mice or knockout mice carrying disrupted genes. In transgenic mice, the choice of the promoter determines whether a specific gene is over-expressed either in all tissues where it is normally expressed or in a restricted manner or at sites where it is normally not expressed. Selective inhibition of an endogenous gene can also be achieved in transgenic mice by producing an antisense RNA or a ribozyme directed against that gene in specific tissues. The transgenic mouse approach has been applied for many genes to develop animal models for NIDDM (review in [7]). The gene targeting approach allows one to generate knockout mice in which specific genes can be disrupted (review in [8]). This approach also makes it possible to introduce point mutations in a gene of interest. The availability of various transgenic and knockout mice

Address for offprints: R.L. Joshi, INSERM U257, 24, rue du Faubourg Saint-Jacques, 75014 Paris, France

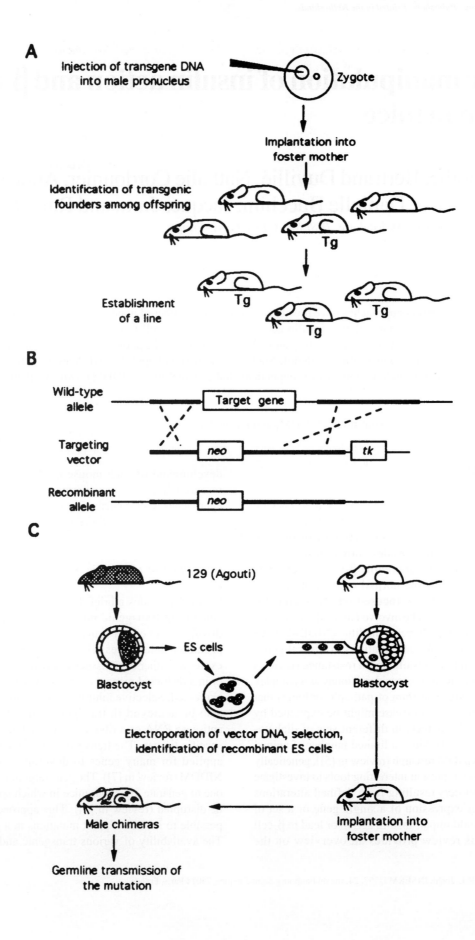

A

Injection of transgene DNA
into male pronucleus

Zygote

Implantation into
foster mother

Identification of transgenic
founders among offspring

Tg

Establishment
of a line

Tg

Tg

Tg

B

Wild-type
allele

Target gene

Targeting
vector

neo

tk

Recombinant
allele

neo

C

129 (Agouti)

Blastocyst

ES cells

Blastocyst

Electroporation of vector DNA, selection,
identification of recombinant ES cells

Male chimeras

Implantation into
foster mother

Germline transmission of
the mutation

makes possible the breeding of these two kinds of mice to generate mice in which the expression of disrupted genes could be reconstituted in specific tissues or the effects of gene mutations could be studied in the absence of endogenous gene expression. Many efforts are currently being devoted to perform more tedious tissue-specific gene disruptions with the use of site specific recombination systems such as the Cre/loxP system from the bacteriophage P4. The use of such a system is a two-step process. Firstly, on one hand loxP sequences are inserted by gene targeting on both sides of the DNA sequence, to be later excised, in the target gene without affecting the gene activity and on the other hand transgenic mice are generated which express the Cre recombinase gene under the control of a tissue-specific promoter. Secondly, breeding of these two kinds of mice generates animals in which the target gene is disrupted in the specific tissue upon excision of the DNA sequence in between the loxP sites by the Cre recombinase. However, a limitation of this approach resides in the fact that the excision by the Cre recombinase is not very efficient and attempts are currently being made to improve the activity of the enzyme by site directed mutagenesis. Recently, application of the gene targeting approach has revived interest in mice as animal models for diabetes. Table 1 presents the list of proteins encoded by genes for which transgenic and/or gene targeting approaches have already been applied with the aim of manipulating insulin action or β-cell function.

Fig. 1. Transgenic and gene targeting approaches in the mouse. (A) For obtention of transgenic mice, microinjected zygotes are reimplanted into foster mothers and founder transgenic animals in the progeny are identified by testing for integration of the transgene by dot-blot, PCR or Southern blot analysis using tail DNA. The transgenic F1 animals are obtained by mating these founders with non-transgenic animals and are used for testing transgene expression and for establishing a line. (B, C) The generation of mice carrying a disrupted gene is a multistep process. The replacement vector designed to inactivate a target gene contains *neo* (neomycin phosphotransferase gene) flanked by regions of 5′- and 3′-homologies as well as *tk* (thymidine kinase gene from herpes simplex virus type 1). The vector DNA is electroporated into male embryonic stem (ES) cells obtained using blastocysts from mouse strain 129 carrying the dominant Agouti fur coat color gene. After positive and negative selection in the presence of G418 and Ganciclovir, respectively, ES cell clones carrying one recombinant allele are identified by Southern blot analysis using appropriate enzymes and probes or by PCR using appropriate primers. Such cells are injected into blastocysts (e.g. from strain C57BL/6) which are reimplanted into foster mothers and generate chimeras (composite fur coat color) at term. Germline transmission of the mutation is obtained through male chimeras which give agouti pups in the progeny when crossed with C57BL/6 females for example. Heterozygous mutants among agouti animals are identified by genotyping and intercrossed to produce homozygous mutants.

Table 1. Genetic manipulation in the mouse of genes coding for proteins involved in insulin action or in β-cell function

Proteins involved in insulin action	Reference #
– The insulin receptor (IR)	6, 9–16
– Insulin receptor substrate-1 (IRS-1)	21, 22
– Hexokinase II	23
– GLUT4/GLUT1	24–41
– Phosphoenolpyruvate carboxykinase (PEPCK)	42
– Glycogen synthase (GS)	43
– Glucokinase (GK)	44
– $G_{i\alpha2}$	45
Proteins relevant to β-cell function	
– Insulin	52–57
– Glucokinase (GK)	58–65
– GLUT2	66, 67
– Islet amyloid polypeptide (IAPP)	68–71

Effects of manipulating insulin action in the mouse

Insulin receptor

The first transgenic mice expressing human IR cDNA in skeletal muscle did not present with any significant changes in basal glucose and insulin levels and the receptor number in the skeletal muscle was only moderately increased [9]. However, plasma glucose and insulin levels were affected after intraperitoneal glucose injections. No glucose intolerance was observed in other transgenic mice expressing human IR cDNA encoding a tyrosine kinase-deficient form, again probably due to limited expression of the transgene [10]. More recently, insulin-dependent glucose uptake and metabolism in skeletal muscle were shown to be impaired in some of the mice overexpressing human IR cDNA encoding another tyrosine kinase-deficient dominant negative form of IR in skeletal and cardiac muscle [11, 12], due to impaired insulin-stimulated IR tyrosine kinase activity leading to reduced phosphorylation of IRS-1 [13]. This resulted in defective activation of PI 3-kinase, MAP kinase or p90[rsk] [13, 14]. However, activation of p70[S6k] and glycogen synthase were unaffected [14]. Finally, production of the extracellular domain of human IR in a soluble form capable of partially sequestering circulating insulin resulted in altered glucose homeostasis in such transgenic mice [15].

More recently, IR-deficient mice have been generated by targeted disruption of the corresponding gene [6, 16]. One strategy consisted in the insertion of translation stop codons in exon 4 and another strategy deleted part of the IR gene around exon 2. Heterozygous animals showed normal glucose tolerance and were fertile. Homozygous IR-deficient pups were near normal at birth but developed, soon after suckling, metabolic disorders such as hyperglycemia and elevated plasma triglyceride levels, which led to diabetes mellitus with ketoacidosis and hepatic steatosis. The

164

Table 2. Phenotype of insulin receptor-deficient mice

Normal at birth

Disorders developed soon after suckling:
    Hyperglycemia
    Hyperinsulinemia
    Hyperlipidemia
    Diabetes mellitus with ketoacidosis
    Liver steatosis
    Growth retardation
    Skeletal muscle hypotrophy
    Reduced brown and white fat deposits

Neonatal lethality

Adapted from [6, 16].

hyperglycemia also resulted in hyperinsulinemia. Absence of insulin signaling in the liver resulted in reduced glycogen content. Marked postnatal skeletal muscle hypotrophy, reduced fat tissue deposits and growth retardation were observed, and IR-deficient pups finally died within a week after birth. Table 2 provides a summary of the phenotype of IR-deficient mice.

Insulin and insulin-like growth factors (IGF-1/IGF-2) and their receptors are structurally very similar and lead to some common as well as specific biological effects [17]. Mice deficient for IGF-1 or IGF-2 or the IGF-1 receptor have already been obtained by targeted disruption of the corresponding genes [18–20]. Since these different ligands are known to bind to and activate the heterologous receptor, breeding of these various mutant mice will allow examination of the question of whether and to what extent insulin and IGF-1 receptors and their ligands can substitute for each other.

Insulin receptor substrate-1
No major metabolic disorders were observed in insulin receptor substrate-1 (IRS-1)-deficient mice generated by targeted disruption of the corresponding gene [21, 22]. However, IRS-1 deficiency resulted in both intra-uterine and postnatal growth retardation. Therefore, IRS-1 appears to be essential in achieving full mitogenic effects of the insulin receptor and/or of the IGF-1 receptor and cannot be entirely substituted by other homologous docking proteins of insulin signaling.

Hexokinase II
No metabolic alterations were detected in transgenic mice overexpressing the human hexokinase II gene in skeletal muscle [23]. Oral glucose tolerance and intravenous insulin tolerance tests did not show any abnormality and insulin and lactate levels in the serum of these mice remained unaffected.

GLUT4 and GLUT1
A number of transgenic mice have been generated which overexpress the human, mouse or rat GLUT4 gene in skeletal

muscle and/or in adipose tissue. The efficiency of insulin action in the whole body could be improved by overproduction of GLUT4 [24–31]. Such mice developed an obesity when fed with a high fat diet but showed proper glycemic control [32]. In the same line, hyperglycemia in diabetic *db/db* mice, which carry a mutated leptin receptor gene, could be reduced by overproduction of GLUT4 and their glycemic control was also improved [33]. Targeted overexpression of the GLUT4 gene in the fat tissue resulted in increased adiposity in transgenic mice [34, 35]. Targeted overproduction of GLUT4 in skeletal muscle resulted in glycogen accumulation [36, 37], and that of GLUT1 increased lactate and glycogen contents in the cells and reduced blood glucose levels [30–40].

Recently, GLUT4-deficient mice were obtained by targeted disruption of the corresponding gene [41]. Surprisingly, these mice did not develop diabetes as severe as one might have expected. A mild fasting hypoglycemia and slight postprandial hyperglycemia were observed only in male animals, although insulin resistance existed in both males and females as evidenced by postprandial hyperinsulinemia and impaired glucose tolerance tests. Lipid metabolism in such mice was very much affected leading to raised serum levels of lactate, free fatty acids and β-hydroxybutyrate in the fasting state. Finally, GLUT4-deficiency resulted in growth retardation, reduced fat tissue deposits and cardiac hypotrophy with reduced longevity (5–6 months) [41].

Phosphoenolpyruvate carboxykinase
Reduced hepatic glycogen storage was observed in transgenic mice overproducing phosphoenolpyruvate carboxykinase (PEPCK) in the liver [42]. These animals presented with fasting hyperglycemia and showed altered glucose tolerance. GLUT4 gene expression was also downregulated in muscle of such mice.

Glycogen synthase
Increased glycogen accumulation was observed in skeletal muscle of transgenic mice overexpressing the rat glycogen synthase cDNA encoding a constitutively activated form of the enzyme in spite of normal glucose transport into these cells [43]. These results indicate that activation of glycogen synthase alone can be sufficient to stimulate glycogen synthesis and lead to increased levels of tissue glycogen [43].

Glucokinase
Treatment of mice with streptozotocin is known to result in β-cell destruction and lead to diabetes mellitus with elevated levels of blood glucose, ketone bodies, triglycerides and free fatty acids. Since liver glucokinase (GK) activity under such conditions is low, transgenic mice overexpressing the rat GK cDNA were produced [44]. Interestingly, strepto-zotocin-treatment of such transgenic mice did not lead to the diabetic alterations that are normally seen, indicating that

the return of glucose uptake by the liver is sufficient to normalize glucose, lipid and ketone body metabolism in the body [44].

$G_{i\alpha2}$

Insulin action was impaired in transgenic mice in which synthesis of the G-protein subunit, $G_{i\alpha2}$, was downregulated in adipose tissue and liver by production of antisense RNA directed against $G_{i\alpha2}$ mRNA [45]. This resulted in hyper-insulinemia, impaired glucose tolerance and resistance to insulin affecting, for example, glucose transporter activity and recruitment, counterregulation of lipolysis and stimulation of glycogen synthesis. Decreased $G_{i\alpha2}$ reduced IRS-1 phosphorylation as a result of increased activities of certain protein tyrosine phosphatases [45].

Manipulating adipose tissue

Genetic manipulation of brown and/or white fat tissues has recently been achieved in transgenic mice using at least three different approaches: (i) specific expression of the gene encoding the diphtheria toxin A chain in white or brown fat tissue [46–48], (ii) overexpression of the glycerol 3-phosphate dehydrogenase gene [49] and (iii) targeted disruption of the gene encoding $R_{II\beta}$, one of the regulatory subunits of cAMP-dependent protein kinase A [50]. Such mice will be useful models for investigating the contribution of fat tissue in glucose homeostasis and the obesity associated insulin resistance leading to NIDDM.

Effects of creating pancreatic β-cell dysfunction in the mouse

As mentioned above, proper β-cell function is obviously crucial for the normal control of glucose homeostasis. In addition to insulin resistance, NIDDM patients also present with β-cell defects (review in [51]). In this respect, a number of attempts have been made to generate metabolic disorders in the mouse by altering β-cell function.

Insulin

Several attempts were made to overexpress the human insulin gene in β-cells of transgenic mice. Normal insulin levels and glucose homeostasis were reported for the first transgenic mice that were generated [52, 53]. In spite of at least a 2-fold increase in total insulin mRNA, the levels of serum and pancreatic insulin were normal in these mice indicating post-transcriptional controls for insulin synthesis [54]. Two transgenic mouse lines carrying multiple copies of the human insulin gene were later reported which became hyperinsulinemic. Their glycemia was normal but they showed glucose intolerance [55]. Transgenic mice producing mutated forms of human insulin were also generated and also

had normal glycemia. It was shown that proinsulin carrying a HisB10 to Asp mutation was secreted at high levels via the unregulated constitutive pathway [56]. Constitutive proinsulin release was also observed for SerB9 to Asp and ThrB27 to Glu mutations [57].

Glucokinase

The GK activity in β-cells of transgenic mice was reduced using two strategies: (i) production an antisense RNA encompassing a GK gene-specific ribozyme element [58] or (ii) production of an antisense RNA to the GK gene [59]. Plasma glucose and insulin levels were normal in the first kind of mice although glucose-induced insulin secretion from in situ-perfused pancreas was reduced. Glucose intolerance was observed in the second type of mice although it varied depending on the mouse strains. On the other hand, increased insulin secretion and reduced blood glucose levels were observed in transgenic mice overexpressing yeast hexokinase B gene in β-cells [60]. Such mice showed significant depletion of insulin in the pancreas [61]. The breeding of these mice with transgenic mice made diabetic by overexpression of a chicken calmodulin gene in β-cells [62] transiently improved their diabetic symptoms. However, this protection was short-lived and did not protect against insulin depletion of β-cells.

More recently, targeted disruption of the mouse GK gene has been achieved [63–65]. Hyperglycemia with mild diabetes and glucose intolerance were observed in heterozygous animals in which the capacity to secrete insulin in response to glucose was reduced. Glucose production by the liver of such mice persisted even under hyperglycemic and hyperinsulinemic conditions and their hepatic glycogen stores were reduced. On the other hand, homozygous GK-deficient pups developed extreme hyperglycemia with severe diabetes. The cholesterol and triglycerides levels in the serum were raised and the glycogen content in the liver was reduced. GK-deficient pups developed hepatic steatosis, were growth retarded and survived only 3–5 days after birth. Interestingly, this lethal phenotype could be rescued upon reconstitution of GK activity only in β-cells of GK-deficient pups using transgenic mice expressing the rat GK cDNA in β-cells. The levels of triglycerides and cholesterol were normalized and 50% animals had normal glucose levels. One study reported that targeted disruption of GK gene results in embryonic lethality of homozygous mutants [65]. This observation is at variance with the phenotype reported by two other groups described above.

GLUT2

Hyperglycemia was observed in transgenic mice producing antisense RNA to GLUT2 in β-cells and in which GLUT2 gene expression could be reduced by 80% at the protein level [66]. Such mice present with abnormal insulin secretion in response to glucose and had impaired glucose tolerance tests. In contrast, neither any abnormality of glucose homeostasis nor any defect in insulin secretion were reported in 2 month-old transgenic animals expressing the human oncogene [Val^{12}]HRAS in β-cells and which also resulted in lower GLUT2 gene expression in β-cells [67].

Islet amyloid polypeptide
Several transgenic mice expressing the human or rat islet amyloid polypeptide (IAPP) in β-cells were produced in which circulating IAPP levels were increased [68–71]. The glycemia and insulinemia in these mice remained unaffected. In some studies, no amyloid deposits, often seen in pathological states, were detected although one study reported accumulation of nonfibrillar human (but not rat) IAPP mass in perivascular spaces [71].

Perspectives
It appears that both insulin action and β-cell function can be manipulated in the mouse by altering the expression of many genes. Systemic application of transgenic and gene targeting approaches to alter the structure or expression of a number of other potential diabetogenes can allow to further develop mouse models in which metabolic alterations resulting from polygenic disorders could be reconstituted. The study of such mouse models will shed light on the molecular basis of complex pathophysiologies seen in various kinds of NIDDM in humans. Such mutant mice can also be very interesting for testing therapeutic strategies aimed at correcting some of the metabolic disorders that would have been created by genetic modification of the mouse genome.

Acknowledgements

We thank Drs. P. De Meyts, E. Karnieli, M.D. Lane, D.E. Moller, O. Pedersen, J.E. Pessin, A.R. Shuldiner, B.M. Spiegelman and R. Taylor for having sent us reprints or preprints of work carried out in their laboratories. We apologize for not being able to cite original research work not dealing with transgenic or knock-out mice.

This work was supported by grants from Association Française contre les Myopathies (AFM), Association pour la Recherche Contre le Cancer (ARC), Fondation de France and Lilly-Alfediam.

References

1. Moller DE, Bjorbaek C, Vidai-Pulg A: Candidate genes for insulin resistance. Diabetes Care 19: 396–400, 1996
2. Shuldiner AR, Silver KD: Candidate genes in non-insulin dependent diabetes mellitus. In: D LeRoith, JM Olelfsky, SI Taylor, (eds). Diabetes Mellitus: A Fundamental and Clinical Text. Lippincott, in press
3. Kahn CR: Insulin action, diabetogenes, and the cause of type II diabetes. Diabetes 43: 1066–1084, 1994
4. De Meyts P: The diabetogenes concept of NIDDM. Adv Exp Med Biol 334: 89–100, 1993
5. Shafrir E: Animal models of non-insulin dependent diabetes. Diabetes Metab Rev 8: 179–208, 1992
6. Joshi RL, Lamothe B, Cordonnier N, Mesbah K, Monthioux E, Jami J, Bucchini D: Targeted disruption of the insulin receptor gene in the mouse results in neonatal lethality. EMBO J 15: 1542–1547, 1996
7. Moller DE: Transgenic approaches to the pathogenesis of NIDDM. Diabetes 43: 1394–1401, 1994
8. Bronson SK, Smithies O: Altering mice by homologous recombination using embryonic stem cells. J Biol Chem 269: 27155–27158, 1994
9. Benecke H, Flier JS, Rosenthal N, Siddle K, Klein HH, Moller DE: Muscle-specific expression of human insulin receptor in transgenic mice. Diabetes 42: 206–212, 1993
10. Nishiyama T, Shirotani T, Murakami T, Shimada F, Todaka M, Saito S, Hayashi H, Noma Y, Shima K, Makino H, Shichiri M, Miyazaki JI, Yamamura KI, Ebina Y: Expression of the gene encoding the tyrosine kinase-deficient human insulin receptor in transgenic mice. Gene 141: 187–192, 1994
11. Chang PY, Benecke H, Le Marchand-Brustel Y, Lawitts J, Moller DE: Expression of a dominant-negative mutant human insulin receptor in the muscle of transgenic mice. J Biol Chem 269: 16034–16040, 1994
12. Moller DE, Chang PY, Yaspelkis BB III, Flier JS, Wallberg-Henriksson H, Ivy JL: Transgenic mice with muscle-specific insulin resistance develop increased adiposity, impaired glucose tolerance, and dyslipidemia. Endocrinol, in press
13. Chang PY, Goodyear W, Benecke H, Markuns JS, Moller DE: Impaired insulin signaling in skeletal muscles from transgenic mice expressing kinase-deficient insulin receptors. J Biol Chem 270: 12593–12600, 1995
14. Chang PY, Le Marchand-Brustel Y, Cheatham LA, Moller DE: Insulin stimulation of mitogen-activated protein kinase, p90^{rsk}, and p70 S6 kinase in skeletal muscle of normal and insulin-resistant mice. J Biol Chem 270: 29928–29935, 1995
15. Schaefer EM, Viard V, Morin J, Ferré P, Pénicaud L, Ramos P, Maika SD, Ellis L, Hammer RE: A new transgenic mouse model of chronic hyperglycemia. Diabetes 43: 143–153, 1994
16. Accili D, Drago J, Lee EJ, Johnson MD, Cool MH, Salvatore P, Asico LD, José PA, Taylor SI, Westphal H: Early neonatal death in mice homozygous for a null allele of the insulin receptor gene. Nature Genet 12: 106–109, 1996
17. De Meyts P, Wallach B, Christoffersen CT, Urso B, Gronskov K, Latus W, Yakushiji F, Ilondo MM, Shymko RM: The insulin-like growth factor-I receptor. Horm Res 42: 152–169, 1994
18. DeChiara TM, Efstratiadis A, Robertson EJ: A growth-deficiency phenotype in heterozygous mice carrying an insulin-like growth factor II gene disrupted by targeting. Nature 345: 78–80, 1990
19. Liu JP, Baker J, Perkins AS, Robertson EJ, Efstratiadis A: Mice carrying null mutations of the genes encoding insulin-like growth factor I (igf-1) and type 1 IGF receptor (igf-1r). Cell 75: 59–72, 1993
20. Powell-Braxton L, Hollingshead P, Warburton C, Dowd M, Pitts-Meek S, Dalton D, Gillett N, Stewart TA: IGF-1 is required for normal embryonic growth in mice. Genes Dev 7: 2609–2617, 1993
21. Tamemoto H, Kadowaki T, Tobe K, Yagi T, Sakura H, Hayakawa T, Terauchi Y, Ueki K, Kaburaji Y, Satoh S, Sekihara H, Yoshioka S, Horikoshi H, Furuta Y, Ikawa Y, Kasuga M, Yazaki Y, Aizawa S: Insulin resistance and growth retardation in mice lacking insulin receptor substrate-1. Nature 372: 182–186, 1994
22. Araki E, Lipes MA, Patti ME, Brüning JC, Haag B III, Johnson RS, Kahn CR: Alternative pathway of insulin signaling in mice with targeted disruption of the IRS-1 gene. Nature 372: 186–190, 1994

23. Chang PY, Jensen J, Printz RL, Granner DK, Ivy JL, Moller DE: Over-expression of hexokinase II in transgenic mice: Evidence that increased phosphorylation augments muscle glucose uptake. J Biol Chem, in press

24. Ezaki O, Flores-Riveros JR, Kaestner KH, Gearhart J, Lane MD: Regulated expression of an insulin-responsive glucose transporter (GLUT4) minigene in 3T3-L1 adipocytes and transgenic mice. Proc Natl Acad Sci USA 90: 3348–3352, 1993

25. Olson AL, Liu ML, Moye-Rowley WS, Buse JB, Bell GI, Pessin JE: Hormonal/metabolic regulation of the human GLUT4/muscle-fat facilitative glucose transporter gene in transgenic mice. J Biol Chem 268: 9839–9846, 1993

26. Deems RO, Evans JL, Deacon RW, Honer CM, Chu DT, Bürki K, Fillers WS, Cohen DK, Young DA: Expression of human GLUT4 in mice results in increased insulin action. Diabetologia 37: 1097–1104, 1994

27. Marshall BA, Mueckler MM: Differential effects of GLUT1 or GLUT4 overexpression on insulin responsiveness in transgenic mice. Am J Physiol 267: E738–E744, 1994

28. Treadway JL, Hargrove DM, Nardone NA, McPherson RK, Russo JF, Milici AJ, Stukenbrok HA, Gibbs EM, Stevenson RW, Pessin JE: Enhanced peripheral glucose utilization in transgenic mice expressing the human GLUT4 gene. J Biol Chem 269: 29956–29961, 1994

29. Hansen PA, Gulve EA, Marshall BA, Gao J, Pessin JE, Holloszy JO, Mueckler M: Skeletal muscle glucose transport and metabolism are enhanced in transgenic mice overexpressing the GLUT4 glucose transporter. J Biol Chem 270: 1679–1684, 1995

30. Ikemoto S, Thompson KS, Itakura H, Lane MD, Ezaki O: Expression of an insulin responsive glucose transporter (GLUT4) minigene in transgenic mice: Effect of exercise and role in glucose homeostasis. Proc Natl Acad Sci USA 92: 865–869, 1995

31. Ren JM, Marshall BA, Mueckler MM, McCaleb M, Amatruda JM, Shulman GI: Overexpression of GLUT4 protein in muscle increase basal and insulin-stimulated whole body glucose disposal in conscious mice. J Clin Invest 95: 429–432, 1995

32. Ikemoto S, Thompson KS, Takahashi M, Itakura H, Lane MD, Ezaki O: High fat diet induced hyperglycemia: Prevention by low level expression of glucose transporter (GLUT4) minigene in transgenic mice. Proc Natl Acad Sci USA 92: 3096–3099, 1995

33. Gibbs EM, Stock JL, McCoid SC, Stukenbrok HA, Pessin JE, Stevenson RW, Milici AJ, McNeish JD: Glycemic improvement in diabetic db/db mice by overexpression of the human insulin-regulatable glucose transporter (GLUT4). J Clin Invest 95: 1512–1518, 1995

34. Shepherd PR, Gnudi L, Tozzo E, Yang H, Leach F, Kahn BB: Adipose cell hyperplysia and enhanced glucose disposal in transgenic mice overexpressing GLUT4 selectively in adipose tissue. J Biol Chem 268: 22243–22246, 1993

35. Gnudi L, Tozzo E, Shepherd PR, Bliss JL, Kahn BB: High level overexpression of glucose transporter-4 driven by an adipose-specific promoter is maintained in transgenic mice on a high fat diet, but does not prevent impaired glucose tolerance. Endocrinol 136: 995–1002, 1995

36. Tsao TS, Burcelin R, Katz EB, Huang L, Charron MJ: Enhanced insulin action due to targeted GLUT4 overexpression exclusively in muscle. Diabetes 45: 28–36, 1996

37. Leturque A, Loizeau M, Vaulont S, Salminen M, Girard J: Improvement of insulin action in diabetic transgenic mice selectively overexpressing GLUT4 in skeletal muscle. Diabetes 45: 23–27, 1996

38. Marshall BA, Ren JM, Johnson DW, Gibbs EM, Liliquist JS, Soeller WC, Holloszy JO, Mueckler M: Germline manipulation of glucose homeostasis via alteration of glucose transporter levels in skeletal muscle. J Biol Chem 268: 18442–18445, 1993

39. Ren JM, Marshall BA, Gulve EA, Gao J, Johnson DW, Holloszy JO, Mueckler M: Evidence from transgenic mice that glucose transport is rate-limiting for glycogen deposition and glycolysis in skeletal muscle. J Biol Chem 268: 16113–16115, 1993

40. Gulve EA, Ren JM, Marshall BA, Gao J, Hansen PA, Holloszy JO, Mueckler M: Glucose transport activity in skeletal muscles from transgenic mice overexpressing GLUT1. J Biol Chem 269: 18366–18370, 1994

41. Katz EB, Stenbit AK, Hatton K, DePinho R, Charron MJ: Cardiac and adipose tissue abnormalities but not diabetes in mice deficient in GLUT4. Nature 377: 151–155, 1995

42. Valera A, Pujol A, Pelegrin M, Bosch F: Transgenic mice overexpressing phosphoenolpyruvate carboxykinase develop non-insulin-dependent diabetes mellitus. Proc Natl Acad Sci USA 91: 9151–9154, 1994

43. Manchester J, Skurat AV, Roach P, Hauschka SD, Lawrence JR JC: Increased glycogen accumulation in transgenic mice overexpressing glycogen synthetase in skeletal muscle. Proc Natl Acad Sci USA 93: 10707–10711, 1996

44. Ferre T, Pujol A, Riu E, Bosch F, Valera A: Correction of diabetic alterations by glucokinase. Proc Natl Acad Sci USA 93: 7225–7230, 1996

45. Moxham CM, Malbon CC: Insulin action impaired by deficiency of the G-protein subunit $G_{i\alpha2}$. Nature 379: 840–844, 1996

46. Ross SR, Graves RA, Spiegelman BM: Targeted expression of a toxin gene to adipose tissue: Transgenic mice resistant to obesity. Genes Dev 7: 1318–1324, 1993

47. Lowell BB, Susulic VS, Hamann A, Lawitts JA, Himms-Hagen J, Boyer BB, Kozak LP, Flier JS: Development of obesity in transgenic mice after genetic ablation of brown adipose tissue. Nature 366: 740–742, 1993

48. Hamann A, Benecke H, Le Marchand-Brustel Y, Susulic VS, Lowell BB, Flier JS: Characterization of insulin resistance and NIDDM in transgenic mice with reduced brown fat. Diabetes 44: 1266–1273, 1995

49. Kozak LP, Kozak UC, Clarke GT: Abnormal brown and white fat development in transgenic mice overexpressing glycerol 3-phosphate dehydrogenase. Genes Dev 5: 2256–2264, 1991

50. Cummings DE, Brandon EP, Planas JV, Motamed K, Idzerda RL, McKnight GS: Genetically lean mice result from targeted disruption of the $R_{II\beta}$ subunit of protein kinase A. Nature 382: 622–626, 1996

51. Taylor SI, Accili D, Imai Y: Insulin resistance or insulin deficiency. Which is the primary cause of NIDDM? Diabetes 43: 735–740, 1994

52. Bucchini D, Ripoche M-A, Stinnakre M-G, Desbois P, Lorès P, Monthioux E, Absil J, Lepesant J-A, Pictet R, Jami J: Pancreatic expression of human insulin gene in transgenic mice. Proc Natl Acad Sci USA 83: 2511–2515, 1986

53. Selden RF, Skoskiewicz MJ, Howie KB, Russell PS, Goodman HM: Regulation of human insulin gene expression in transgenic mice. Nature 321: 525–528, 1986

54. Schnetzler B, Murakawa G, Abalos D, Halban P, Selden R: Adaptation to supraphysiologic levels of insulin gene expression in transgenic mice: evidence for the importance of post-transcriptional regulation. J Clin Invest 92: 272–280, 1993

55. Marban SL, DeLoia JA, Gearhart JD: Hyperinsulinemia in transgenic mice carrying multiple copies of the human insulin gene. Develop Genet 10: 356–364, 1989

56. Carroll RJ, Hammer RE, Chan SJ, Swift HH, Rubenstein AH, Steiner DF: A mutant human proinsulin is secreted from islets of Langherans in increased amounts via an unregulated pathway. Proc Natl Acad Sci USA 85: 8943–8947, 1988

57. Ma YH, Lorès P, Wang J, Jami J, Grodsky GM: Constitutive (pro)insulin release from pancreas of transgenic mice expressing monomeric insulin. Endocrinol 136: 2622–2630, 1995

58. Efrat S, Leiser M, Wu YJ, Fusco-DeMane D, Emran OA, Surana M, Jetton TL, Magnuson MA, Weir G, Fleischer N: Ribozyme-mediated attenuation of pancreatic β-cell glucokinase expression in transgenic mice results in impaired glucose-induced insulin secretion. Proc Natl Acad Sci USA 91: 2051–2055, 1994

168

59. Ishihara H, Tashiro F, Ikuta K, Asano T, Katagiri H, Inukai K, Kikuchi M, Yazaki Y, Oka Y, Miyazaki JI: Inhibition of pancreatic β-cell glucokinase by antisense RNA expression in transgenic mice: mouse strain-dependent alteration of glucose tolerance. FEBS Lett 371: 329–332, 1995

60. Epstein PN, Boschero AC, Atwater I, Cai X, Overbeek PA: Expression of yeast hexokinase in pancreatic β-cells of transgenic mice reduces blood glucose, enhances insulin secretion, and decreases diabetes. Proc Natl Acad Sci USA 89: 12038–12042, 1992

61. Voss-McCowan ME, Bonner-Weir S, Klevay LM, Epstein PN: A yeast hexokinase transgene decreases pancreatic insulin and transiently reduces diabetes. 1: 103–111, 1993

62. Epstein PN, Overbeek PA, Means AR: Calmodulin-induced early-onset diabetes in transgenic mice. Cell 58: 1067–1073, 1989

63. Grupe A, Hultgren B, Ryan A, Ma YH, Bauer M, Stewart TA: Transgenic knockouts reveal a critical requirement for pancreatic β-cell glucokinase in maintaining glucose homeostasis. Cell 83: 69–78, 1995

64. Terauchi Y, Sakura H, Yasuda K, Iwamoto K, Takahashi N, Ito K, Kasai H, Suzuki H, Ueda O, Kamada N, Jishage K, Komeda K, Noda M, Kanazawa Y, Taniguchi S, Miwa I, Akanuma Y, Kodama T, Yazaki Y, Kadowaki T: Pancreatic β-cell-specific targeted disruption of glucokinase gene. J Biol Chem 270: 30253–30256, 1995

65. Bali D, Svetlanov A, Lee HW, Fusco-DeMane D, Leiser M, Li B, Barzilai N, Surana M, Hou H, Fleischer N, DePinho R, Rossetti L, Efrat S: Animal model for maturity-onset diabetes of the young generated by disruption of the mouse glucokinase gene. J Biol Chem 270: 21464–21467, 1995

66. Valera A, Solanes G, Fernandez-Alvarez J, Pujol A, Ferrer J, Asins G, Gomis R, Bosch F: Expression of GLUT2 antisense RNA in β-cells of transgenic mice leads to diabetes. J Biol Chem 269: 28543–28546, 1994

67. Tal M, Wu YJ, Leiser M, Surana M, Lodish H, Fleischer N, Weir G, Efrat S: [Val^{12}]HRAS downregulates GLUT2 in β-cells of transgenic mice without affecting glucose homeostasis. Proc Natl Acad Sci USA 89: 5744–5748, 1992

68. Yagui K, Yamaguchi T, Kanatsuka A, Shimada F, Huang CI, Tokuyama Y, Ohsawa H, Yamamura KI, Miyazaki JI, Mikata A, Yoshida S, Makino H: Formation of islet amyloid fibrils in beta-secretory granules of transgenic mice expressing human islet amyloid polypeptide/amylin. Eur J Endocrinol 132: 487–496, 1995

69. Fox N, Schrementi J, Nishi M, Ohagi S, Chan SJ, Heisserman JA, Westermark GT, Leckström A, Westermark P, Steiner DF: Human islet amyloid polypeptide transgenic mice as a model of non-insulin-dependent diabetes mellitus (NIDDM). FEBS Lett 323: 40–44, 1993

70. Höppener JW, Oosterwijk C, van Hulst KL, Verbeek JS, Capel PJ, de Koning EJ, Clark A, Jansz HS, Lips CJ: Molecular physiology of the islet amyloid polypeptide (IAPP)/amylin gene in man, rat, and transgenic mice. J Cell Biochem 55: 39–53, 1994

71. de Koning EJP, Höppener JWM, Verbeek JS, Oosterwijk C, van Hulst KL, Baker CA, Lips CJM, Morris JF, Clark A: Human islet amyloid polypeptide accumulates at similar sites in islets of transgenic mice and humans. Diabetes 43: 640–644, 1994

Molecular and Cellular Biochemistry **182**: 169–175, 1998.
© 1998 *Kluwer Academic Publishers. Printed in the Netherlands.*

TNF-α and insulin resistance: Summary and future prospects

Pascal Peraldi and Bruce Spiegelman

Dana Farber Cancer Institute, Harvard Medical School, Boston, MA, USA

Abstract

While the causes of obesity remain elusive, the relationship between obesity and insulin resistance is a well-established fact [1]. Insulin resistance is defined as a smaller than normal response to a certain dose of insulin, and contributes to several pathological problems of obese patients such as hyperlipidemia, arteriosclerosis and hypertension. Several pieces of evidence indicate that the cytokine tumor necrosis factor a (TNF-α) is an important player in the state of insulin resistance observed during obesity. In this review we will try to summarize what is known about the function of TNF-α in insulin resistance during obesity and how TNF-α interferes with insulin signaling. (Mol Cell Biochem **182**: 169–175, 1998)

Key words: insulin resistance, TNF-α, obesity, insulin signaling, adipocyte, IRS-1

TNF-α: the cytokine

TNF-α was originally identified as an endogenous factor, induced after inflammation or mitogenic stimulus, that kills tumor cells [2–4]. TNF-α is produced mainly by macrophages, but also by other cell types such as T cells and fibroblasts [5]. Roughly one third of tumor cells are sensitive to the cytotoxic effect of TNF-α while non-tumorogenic cells are generally resistant to TNF-α. This cytokine also induces several pathological states such as the endotoxic shock observed after infection by gram-negative bacteria, fever, anorexia etc. TNF-α is also known to induce the synthesis of other proinflammatory molecules such as interleukin 1 and 6, interferon γ, prostaglandines, PDGF [6, 7].

TNF-α is produced as a pre-protein of 26 kDa bound to the cell membrane, probably as a trimer [8]. This membrane-bound protein is active, but only as an autocrine and paracrine factor. A cleavage releases the circulating form of TNF-α, a homotrimer of 51 kDa. Metalloprotease inhibitors have been produced to inhibit the maturation of TNF-α and have been used successfully to protect mice against a lethal dose of endotoxin [9, 10].

TNF-α: the receptors

Two TNF receptors have been identified and named according to their molecular mass, p55TNFR (also called TNFR1) and p75TNFR (or TNFR2). Both receptors are expressed in virtually all tissues but with different ratios [7, 11]. TNF receptors are expressed as a monomer at the cell surface; the ligand-induced homotrimerisation of the receptor triggers TNF-α signaling. Both receptors can be found as soluble proteins after proteolytic cleavage. These soluble receptors can bind TNF-α and their concentration is increased during bacterial infection. The function of these soluble receptors is thought to modulate the available circulating concentration of TNF-α.

p55TNFR and p75TNFR are totally unrelated proteins outside their ligand binding domains, suggesting that they control different intracellular signaling events [12]. So far, p55TNFR seems to be responsible for the majority of the biological effects of TNF-α. However, p75TNFR does possess signaling mechanisms. Indeed, p75TNFR activates the transcription factor NF-κB, and under certain conditions can mediate cell death. Moreover, according to the ligand passing idea, p75TNFR can increase locally TNF-α

Address for offprints: B. Spiegelman, Dana Farber Cancer Institute, Harvard Medical School, 44 Binney Street, Boston, MA 02115, USA

concentrations. Indeed, p75TNFR binds TNF-α with a higher affinity and a faster dissociation rate than p55TNFR (Kd of 100 pM vs. 500 pM and $t_{1/2}$ of 10 min. versus 3 h). Therefore, at low TNF-α concentration, p75TNFR binds TNF-α but releases it quickly and makes it potentially available for p55TNFR [13].

TNF-α: the signaling

TNF receptors are devoid of any intrinsic catalytic activity, and are not coupled to GTP binding proteins. However, TNF receptors activate several signaling pathways, such as the MAP kinase cascade, protein kinase C activation, sphingomyelinase activities, etc. [7].

This last two years have provided us with several insights in the early biological events triggered by TNF-α. Using fusion proteins and two-hybrid screen several laboratories have isolated proteins through their ability to bind to the cytoplasmic domain of p55TNFR and p75TNFR (Fig. 1). Indeed, the early steps of TNF-α signaling are totally dependent upon protein-protein interactions, in the absence of any phosphorylation mechanisms. TRAF1 and TRAF2 (TNF receptor associated protein) form homo- or heterodimers, through their 'TRAF' domains and bind to p75TNFR [14]. Only TRAF2 seems to bind directly to p75TNFR upon TNF-α treatment. TRAFs are sequestered in the cytosol by the ITRAF (inhibitor of TRAF) that maintained the TRAFs in an inactive form (i.e. unable to bind p75TNFR). TRAF proteins are also associated to the cIAP (inhibitor of apoptosis) through a domain of TRAF different from the one binding the I-TRAF [15, 16]. TRAF2 appears to be important for p75TNFR mediated NF-κB activation, but the function of TRAF1, and cIAP is still unknown.

After binding of TNF-α to p55TNFR, p55 binds to TRADD (p55TNF receptor associated death domain) and FAN (Factor associated with N-SMase activation). FAN binds to the NSD (N-sphingomyelinase domain) domain of p55TNFR and is responsible for the activation of the neutral sphingomyelinase [17]. TRADD binds to the 'death domain' of p55TNFR and is itself associated with FADD (Fas associated death domain protein), TRAF-2 and RIP [18–21]. The N-terminal domain of FADD binds to the protease MACH (also called FLICK) which is proposed to be a direct activator of the proteolysis cascade responsible of apoptosis [22, 23]. RIP is a serine/threonine kinase which is also involved in cell death [20, 21]. Therefore TRADD controls two different pathways; apoptosis (through its association with FADD and RIP) and NF-κB activation (through its association with TRAF-2). However, several reports indicate that these two pathways are to some extent antagonistic. Indeed, the transcription factor NF-κB protects cells from apoptosis induced by TNF-α but also from stress such as ionizing radiation and IL-1 [24–27]. The cell-specific effect of TNF-α on cell growth or apoptosis is therefore possibly regulated by the amount of NF-κB activation versus RIP/FADD activation.

TNF-α and glucose homeostasis

Although the function of TNF-α in the development of insulin resistance during obesity has been indicated recently by experiments performed in our laboratory, several earlier observations suggested that TNF-α affects glucose and lipid metabolism. In rat and human, TNF-α injection induces an increase in the concentration of plasma triglyceride and very low density lipoproteins [28, 29]. This hyperlipidemia is thought to be due to an increase in hepatic lipogenesis and lypolysis. TNF-α and other cytokines (such as interleukin 1 and interferon γ), affect glucose homeostasis in several tissues mainly by increasing non insulin dependent glucose transport through synthesis of the glucose transporter Glut-1, and by decreasing insulin stimulated glucose transport [29, 30]. In adipocytes, high concentration of TNF-α decreases the expression of lipogenic enzymes and can also induce adipocytes dedifferentiation [31]. Finally, an insulin resistant state is observed during certain cancers, infections, and trauma, such as burn injuries, conditions in which high level of circulating TNF-α has been detected [32–34].

TNF-α production during obesity

The first evidence for a function of TNF-α in the state of insulin resistance observed during obesity was provided by the observation that adipocytes of obese animals overexpress TNF-α [35, 36]. This overexpression of TNF-α messenger was observed in multiple models of rodent obesity such as fa/fa rats and ob/ob, tub/tub, KKAY mice and in a strain of transgenic mice lacking brown adipose tissue that develops spontaneously obesity [37]. In these rodents an increase in TNF-α messenger in the adipose tissue, as well as an increase in circulating TNF-α was observed. However, circulating levels were low in absolute terms.

An increase in TNF-α messenger is also observed in adipocytes from obese humans. TNF-α expression in humans is in positive correlation with the degree of obesity (body mass index) and of hyperinsulinemia, and in negative correlation with lipoprotein lipase activity in the adipose tissue [38, 39]. Moreover, after a weight reduction program (which is known to improve insulin sensitivity), a decrease in the level of TNF-α expression was observed. TNF-α overexpression is not completely restricted to fat, since by using RT-PCR TNF-α was also found to be expressed in skeletal muscle and the heart, although at lower levels than

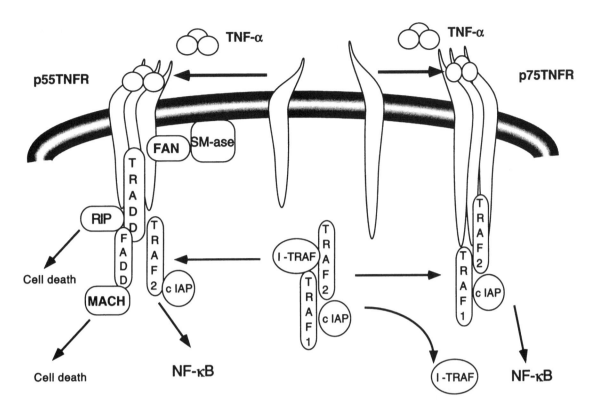

Fig. 1. TNF-α induces the trimerization of its receptors. This oligomerization induces the association of several intracellular proteins with the receptors and triggers TNF-α signaling.

those seen in adipocytes [40]. As muscle is the main post-prandial glucose utilization site, TNF-α expression in muscle could theoretically play an important role in the development of insulin resistance.

TNF-α and insulin resistance during obesity

Experiments neutralizing TNF-α (using immunoadhesins) in hyperinsulinemic, euglycemic clamps of obese animals have shown that this cytokine was causally involved in this syndrome of insulin resistance [35]. Indeed, fa/fa obese rats in which TNF-α is neutralized are more sensitive to insulin than untreated animals. This is due to a 2–3 fold increase in peripheric glucose uptake. On the other hand, hepatic glucose production was not affected. In non-clamped obese animals, three days of TNF-α neutralization results in decrease glucose, insulin and the circulating free fatty acid levels close to those observed in lean animals [41].

Additional evidence for a role of TNF-α in the development of insulin resistance has recently been provided by the study of the knock-out mice for the adipocyte-specific fatty acid binding protein (aP2). Indeed, these animals develop obesity as usual on a high fat diet, but they are the first known model in which obesity is not associated with development of

insulin resistance or diabetes [42]. Interestingly, TNF-α overexpression in adipocytes was not observed, consistent with the notion that TNF-α has a function for the development of insulin resistance during obesity. Furthermore, these data prove that aP2 is linked to the state of insulin resistance observed during obesity, probably through the influence of fatty acid flux on the expression of TNF-α.

TNF-α and insulin signaling

Several laboratories have reported that the tyrosine kinase activity of the insulin receptor is decreased during obesity in muscle and fat. As the enzymatic activity of the insulin receptor is necessary for all of the known biological function of this hormone, this decrease is likely to be an important cause of the state of insulin resistance in obesity.

Neutralization of TNF-α in obese fa/fa rats restores the tyrosine kinase activity in fat and muscle, as well as the insulin-induced phosphorylation of IRS-1 to levels comparable to the ones observed in lean animals [41]. In lean animals, neutralization of TNF-α had no effect. These data suggest that TNF-α induces insulin resistance during obesity by interfering on the tyrosine kinase activity of the insulin receptor. Indeed, *in vitro* experiments have shown that

TNF-α treatment of adipocyte in culture induces a state of 'insulin resistance' [43]. This is due to an inhibition of the tyrosine kinase activity of the insulin receptor leading to a decrease to all the biological function of insulin such as the insulin-induced tyrosine phosphorylation of IRS-1 and glucose transport. While TNF-α induces a decrease of the expression of insulin receptor, IRS-1 and Glut-4 at high concentration [44], specific inhibition of the tyrosine kinase of the insulin receptor could be seen at concentration where insulin receptor, IRS-1, and the glucose transporter Glut-4 expression was unchanged. Since the concentration of circulating TNF-α during obesity are low (90 pg/ml in fa/fa rats, and undetectable in obese humans), it is likely that TNF-α induces insulin resistance by interfering with the tyrosine kinase of the insulin receptor and not by interfering with the expression of proteins, a process which need higher TNF-α concentration.

TNF-α-induced insulin receptor inhibition

The ability to inhibit insulin signaling by TNF-α in cell culture has facilitated the study of the mechanisms involved in this process. Several reports have shown that TNF-α interferes with insulin signaling in various cell lines such as hepatocytes, fibroblasts and myeloid cells [45–47].

Several steps of the signaling in the interaction between TNF receptors and insulin receptor have been elucidated (Fig. 2). IRS-1 appears to be a key molecule in this inter- action. IRS-1 is one of the direct substrates of the insulin receptor and is necessary for several of the biological function of insulin [48]. The tyrosine phosphorylation of IRS-1 induce the binding of several SH2 domain containing proteins. These associations induce an activation of the protein (for example the PI3 kinase) or modify the compart- mentalization of proteins, bringing them close to their substrates (for Grb2-Sos for example).

Treatment of adipocytes or hepatocytes with TNF-α induces an increase in the serine phosphorylation of IRS-1 [49, 50]. This phosphorylation is an important event since this modified form of IRS-1 act as an inhibitor of the insulin receptor *in vitro* [49]. This inhibition is dependent upon the phosphorylation of IRS-1 since dephosphorylation of IRS-1 causes it to lose its inhibitory activity. This mechanism probably happens in intact cells since TNF-α does not interfere with insulin receptor phosphorylation in myeloid cells that lack IRS-1 (32D cells). However, if IRS-1 is ectopically expressed in the same cells, insulin receptor and IRS-1 tyrosine phosphorylation become highly sensitive to TNF-α, indicating that IRS-1 is an important molecule in TNF-α-mediated inhibition of insulin-signaling. This mechanism may also be responsible for insulin resistance in obesity since IRS-1 from muscle and fat of obese animals

is also an inhibitor of the insulin receptor tyrosine kinase activity. Hence, IRS-1 appears as a dual function protein involved in positive insulin signaling but also involved in a mechanism leading to the inhibition of insulin signaling.

The first steps of TNF-α signaling in insulin resistance have been resolved. It is mainly, if not entirely through stimulating p55TNFR that TNF-α inhibits insulin signaling in cell culture [46]. However, it is plausible that p75TNFR plays a role in obesity. Indeed, as previously described the concentration of circulating TNF-α during obesity are rather low, and one of the function of p75TNFR is to concentrate its ligand to make it available for p55TNFR by the ligand passing mechanism. Hence, although the signaling function of p75TNFR does not seem to be necessary for insulin resistance its 'concentrating' function could be important to allow p55TNFR to signal.

The p55TNFR stimulates a neutral sphingomyelinase which hydrolyzes sphingomyelin into ceramide and choline [51]. Ceramides activates directly several enzymes such as PKC-ζ [52, 53], a ceramide-activated kinase which phosphorylates and activates Raf-1 [54], and a ceramide-activated phosphatase which belongs to the phosphatase 2A family [55]. Synthetic cell permeant ceramides and exogenous sphingomyelinase mimic the effect of TNF-α on insulin signaling in cell culture [46]. Like TNF-α, these molecules convert IRS-1 into an inhibitor of insulin receptor, suggesting that stimulation of sphingomyelinase and production of ceramides are the first step of TNF-α signaling leading to insulin signaling inhibition.

The questions

A complete understanding of the function of TNF-α during obesity is an important task since TNF-α production and signaling could be a potential therapeutic target for the treatment of insulin resistance during obesity. Indeed, several questions remains to be addressed:

– What is the physiological component of obesity that triggers the production of TNF-α by adipocytes? TNF-α overexpression is observed in all the models of obesity, and in adipocytes (and muscles to a lower extent). Therefore is it unlikely that TNF-α overexpression is linked proximally to a particular genetic defect. It is more likely that a common modification of the hormonal or the metabolic balance in obesity is involved. As aP2 knock out mice develop obesity without insulin resistance and without TNF-α production, it is likely that the aP2 protein plays a role in TNF-α synthesis [42]. The only known function of the aP2 protein is to bind fatty acids, suggesting that fatty acids are good candidates to regulate the level of TNF-α during obesity.

– What is the tissue basis of TNF-α actions? Circulating concentration of TNF-α are very low. It is therefore unlikely that TNF-α could act as an endocrine factor in obesity. In

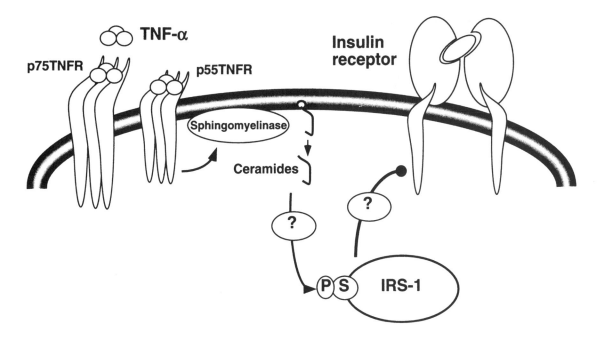

Fig. 2. By binding to p55 TNFR and activating sphingomyelinase (which produces ceramides) TNF-α increases IRS-1 serine phosphorylation. This phosphorylated IRS-1 act as an inhibitor of the insulin receptor by an unknown mechanism.

adipose tissue TNF-α could act as an autocrine and/or paracrine factor. The state of insulin resistance in muscle is probably due to a paracrine action of TNF-α provided by the fat, and/or an autocrine function of TNF-α produced by muscle itself. Of interest would be to treat animals with inhibitors of TNF-α processing. If the function of TNF-α is only autocrine these inhibitors could be without influence on insulin resistance, however if the function is paracrine the inhibitors should improve insulin sensitivity.

– By what mechanism(s) does serine-phosphorylated IRS-1 inhibit insulin receptor activity? Several hypotheses are possible. It is possible that the affinity of serine phosphorylated IRS-1 for insulin receptor is modified, and inhibits insulin receptor activity by steric hindrance. On the other hand, IRS-1 is known to be a docking protein. Therefore, it is conceivable that IRS-1 associates with an inhibitor of insulin receptor upon TNF-α treatment. This inhibitor (a tyrosine phosphatase for example) could be activated after IRS-1 binding, or physically brought to the insulin receptor by IRS-1. Of interest would also be to identify the kinase stimulated by TNF-α that phosphorylates IRS-1, since an inhibitor of this kinase could uncouple TNF-α binding from insulin resistance.

The future

If the observations performed in obese animals can be extended to human obesity, inhibition of TNF-α synthesis

or signaling in adipocytes and muscle could provide a novel way to increase insulin sensitivity in the obese patient. So far only one report has been published concerning the neutralization of TNF-α in a population of obese patients and the results have been disappointing [56]. However, this study has been performed in a population of adults with established diabetes. These patients had hyperglycemia and elevated free fatty acid level, both being known to induce a state of insulin resistance by themselves [57, 58]. In addition, this population was not examined for the production of TNF-α. It has been shown that in very obese people TNF-α production tends to go down, the insulin resistance state being maintained in all likelihood by hyperglycemia and high free fatty acid level. Therefore it is difficult to draw any conclusion on the function of TNF-α in the development of insulin resistance based only on these data. Of interest will be to study the effect of TNF-α neutralization in a population of insulin resistant obese subjects, having not yet developed diabetes. To be fully efficient in humans the neutralization should happen in the adipose tissue, in order to inhibit the autocrine and paracrine effect of TNF-α production by the fat. Ideally, the molecule used should be relatively stable and should diffuse through the blood vessels to reach the adipocytes and/or muscle tissue. Theoretically, three strategies are possible: inhibiting the binding of TNF-α to p55TNFR, inhibiting TNF-α production in fat and interfering with TNF-α signaling. As previously described, the first strategy has been used in obese rats by neutralizing circulating TNF-α. However, it requires frequent injection of antibody and this approach does not affect the autocrine and paracrine effect of TNF-α, which could have a more important function in obese humans who have lower circulating level of TNF-α. Theoretically, inhibition of ceramide production during obesity could reverse insulin resistance, however to our knowledge no specific inhibitors of sphingomyelinase have been produced. We still need a better understanding of the mechanisms of TNF-α production during obesity and the signaling pathways of TNF-α leading to insulin resistance to develop specific drugs. Neutralization of the effect of TNF-α could increase glucose tolerance of obese people and delay or even thwart the development of diabetes.

References

1. Sigal RJ, Warram JH: The interaction between obesity and diabetes. Curr Opin Endocrinol Diab 3: 3–9, 1996

2. Pennica D, Nedwin GE, Hayflick JS, Seeburg PH, Dernick R, Palladino MA, Kohr WJ, Aggarwal BB, Goeddel DV: Human tumor necrosis factor: Precursor structure, expression and homology to lymphotoxin. Nature 312: 724–727, 1984

3. Gray PW, Aggarwal BB, Benton CV, Bringman TS, Henzel WJ, Jarret JA, Leung DW, Moffat BNP, Svedersky LP, Palladino MA, Nedwin GE: Cloning and expression of cDNA for human lymphotoxin, a lymphokine with tumor necrosis activity. Nature 312: 721–724, 1984

4. Fiers W: Tumor necrosis factor. FEBS Lett 285: 199–212, 1991

5. Vassalli P: The pathophysiology of tumor necrosis factors. Annu Rev Immunol 10: 411–452, 1992

6. Tracey KJ, Cerami A: Tumor necrosis factor, and other cytokines and disease. Annu Rev Cell Biol 9: 317–343, 1993

7. Vandenabeele P, Declercq W, Beyaert R, Fiers W: Two tumor necrosis factor receptors: Structure and function. Trends Cell Biol 5: 392–399, 1995

8. Kriegler M, Perez C, DeFay K, Albert I, Lu S: A novel form of TNF/cachectin is a cell surface cytotoxic transmembrane protein: Ramifications for the complex physiology of TNF. Cell 53: 45–53, 1988

9. Mohler K, Sleath P, Fitzner J, Cerretti D, Alderson M, Kerwar S, Torrance D, Otten-Evans C, Greenstreet T, Weerawarna K, Al E: Protection against a lethal dose of endotoxin by an inhibitor of tumour necrosis factor processing. Nature 370: 218–220, 1994

10. Gearing A, Beckett P, Christodoulou M, Churchill M, Clements J, Davidson A, Drummond A, Galloway W, Gilbert R, Gordon J, Al E: Processing of tumour necrosis factor-alpha precursor by metalloproteinases. Nature 370: 555–558, 1994

11. Bazzoni F, Beutler B: How do tumor necrosis factor receptors work? J Inflam 45: 221–238, 1995

12. Tartaglia LA, Weber RF, Figari IS, Reynolds C, Palladino JMA, Goeddel DV: The two different receptors for tumor necrosis factor mediates distinct cellular responses. Proc Natl Acad Sci USA 88: 9292–9296, 1991

13. Tartaglia L, Pennica P, Goeddel D: Ligand passing: The 75-kDa tumor necrosis factor (TNF) receptor recruits TNF for signaling by the 55-kDa TNF receptor. J Biol Chem 268: 18542–18548, 1993

14. Rothe M, Wong SC, Henzel WJ, Goeddel DV: A novel family of putative signal transducers associated with the cytoplasmic domain of the 75 kDa tumor necrosis factor receptor. Cell 78: 681–692, 1994

15. Rothe M, Pan M, Henzel W, Ayres T, Goeddel D: The TNFR2-TRAF signaling complex contains two novel proteins related to baculoviral inhibitor of apoptosis proteins. Cell 83: 1243–1252, 1995

16. Rothe M, Xiong J, Shu H-B, Williamson K, Goddard A, Goeddel DV: I-TRAF is a novel TRAF-interacting protein that regulates TRAF-mediated signal transduction. Proc Natl Acad Sci USA 93: 8241–8246, 1996

17. Adam-Klages S, Adam D, Wiegman K, Struve S, Kolanus W, Schneider-Mergener J, Kronke M: FAN, a novel WD-repeat protein, couples the p55TNF-receptor to neutral sphingomyelinase. Cell 86: 937–947, 1996

18. Hsu H, Xiong J, Goeddel D: The TNF receptor 1-associated protein TRADD signals death and NF-κB activation. Cell 81: 495–504, 1995

19. Hsu H, Shu HB, Pan MG, Goeddel DV: TRADD-TRAF2 and TRADD-FADD interactions define two distinct TNF receptor 1 signaling transduction pathways. Cell 84: 299–308, 1996

20. Stanger B, Leder P, Lee T, Kim E, Seed B: RIP: A novel protein containing a death domain that interacts with Fas/APO-1 (CD95) in yeast and causes cell death. Cell 81: 513–523, 1995

21. Hsu H, Huang J, Shu HB, Baichwal V, Goeddel DV: TNF-dependent recruitment of the protein kinase RIP to the TNF receptor-1 signaling complex. Immunity 4: 387–396, 1996

22. Boldin MP, Goncharov TM, Goltsev YV, Wallach D: Involvement of MACH, a novel MORT1/FADD-interacting protease, in FAS/APO-1 and TNF receptor induced cell death. Cell 85: 803–815, 1996

23. Muzio M, Chinnaiyan A, Kischkel F, O'Rourke K, Shevchenko A, Ni J, Scaffidi C, Bretz J, Zhang M, Gentz R, Mann M, Krammer P, Peter M, Dixit V: FLICK, a novel FADD-homologous ICE/CED-3-like protease, is recruited to the CD95 (Fas/APO-1) death-inducing signaling complex. Cell 85: 817–827, 1996

24. Wang C-Y, Mayo MW, Baldwin AS: TNF- and cancer therapy-induced apoptosis: potential by inhibition of NF-κB. Science 274: 784–786, 1996

25. Beg AA, Baltimore D: An essential role for NF-κB in preventing TNF-α-induced cell death. Science 274: 782–784, 1996

26. Van Antwerp DJ, Martin SJ, Kafri T, Green DR, Verma IM: Supression of TNFα-induced apoptosis by NF-κb. Science 274: 787–789, 1996

27. Liu Z-g, Hsu H, Goeddel DV, Karin M: Dissection of TNF receptor 1 effector functions: JNK activation is not linked to apoptosis while NF-κB activation prevents cell death. Cell 87: 565–576, 1996

28. Van Dongen CJ, Zwiers H, Gispen WH: Purification and partial characterization of the phosphatidylinositol 4-phosphate kinase from rat brain. Biochem J 223: 197–203, 1984

29. Lang CH, Dobrescu C, Bagby GJ: Tumor necrosis factor impairs insulin action on peripheral glucose disposal and hepatic glucose output. Endocrinology 130: 43–52, 1992

30. Grunfeld C, Feingold K: The metabolic effect of tumor necrosis factor and other cytokines. Biotherapy 3: 143–158, 1991

31. Torti FM, Dieckmann B, Beutler B, Cerami A, Ringold GM: A macrophage factor inhibits adipocyte gene expression: An in vitro model for cachexia. Science 229: 867–869, 1985

32. Grunfeld C, Feingold K: Metabolic disturbance and wasting in the acquired immunodeficiency syndrome. N Engl J Med 327: 329–337, 1992

33. Copeland GP, Leinster SJ, Davis JC, Hipkin LJ: Insulin resistance in patient with colorectal cancer. Br J Surg 74: 1031–1035, 1987

34. Marano M, Moldawer L, Fong Y, Al E: cachectin/TNF production in experimental burns and pseudonomas infection. Arch Surg 123: 1383–1388, 1988

35. Hotamisligil GS, Shargill NS, Spiegelman BM: Adipose expression of tumor necrosis factor-α: Direct role in obesity-linked insulin resistance. Science 259: 87–91, 1993

36. Hofmann C, Lorenz K, Braithwaite SS, Colca JR, Palazuk BJ, Hotamisligil GS, Spiegelman BM: Altered gene expression for tumor necrosis factor-α and its receptors during drug and dietary modulation of insulin resistance. Endocrinology 134: 264–270, 1994

37. Lowell BB, Susulic S, Hamann A, Lawitts JA, Himms H-J, Boyer BB, Kozak LP, Flyers JS: Development of obesity in transgenic mice after genetic ablation of brown adipose tissue. Nature 366: 740–742, 1993

38. Hotamisligil GS, Arner P, Caro JF, Atkinson RL, Spiegelman BM: Increased adipose tissue expression of tumor necrosis factor-α in human obesity and insulin resistance. J Clin Invest 95: 2409–2415, 1995

39. Kern P, Saghizadeh M, Ong J, Bosch R, Deem R, Simsolo R: The expression of tumor necrosis factor in human adipose tissue. Regulation by obesity, weight loss, and relationship to lipoprotein lipase. J Clin Invest 95: 2111–2119, 1995

40. Saghizadeh M, Ong JM, Garvey TW, Henry R, Kern PA: The expression of TNF-α by human muscle. J Clin Invest 97: 1111–1116, 1996

41. Hotamisligil GS, Budavari A, Murray D, Spiegelman BM: Reduced tyrosine kinase activity of the insulin receptor in obesity-diabetes. J Clin Invest 94: 1543–1549, 1994

42. Hotamisligil GS, Johnson RS, Distel RJ, Ellis R, Papaioannou VE, Spiegelman BM: Uncoupling of obesity from insulin resistance through a targeted mutation in aP2, the adipocyte fatty acid binding protein. Science 274: 1377–1379, 1996

43. Hotamisligil GS, Murray DL, Choy LN, Spiegelman BM: Tumor necrosis factor α inhibits signaling from the insulin receptor. Proc Natl Acad Sci 91: 4854–4858, 1994

44. Stephens J, Pekala P: Transcriptional repression of the C/EBP-alpha and GLUT4 genes in 3T3-L1 adipocytes by tumor necrosis factor-alpha. Regulations is coordinate and independent of protein synthesis. J Biol Chem 267: 13580–13584, 1992

45. Feinstein R, Kanety H, Papa MZ, Lunenfeld B, Karasik A: Tumor necrosis factor-α suppresses insulin-induced tyrosine phosphorylation of insulin receptor and its substrates. J Biol Chem 268: 26055–26058, 1993

46. Peraldi P, Hotamisligil GS, Buurman WA, White MF, Spiegelman BM: Tumor necrosis factor (TNF)-α inhibits insulin signaling through stimulation of the p55 TNF receptor and activation of sphingomyelinase. J Biol Chem 271: 13018–13022, 1996

47. Kroder G, Bossenmayer B, Kellerer M, Capp E, Stoyanov B, Muhlhofer A, Berti L, Horikoshi H, Ullrich A, Haring H: Tumor necrosis factor α and hyperglycemia-induced insulin resistance. J Clin Invest 97: 1471–1477, 1996

48. White MF, Kahn CR: The insulin signaling system. J Biol Chem 269: 1–4, 1994

49. Hotamisligil GS, Peraldi P, Budavari A, Ellis R, White MF, Spiegelman BM: IRS-1-mediated inhibition of insulin receptor tyrosine kinase activity in TNF-α-and obesity-induced insulin resistance. Science 271: 665–668, 1996

50. Kanety H, Feinstein R, Papa M, Hemi R, Karasik A: Tumor necrosis factor-α-induced phosphorylation of insulin receptor substrate-1 (IRS-1). J Biol Chem 270: 23780–23784, 1995

51. Kolesnick R, Golde D: The sphingomyelin pathway in tumor necrosis factor and interleukin-1 signaling. Cell 77: 325–328, 1994

52. Lozano J, Berra E, Municio M, Diaz-Meco M, Dominguez I, Sanz L, Moscat J: Protein kinase C isoform is critical for kappa B-dependent promoter activation by sphingomyelinase. J Biol Chem 269: 19200–19202, 1994

53. Muller G, Ayoub M, Storz P, Rennecke J, Fabbro D, Pfizenmaier K: PKC ζ is a molecular switch in signal transduction of TNF-α bifunctionally regulated by ceramide and arachidonic acid. EMBO J 14: 1156–1165, 1995

54. Yao B, Zhang Y, Delikat S, Mathias S, Basu S, Kolesnick R: Phosphorylation of Raf by ceramide-activated protein kinase. Nature 378: 307–310, 1995

55. Dobrowsky R, Hannun Y: Ceramide stimulates a cytosolic protein phosphatase. J Biol Chem 267: 5048–5051, 1992

56. Ofei F, Hurel S, Newkirk J, Sopwith M, Taylor R: Effects of an engineered human anti-TNF-α antibody (CPD571) on insulin sensitivity and glycemic control in patients with NIDDM. Diabetes 45: 881–885, 1996

57. Boden G, Chen X, Ruiz J, White JV, Rosseti L: Mechanisms of fatty acid-induced inhibition of glucose uptake. J Clin Invest 93: 2438–2446, 1994

58. Muller HK, Kellerer B, Ermel A, Muhlhofer B, Obermaier-Kusser B, Vogt B, Haring HC: Prevention by protein kinase C inhibitors of glucose-induced insulin receptor tyrosine kinase resistance in rat fat cells. Diabetes 40: 1440–1448, 1991

Molecular and Cellular Biochemistry **182**: 177–184, 1998.

Membrane glycoprotein PC-1 and insulin resistance

Ira D. Goldfine, Betty A. Maddux, Jack F. Youngren, Lucia Frittitta,
Vincenzo Trischitta and G. Lynis Dohm
Division of Diabetes and Endocrine Research, San Francisco, CA, USA

Abstract

Peripheral resistance to insulin is a major component of non-insulin dependent diabetes mellitus. Defects in insulin receptor tyrosine kinase activity have been demonstrated in several tissues from insulin resistant subjects, but mutations in the insulin receptor gene occur in only a small fraction of cases. Therefore, other molecules that are capable of modulating the function of the insulin receptor are likely candidates in the search for the cellular mechanisms of insulin resistance. We have isolated an inhibitor of insulin receptor tyrosine kinase activity from cultured fibroblasts of an insulin resistant NIDDM patient and identified it as membrane glycoprotein PC-1. Subsequently we have demonstrated that expression of PC-1 is elevated in fibroblasts from other insulin resistant subjects, both with and without NIDDM. Studies in muscle, the primary site for insulin-mediated glucose disposal, have shown that the levels of PC-1 in this tissue are inversely correlated to insulin action both *in vivo* and *in vitro*. Transfection of PC-1 into cultured cells has confirmed that overexpression of PC-1 can produce impairments in insulin receptor tyrosine kinase activity and the subsequent cellular responses to insulin. Preliminary data suggests a direct interaction between PC-1 and the insulin receptor. However, the mechanisms whereby PC-1 inhibits insulin receptor signaling remain to be determined. (Mol Cell Biochem **182**:177–184, 1998)

Key words: membrane glycoprotein, diabetes mellitus, hyperglycemia, insulin receptor, tyrosine kinase

Introduction

Non insulin dependent diabetes mellitus (NIDDM) is a major health problem in the United States affecting approximately 5% of the population. Development of NIDDM involves both insulin secretory abnormalities and insulin resistance in peripheral tissues [1–4]. In NIDDM patients, insulin resistance both precedes and contributes to the development of the disease [5, 6]. In non-diabetic individuals, increased secretion of insulin is sufficient to maintain normoglycemia, although the resultant hyperinsulinemia is believed to play a major role in causing hypertension, coronary artery disease and, hyperlipidemia [1].

There are numerous factors which can contribute to insulin resistance in an individual [1, 2, 4]. Data from ethnic, family, and longitudinal studies suggest that insulin resistance can be inherited, and is an intrinsic feature of many NIDDM patients [7–9]). There also are extrinsic causes by which insulin resistance can develop in an individual.

Obesity and inactivity are common contributors to insulin resistance [10, 11]. In addition, an altered hormonal milieu (i.e. pregnancy, an excess of growth hormone, gluco-corticoids, catecholamines etc.) can induce peripheral insulin resistance [4, 12, 13]. It is not clear whether the various contributors to insulin resistance act through similar or distinct mechanisms, and in the majority of individuals the molecular basis of the resistance to insulin is unknown.

The cellular response to insulin is mediated through the insulin receptor (IR), which is a tetrameric protein consisting of two identical extracellular alpha subunits that bind the hormone and two identical transmembrane beta subunits that have intracellular tyrosine kinase activity (Fig. 1). The IR is highly homologous to the receptor for IGF-1 (IGF-1-R) [14]. When insulin binds to the IR alpha subunit, the beta subunit tyrosine kinase domain is activated, and insulin action ensues [15, 16]. When insulin activates the receptor, the beta subunit is autophosphorylated at the juxtamembrane domain (tyr 960), the kinase domain (tyrs 1146, 1150 and 1151) and the

Address for offprints: I.D. Goldfine, Division of Diabetes and Endocrine Research, 1600 Divisadero Street, UCSF Box 1616, San Francisco, CA 94143-1616, USA

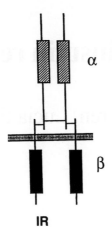

Fig. 1. Subunit structure of the insulin receptor.

C-terminal domain (tyr 1316 and 1322) [16]. Subsequently, endogenous substrates including IRS-1, IRS-2 and SHC are tyrosine phosphorylated [16, 17]. These phosphorylated substrates act as docking molecules to activate SH2 domain molecules including: GRB-2 which activates the ras pathway; the p85 subunit of PI-3kinase; protein tyrosine phosphatase PTP2/SYP, PLCy/NCK, and others [16, 17].

Various groups, including our own, have reported that IR tyrosine kinase activity, but not IGF-1-R kinase activity, is impaired in muscle, fat, fibroblasts and other tissues of patients with NIDDM (for reviews see [3, 18]). In addition, IR tyrosine kinase activity is impaired in tissue from insulin resistant, non-diabetic subjects, both lean and obese [19–21]. In addition, defects in IR tyrosine kinase activity have been reported in the hyperandrogenic, insulin-resistant condition of polycystic ovary syndrome [22]. Abnormalities in the IR gene do not appear to be the major cause of the decreased IR tyrosine kinase activity in the vast majority of patients examined [23]. Rather, the defect(s) in IR signaling capacity that occurs in insulin resistant subjects most likely involves molecules, apart from the receptor, that are capable of modulating its function. One potential regulatory molecule is PC-1.

Characteristics of PC-1

PC-1 is a class II transmembrane glycoprotein that is located both on plasma membrane and in the endoplasmic reticulum (Fig. 2). PC-1 is the same protein as liver nucleotide pyrophosphatase/alkaline phosphodiesterase 1 [24–32]. PC-1 is also related to the tumor mobility factor autotaxin, brain specific phosphodiesterase PD-Iα, and the neuroprotein gp130 RB13-6 [33–35]. PD-Iα and autotaxin may be the product of the same gene [33]. PC-1 has been reported to be expressed in plasma and intracellular membranes of:

plasma cells, placenta, the distal convoluted tubule of the kidney, ducts of the salivary gland, epididymis, proximal part of the *vas deferens*, chondrocytes, dermal fibroblasts, skeletal muscle, and adipose tissue [31, 36, 37]. PC-1 exists as a homodimer of 230–260 kDa; the reduced form of the protein has a molecular size of 115–135 kDa, depending on the cell type. Human PC-1 is predicted to have 873 amino acids, but the translation start site has been controversial [27, 29]. Human and mouse PC-1 are similar; there is an 87% nucleotide and 77% amino acid sequence homology between the two proteins [27, 29].

Human PC-1 maps to the chromosome location, 6q22-6q23 [27, 29]. Preliminary studies have been carried out on the exon structure of PC-1 by Goding and colleagues [38]. They have suggested that PC-1 has at least 6 exons and report that: exon 1 encodes the cytoplasmic tail and the proximal part of the transmembrane region; exon 2 encodes the remainder of the transmembrane region and a short extracellular sequence; exons 3 and 4 encode the cysteine-rich repeats noted in the cDNA sequence; and exon 5 contains the proteolytic cleavage site (pro-ala).

On the basis of studies with monoclonal antibodies it has been suggested that a portion of PC-1 is inserted into the membrane such that there is a small cytoplasmic amino terminus, and a larger extracellular carboxyl terminus [27, 29] (Fig. 2). In addition, a soluble form of PC-1 has been reported [38]. It has been proposed that this soluble form derives from the membrane bound PC-1 form by proteolytic cleavage of a pro-ala sequence. There is an extracellular cysteine-rich domain adjacent to the plasma membrane, and 10 potential N-glycosylation sites. PC-1 has an ATP binding site, GXGXXG...A/GXK, an EF-hand motif found in calcium binding proteins and a somatomedin-like region [27, 29]. The extracellular domain of PC-1 cleaves sugar-phosphate, phosphosulfate, pyrophosphate, and phosphodiesterase linkages [28, 31]. The active enzyme site for phosphodiesterase and pyrophosphatase contains a key threonine residue [27, 29]. We have recently shown that mutation of this residue does not impair the ability of PC-1 to inhibit IR function [39] and thus the phosphodiesterase activity of PC-1 is not necessary for its ability to inhibit the IR. Just under the plasma membrane is a potential tyrosine phosphorylation site. It has been suggested that PC-1 may have threonine-specific protein kinase activity [40], although it does not have sequence homology to the known threonine kinases.

The physiological function of PC-1 is unknown. It has been proposed that PC-1 may belong to an enzyme cascade system working in concert with other ectoenzymes that hydrolyze nucleotides and nucleic acids to nucleosides which are then taken up by cells via a nucleoside transporter [31, 32]. This pathway would allow the salvage of nucleotides from the extracellular fluid, and also allow the uptake of nucleosides by cells that are unable to synthesize purines by the *de novo* pathway.

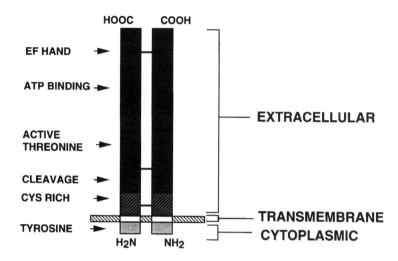

Fig. 2. Schematic diagram of membrane glycoprotein PC-1.

PC-1 and diabetes

Overall, genome-wide scans have been largely unsuccessful in identifying major genes predisposing for common forms of NIDDM. However, several independent studies have reported linkage between the region of the PC-1 gene (chromosome 6q22-q23) and different forms of NIDDM. Stern *et al.* have found a marker related to 2 h glucose levels following oral glucose tolerance tests in Mexican-Americans (D6S290) that is located near the PC-1 gene [41]. Temple *et al.* have found that a gene for neonatal diabetes is localized to 6q22-q23 [42]. Neonatal diabetes is usually resolved in the first 6 months of life, but predisposes the individual to NIDDM [43]. Investigating PC-1 as a candidate gene for NIDDM, Doria *et al.* reported DNA polymorphisms in the region of the PC-1 gene, one of which was twice as frequent in NIDDM patients than controls [44].

Our investigation of PC-1 began when we studied skin fibroblasts derived from a 42-year old woman (MW) who had resistance to insulin [45, 46]. In her cultured cells, insulin stimulation of biological functions such as glucose transport was impaired (Fig. 3). The IR content and insulin binding to her fibroblasts was normal, and, when purified, the IR had normal tyrosine kinase activity [45]. However, her fibroblasts produced a glycoprotein inhibitor of IR tyrosine kinase resulting in decreased whole cell IR tyrosine kinase activity (Fig. 4). The closely related IGF-1 receptor however was not affected. We purified this inhibitor by lectin affinity chromatography followed by ATP-agarose chromatography. Sequence was obtained and found to be identical to the corresponding cyanogen bromide peptides of membrane glycoprotein PC-1 [46]. Western blot analysis and ELISA with antibodies specific for PC-1 indicated that there was an approximate 5–10 fold increase in PC-1 content when

extracts from the patient's cells were compared to those from controls [46] (Fig. 5). Thereafter, we carried out Northern blot analysis of PC-1 mRNA. In mRNA from MW, there was an approximate 5–10 fold increase in the major mRNA species for PC-1 [46].

We next determined whether an elevation in PC-1 was unique to patient MW or was present in cells of others. Accordingly, we studied PC-1 content in fibroblasts from 9 additional insulin resistant NIDDM patients (Fig. 6). Seven of the 9 patients had elevated PC-1. In fibroblasts from patients with increased PC-1, IR tyrosine kinase activity was found to be decreased [46]. Another group has recently reported that PC-1 is elevated in fibroblasts of insulin resistant individuals [47]. These studies suggest therefore that in fibroblasts from NIDDM patients, elevations in PC-1 may play a role in insulin resistance.

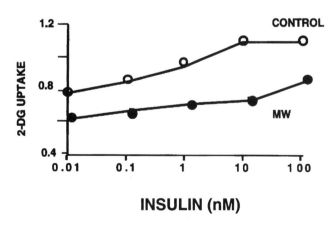

Fig. 3. Impaired insulin action in fibroblasts from patient MW. Cells were incubated 1 h with insulin and 30 min with [^{3}H]2-deoxy-D-glucose (2DG). Uptake was measured and expressed as femtomoles per mg protein.

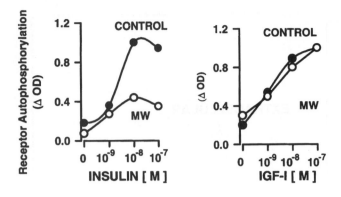

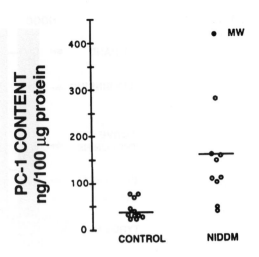

Fig. 4. Stimulation of IR (left) and IGF-1-R (right) beta subunit auto-phosphorylation in fibroblasts from patient (MW) and a sex and age-matched control. Fibroblasts were stimulated with insulin or IGF-1, for 5 min. Receptors were solubilized and captured on ELISA plates coated with either an IR antibody or an IGF-1-R antibody. Receptor autophosphorylation was determined by labelling with an anti-phosphotyrosine antibody and subsequent peroxidase readout.

Fig. 6. PC-1 content in dermal fibroblasts from controls and NIDDM patients. PC-1 content in fibroblasts from controls and NIDDM patients was determined by ELISA.

We have recently undertaken clinical studies on the role of PC-1 in insulin resistance. These studies have focused on skeletal muscle, as this is the major tissue for insulin-mediated glucose disposal [48]. In collaboration with the laboratory of Dr. Riccardo Vigneri at the University of Catania, Italy, we examined the role of PC-1 in the 'intrinsic' insulin resistance seen in a subpopulation of lean, non-diabetic subjects. Biopsies of external oblique muscle were obtained from healthy non diabetic, non obese subjects (11 females, 6 males; age, 31 ± 3 years; body mass index (BMI), 24.1 ± 1 kg/m^2; fasting glucose, 91 ± 2 mg/dl) [49]. In this group, there was a wide range of insulin sensitivity as determined by intravenous insulin tolerance tests. There was a significant, inverse correlation between PC-1 content in muscle, and the K value of the insulin tolerance tests ($r = -0.57$, $p = 0.035$) (Fig. 7). Further, PC-1 levels correlated negatively and significantly with the ability of insulin to stimulate muscle IR tyrosine kinase *in vitro* ($r = 0.45$, $p = 0.022$) (Fig. 8). These data suggested that in some individuals, an intrinsic

upregulation of skeletal muscle PC-1 expression may contribute to impaired insulin receptor function and insulin resistance.

In collaboration with Dr. Lynis Dohm at East Carolina University, we examined the role of PC-1 in the insulin resistance that develops secondary to obesity [36]. In this study, muscle biopsies were obtained from subjects during elective abdominal surgery. Subjects represented a wide range of obesity (BMI range 19–90) and insulin sensitivity.

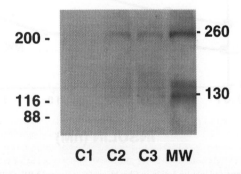

Fig. 5. Western blot analysis of PC-1 in fibroblasts. (Adapted from reference [46], with permission).

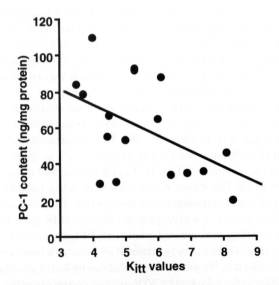

Fig. 7. Correlation between PC-1 content in skeletal muscle of non obese, non diabetic subjects and whole body insulin sensitivity. PC-1 content in solubilized biopsy extracts was determined by radioimmunoassay. Insulin sensitivity determined by intravenous insulin tolerance test and calculated as the rate constant for glucose disappearance (K_{itt}). ($r = -0.57$, $p = 0.035$). (Adapted from reference [37], with permission).

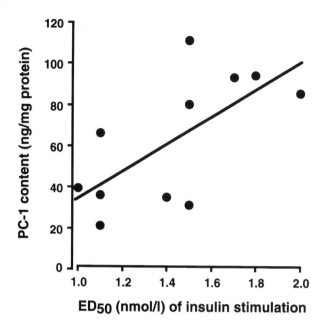

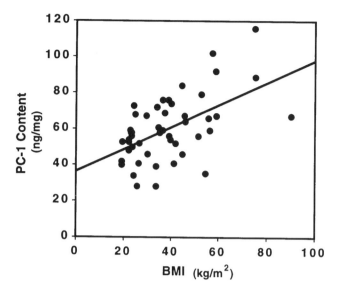

Fig. 9. Correlation between PC-1 in skeletal muscle and subject BMI. PC-1 content of soluble muscle extracts was determined by radioimmunoassay. (Adapted from reference [36], with permission).

vs. 336 ± 45 ng/mg, p = 0.012). In addition, PC-1 content was significantly and inversely correlated with both the K value of insulin tolerance tests and fasting plasma insulin levels (r = –0.5 and –0.58, p = 0.04 and 0.009, respectively).

These studies demonstrated an inverse relationship between PC-1 content and tyrosine kinase activity, but did

Fig. 8. Correlation between PC-1 content and the ED_{50} of *in vitro* insulin stimulation of insulin receptor tyrosine kinase activity in skeletal muscle of non obese, non diabetic subjects. PC-1 content in solubilized biopsy extracts was determined by radioimmunoassay. Solubilized IR were captured on ELISA plates coated with an IR antibody. Immunocaptured receptors were incubated with ^{32}P ATP and the peptide substrate poly Glu 4:Tyr 1 at varying concentrations of insulin (0–100 nM). Substrate phosphorylation was determined by adsorbing the peptide substrate to filter paper and scintillation counting. (r = 0.45, p = 0.022). (Adapted from reference [37], with permission).

PC-1 content of muscle biopsies correlated with BMI in these subjects (r = 0.55, p < 0.001) (Fig. 9), and was inversely related to insulin stimulation of glucose transport in isolated muscle strips (r = –0.58, p = 0.008) (Fig. 10). Multivariate analysis indicated that muscle PC-1 content was an independent predictor of glucose transport, and accounted for 42% of the variance in glucose transport values. These data are consistent with an upregulation of PC-1 expression in obesity that may play a role in the impaired IR tyrosine kinase activity and insulin resistance that occur in the obese state.

While adipose tissue accounts for only a small portion of whole-body, insulin-stimulated glucose disposal [48] there is strong evidence that fat cell insulin contributes to the development of NIDDM through the regulation of hepatic glucose output by plasma free fatty acids [50]. Therefore, we were interested in seeing whether PC-1 might play a role in adipose tissue insulin resistance. Again in collaboration with the laboratory of Dr. Vigneri, we obtained adipose tissue biopsies from and administered intravenous insulin tolerance tests to 20 (13 males, 7 females) healthy, lean, non-diabetic subjects [37]. In this population we found that fat cell membrane content of PC-1 was elevated in insulin resistant vs. insulin sensitive individuals (525 ± 49

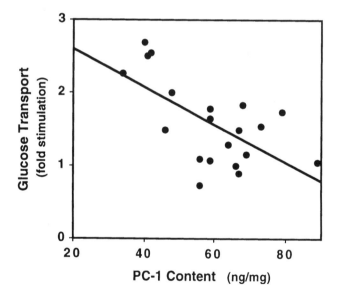

Fig. 10. Correlation between PC-1 content in skeletal muscle and insulin-stimulated glucose transport in incubated muscle strips. PC-1 content of soluble muscle extracts was determined by radioimmunoassay. Glucose transport was measured in muscle strips incubated in the presence and absence of 100 nM insulin. Insulin stimulation of transport was calculated as the fold increase in [^{14}C] 2-deoxyglucose uptake over basal values. (Adapted from reference [37], with permission).

182

not demonstrate causality. Thus, we examined the effect of PC-1 overexpression by transfecting PC-1 into cultured cell lines [39, 46]. For most studies, we employed MCF-7 cells, an insulin sensitive human cell line [39, 46, 51, 52], that normally has low levels of PC-1. These cells were transfected with an expression plasmid containing PC-1 cDNA. Overexpression of this protein in MCF-7 cells, like in fibroblasts overexpressing PC-1, did not alter insulin binding. In MCF-7 cells, PC-1 overexpression induced marked inhibition of insulin stimulated IR tyrosine kinase activity, as assessed by IR autophosphorylation and phosphorylation of the intracellular protein, IRS-1 (Fig. 11) [16, 17]. Various biological effects of insulin were also blunted (see Fig. 12). As in fibroblasts, the related IGF-1-R and EGFR tyrosine kinases were not affected. Thus overexpression of PC-1 induces *in vitro* insulin resistance, and likely plays a direct role in regulating IR function.

Future directions

A major area to be studied further is how PC-1 regulates the IR. Most likely PC-1 modifies IR signaling by either: directly binding to the IR; covalently modifying it by a process such as phosphorylation; or generating an intracellular signal that in turn modifies the IR. The latter possibility is somewhat unlikely since PC-1 has a short intracellular domain and thus may not be involved in signaling. PC-1 may alter IR alpha subunit function, even though it does not influence insulin binding. PC-1 inhibition of the IR is relatively specific since PC-1 does not have effects on IGF-I-R and EGFR autophosphorylation. Since, the IGF-I-R has a very similar beta subunit to the IR, but a different alpha subunit [14], it is more likely that PC-1 interacts with the IR alpha subunit.

It has been reported that PC-1 copurifies with the IR on lectin affinity chromatography [30]. In concert, we have observed that the IR purified by IR monoclonal antibody columns [53] have no detectable PC-1 activity, whereas the IR isolated by an insulin affinity column has both detectable PC-1 activity and relatively decreased IR tyrosine kinase activity. Also, we have preliminary data that purified PC-1 interacts with the IR both *in vivo* and *in vitro*. It is therefore possible that PC-1 binds to the IR alpha subunit and alters its ability to activate the beta subunit.

It is also possible that protein phosphorylation mediates the interaction of PC-1 with the IR. PC-1 has been reported to be a threonine kinase [40], and the IR is modified by serine/threonine phosphorylation [15]. The IR contains multiple serine/threonine phosphorylation sites, and thus is a potential substrate for this putative PC-1 threonine kinase activity [40].

The active site of the PC-1 molecule involved in IR regulation is unknown. We have found that the active phosphodiesterase/pyrophosphatase site of PC-1 does not regulate this function [39]. The ATP, Ca^{2+} binding, and somatomedin-like domains are other candidate sites. These domains need to be investigated to determine whether the action of PC-1 on the IR is modified when the mutant enzyme is expressed in cultured cells.

Little is known about the human PC-1 gene and regulation of its expression. In particular the nature of its promoter and other regulatory elements have not been reported. It has been observed that glucocorticoids raise PC-1 levels in several cell types [31, 54]. We have been able to demonstrate that dexamethasone, a synthetic glucocorticoid, raises PC-1 content in MCF-7 cells and in IM-9 cells. These data suggest therefore the presence of a GRE in the regulatory sequences of the PC-1 gene. In the mouse, PC-1 is linked to the proto-oncogene Myb on mouse chromosome 10 [55]. This region is syntenic with human gene chromosome region 6q where the PC-1 and Myb genes reside. This observation may be important because PC-1 is overexpressed in malignant plasma cells and other tumor cells [55].

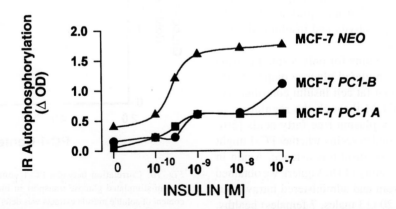

Fig. 11. Insulin stimulation of IR autophosphorylation in transfected MCF-7 cell lines. cells were stimulated with various concentrations of insulin for 5 min. Receptors were solubilized and captured on ELISA plates coated with either an IR antibody or an IGF-1-R antibody. Receptor autophosphorylation was determined by labelling with an antiphosphotyrosine antibody and subsequent peroxidase readout.

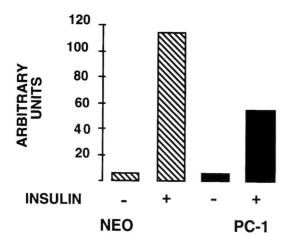

Fig. 12. Insulin stimulation of [³H]thymidine incorporation into DNA in transfected MCF-7 cell lines. Cells were stimulated with 100 nM insulin for 24 h. Incorporation. is expressed as femtomoles per mg protein. (Adapted from reference [46], with permission).

Acknowledgements

This work was funded in part by NIH grant DK-38416. J.F.Y was supported by NHRSA T32AG00212 and NHRSA DK 07418.

References

1. Reaven GM: Role of insulin resistance in human disease. Diabetes 37: 1595–607, 1988
2. Kahn CR: Insulin action, diabetogenes, and the cause of type II diabetes. Diabetes 43: 1066–1084, 1994
3. Kruszynska YT, Olefsky JM: Cellular and molecular mechanisms of non-insulin dependent diabetes mellitus. J Invest Med 44: 423–428, 1996
4. Hamman RF: Genetic and environmental determinants of non-insulin-dependent-diabetes mellitus (NIDDM). Diabet/Metabol Rev 8: 287–338, 1992
5. Lillioja S, Mott DM, Howard BV, Bennett PH, Yki-Järvinen H, Freymond D, Nyomba BL, Zurlo F, Swinburn B, Bogardus C: Impaired glucose tolerance as a disorder of insulin action: longitudinal and cross-sectional studies in Pima Indians. N Engl J Med 318: 1217–1225, 1988
6. Martin BC, Warrarn JH, Krolewski AS, Bergman RN, Soeldner JS, Kahn CR: Role of glucose and insulin resistance in development of type 2 diabetes mellitus: Results of a 25-year follow-up study. Lancet 340: 925–929, 1992
7. Bogardus C, Lillioja S, Nyomba BL, Zurlo F, Swinburn B, Puente AE-D, Knowler WC, Ravussin ER, Mott DM, Bennett PH: Distribution of *in vivo* insulin action in Pima Indians as mixture of three normal distributions. Diabetes 38: 1423–1432, 1989
8. Lillioja S, Mott DM, Zawadzki KK, Young AA, Abbott WGH, Knowler WC, Bennett PH, Moll P, Bogardus C: *In vivo* insulin action is a familial characteristic in nondiabetic Pima Indians. Diabetes 36: 1329–1335, 1987
9. Vaag A, Henriksen JE, Maclsbad S, Holm N, Beck-Nielsen H: Insulin secretion, insulin action, and hepatic glucose production in identical twins discordant for non-insulin-dependent diabetes mellitus. J Clin Invest 95: 690–698, 1995
10. Bogardus C, Lillioja S, Mott DM, Hollenbeck C, Reaven GM: Relationship between degree of obesity and *in vivo* insulin action in man. Am J Physiol 248: E286–E291, 1985
11. Ludvik B, Nolan JJ, Baloga J, Sacks D, Olefsky JM: Effect of obesity on insulin resistance in normal subjects and patients with NIDDM. Diabetes 44: 1121–1125, 1995
12. Brindley DN: Role of glucocorticoids and fatty acids in the impairment of lipid metabolism observed in the metabolic syndrome. Int J Obes Rel Metabol Disorders 19(suppl 1): S69–S75, 1995
13. Moller N, Jorgensen JD, Moller J, Orskov L, Ovesen P, Schmitz 0, Christiansen JS, Orskov H: Metabolic effects of growth hormone in humans. Metabol Clin Exp 44(suppl 4): 33–36, 1995
14. DeMeyts P: The structural basis of insulin and insulin-like growth factor-1 receptor binding and negative cooperativity, and its relevance to mitogenic versus metabolic signaling. Diabetologia 37 (Suppl 2): S135–S148, 1994
15. Goldfine ID: The insulin receptor: Molecular biology and transmembrane signalling. Endocr Rev 8: 235–255, 1987
16. Kahn CR, White MF: The insulin receptor and the molecular mechanism of insulin action. J Clin Invest 82: 1151–1156, 1988
17. White MF: The IRS-1 signaling system. Curr Opin Genet Devel 4: 47–54, 1994
18. Moller DE, Flier JS: Insulin resistance – mechanisms, syndromes, and implications. New Eng J Med 325: 938–948, 1991
19. Caro JF, Sinha MK, Raju SM, Ittoop O, Pories WJ, Flickinger EG, Meelheim D, Dohm GL: Insulin receptor kinase in human skeletal muscle from obese subjects with and without noninsulin dependent diabetes. J Clin Invest 79: 1330–1337, 1987
20. Grasso G, Frittitta L, Anello M, Russo P, Susti G, Trischitta V: Insulin receptor tyrosine kinase activity is altered in both muscle and adipose tissue from nonobese normoglycemic insulin resistant subjects. Diabetologia 38: 55–61, 1995
21. Handberg A, Vaag A, Vinten J, Beck-Nielsen H: Decreased tyrosine kinase activity in partially purified insulin receptors from muscle of young, nonobese first degree relatives of patients with type 2 (non-insulin-dependent) diabetes mellitus. Diabetologia 36: 668–674, 1993
22. Dunaif A, Xia J, Book C-B, Schenker E, Tang Z Excessive insulin receptor serine phosphorylation in cultured fibroblasts and in skeletal muscle: A potential mechanism for insulin resistance in the Polycystic Ovary Syndrome. J Clin Invest 96: 801–810, 1995
23. Taylor SI: Lessons from patients with mutations in the insulin- receptor gene. Diabetes 41: 1473–1490, 1992
24. Yano T, Funakoshi I, Yamashina I: Purification and properties of nucleotide pyrophosphatase. J Biochem 98: 1097–1107, 1985
25. van Driel IR, Goding JW: Plasma cell membrane glycoprotein PC-1. J Biol Chem 262: 4882–4887, 1987
26. Harahap AR, Goding M Distribution of PC-1 in non lymphoid tissues. J Immunol 141: 2317–2320, 1988
27. Buckley MF, Loveland KA, McKinstry WJ, Garson OM, Goding JW: Plasma cell membrane glycoprotein PC-1 cDNA cloning of the human molecule, amino acid sequence, and chromosomal location. J Biol Chem 265: 17506–17511, 1990
28. Rebbe NF, Tong BD, Finley EM, Hickman S: Identification of nucleotide pyrophosphatase/alkaline phosphodiesterase I activity associated with the mouse plasma cell differentiation antigen PC-1. Proc Natl Acad Sci USA 88: 5192–5196, 1991
29. Funakoshi I, Kato H, Horie K, Yano T, Hori Y, Kobayashi H, Inoue T, Suzuki H, Fukui S, Tsukahara M, Kajii T, Yamashina I: Molecular cloning of human nucleotide pyrophosphatase. Arch Biochem Biophys 295: 180–187, 1992
30. Uriarte M, Stalmans W, Hickman S, Bollen M: Phosphorylation and nucleotide-dependent dephosphorylation of hepatic polypeptides related to the plasma cell differentiation antigen PC-1. Biochem J 293: 93–100, 1993

31. Rebbe NF, Tong BID, Hickman S: Expression of nucleotide pyrophosphatase and alkaline phosphodiesterase I activities of PC-1, the murine plasma cell antigen. Mol Immunol 30: 87–93, 1993

32. Yoshida H, Fukui S, Funakoshi I, Yamashina I: Substrate specificity of a nucleotide pyrophosphatase responsible for the breakdown of 3′-phosphoadenosine 5′-phosphosulfate (PAPS) from human placenta. J Biochem 93: 1641–1648, 1983

33. Kawagoe H, Soma O, Goji J, Nishimura N, Narita M, Inazawa J, Nakamura H, Sano K: Molecular cloning and chromosomal assignment of the human brain-type phosphodiesterase I/Nucleotide Pyrophosphotase Gene (PDNP2). Genomics 30: 380–384, 1995

34. Murata J, Lee HY, Clair T, Krutzsch HC, Arestad AA, Sobel ME, Liotta LA, Stracke M: cDNA cloning of the human tumor motility-stimulating protein, autotaxin, reveals a homology with phosphodiesterase. J Biol Chem 269: 30479–30484, 1994

35. Deisler H, Lottspeich F, Rajewsky MF: Affinity purification and cDNA cloning of rat neural differentiation and tumor cell surface antigen gp130 RB13-6 reveals relationship to human and murine PC-1. J Biol Chem 270: 9849–9855, 1995

36. Youngren J, Maddux BA, Sasson S, Sbraccia P, Tapscott EB, Swanson MS, Dohm GL, Goldfine ID: Skeletal muscle content of membrane glycoprotein PC-1 in obesity. Diabetes 45: 1324–1328, 1996

37. Frittitta L, Youngren JF, Sbraccia P, D'Adamo M, Buongiorno A, Vigneri R Goldfine ID, Trischitta V: Increased adipose tissue PC-1 protein content but not TNF-α gene expression is associated to a reduction in whole body insulin sensitivity and insulin receptor tyrosine kinase activity. Diabetologia, March 1997, (in press)

38. Belli AI, van Driel IR, Goding JW: Identification and characterization of a soluble form of the plasma cell membrane glycoprotein PC-1 (5′-nucleotide phosphodiesterase) Endo J Biochem 217: 421–428, 1993

39. Grupe A, Alleman J, Goldfine ID, Sadick M, Stewart T: Inhibition of insulin receptor phosphorylation by PC-1 is not mediated by the hydrolysis of Adenosine triphosphate or the generation of Adenosine. J Biol Chem 270: 22085–22088, 1995

40. Oda Y, Kuo M-D, Huang SS, Huang JS The plasma cell membrane glycoprotein, PC-1, is a threonine specific protein kinase stimulated by acidic fibroblast growth factor. J Biol Chem 266: 16791–16795, 1991

41. Stern MP, Duggurila R, Mitchell BD, Reinhart U, Shivakumar S, Shipman PA, Uresandi OC, Benavides E, Blangero J, O'Connell P: Evidence for linkage of regions on chromosomes 6 and 11 to plasma glucose concentrations in Mexican Americans. Genome Res 6: 724–734, 1996

42. Temple IK, Gardner RJ, Robinson DO, Kilbirige MS, Ferguson AW, Baum JD, Barber JCK, James RS, Shield JPH: Further evidence for an imprinted gene for neonatal diabetes localised to chromosome 6q22-q23. Human Mol Genet 5: 1117–1121, 1996

43. von Muhlendahl KE, Herkennoff H: Longterm course of neonatal diabetes. New Engl J Med 333: 704–708, 1995

44. Doria A, Federman S, Rich SS, Warram JH, Krolewski AS: DNA polymorphisms and mapping on chromosome 6q of plasma-cell antigen 1 (PC-1), an insulin signal inhibitor. Diabetologia 39(suppl 1): A77, 1996

45. Sbraccia P, Goodman PA, Maddux BA, Chen Y-DI, Reaven GM, Goldfine ID: Production of an inhibitor of insulin receptor tyrosine kinase in fibroblasts from a patient with insulin resistance and NIDDM. Diabetes 40: 295–299, 1991

46. Maddux BA, Sbraccia P, Kumakura S, Sasson S, Youngren J, Fisher A, Spencer S, Grupe A, Henzel W, Stewart TA, Reaven GM, Goldfine ID: Membrane Glycoprotein PC-1 and insulin resistance in non-insulin-dependent diabetes. Nature 373: 448–451, 1995

47. Whitehead JP, Humphreys PJ, Maasen JA, Moller DE, Krook A, O'Rahilly S: Increased PC-1 phosphodiesterase activity in patients with postreceptor insulin resistance but not in patients with insulin receptor mutations. Diabetologia [Suppl 1]: I–V, A65, 1995

48. Bonadonna RC, De Fronzo RA: Glucose metabolism in obesity and type 2 diabetes. Diabet Metab 17: 112–135, 1991

49. Frittitta L, Youngren J, Trischitta V, Goldfine ID: Membrane glycoprotein PC-1 content in skeletal muscle of non obese, non diabetic subjects: Relationship to insulin receptor tyrosine kinase activity and whole body insulin sensitivity. Diabetologia 39: 1190–1195, 1996

50. Reaven GM: The fourth musketeer, from Alexander Dumas to Claude Bernard. Diabetologia 38: 3–13, 1995

51. Milazzo G, Giorgino F, Damante G, Sung C, Stampfer MR, Vigneri R Goldfine ID, Belfiore A: Insulin receptor expression and function in human breast cancer cell lines. Cancer Res 52: 3924–3930, 1992

52. Osborne CK, Monaco ME, Lippmann ME: Hormone responsive human breast cancer in long-term tissue culture: Effect on insulin. Proc Natl Acad Sci USA 73(12): 4536–4540, 1976

53. Ebina Y, Ellis L, Jamagin K, Edery M, Graf L, Clauser E, Ou J-H, Maslarz F, Kan YW, Goldfine ID, Roth RA, Rutter WJ: The human insulin receptor cDNA: The structural basis for hormone-activated transmembrane signalling. Cell 40: 747–758, 1985

54. Rousseau GG, Amar-Costesec A, Verhaegen M, Granner DK: Glucocorticoid hormones increase the activity of plasma membrane alkaline phosphodiesterase I in rat hepatoma cells. Proc Natl Acad Sci USA 77: 1005–1009, 1980

55. Buckley MF, Goding JW: Plasma cell membrane glycoprotein gene PC-1 (alkaline phosphodiesterase 1) is linked to the proto-oncogene Myb on mouse chromosome 10. Immunogenetics 36: 199–201, 1992

Molecular and Cellular Biochemistry **182**: 185–191, 1998.
© 1998 *Kluwer Academic Publishers. Printed in the Netherlands.*

Insulin action on protein phosphatase-1 activation is enhanced by the antidiabetic agent pioglitazone in cultured diabetic hepatocytes

Subbiah Pugazhenthi and Ramji L. Khandelwal
Department of Biochemistry, University of Saskatchewan, Saskatoon, Saskatchewan, Canada

Abstract

Effect of the antidiabetic agent pioglitazone on the insulin-mediated activation of protein phosphatase-1 was examined in diabetic hepatocytes. Streptozotocin-induced diabetes in Sprague Dawley rats caused a significant decrease in the activation of glycogen synthase in hepatocytes isolated from these animals. There was an inverse correlation between the *in vivo* hyperglycemic condition and the *in vitro* activation of glycogen synthase in liver cells (r = 0.93, p < 0.001). Long term incubation of diabetic hepatocytes with insulin and dexamethasone caused significant (p < 0.001) improvement in the activation of glycogen synthase activation. When incubated along with hormones, pioglitazone enhanced their action (p < 0.05–0.01). Diabetic hepatocytes were also characterized by 50% decrease in the activity of protein phosphatase-1, the enzyme which dephosphorylates and activates glycogen synthase. Pioglitazone potentiated the acute stimulatory effect of insulin on protein phosphatase-1 in normal hepatocytes but not in diabetic hepatocytes. Long term incubation of diabetic hepatocytes with insulin ameliorated the decrease in the protein phosphatase -1 activity in these cells. This stimulatory long-term effect of insulin was significantly (p < 0.05) enhanced by the antidiabetic agent pioglitazone. (Mol Cell Biochem **182**: 185–191, 1998)

Key words: pioglitazone, protein phosphatase-1, glycogen synthase, diabetes, hepatocytes

Abbreviations: STZ – streptozotocin; DMEM – Dulbecco's modified eagle medium; MAP kinase – mitogen-activated protein kinase; MEK – MAP kinase kinase; NIDDM – non-insulin-dependent diabetes mellitus; PI3-kinase – phosphatidylinositol 3-kinase; PP-1 – protein phosphatase-1

Introduction

Insulin-resistance is a characteristic feature of both non-insulin-dependent and insulin-dependent types of diabetes mellitus [1, 2]. While developing orally effective drugs for the treatment of this disease, one needs to specifically target this defect to enhance insulin action. A group of thiazolidinedione derived antidiabetic agents such as pioglitazone, englitazone and troglitazone have been shown to possess such insulin potentiating properties [3–5]. They are structurally different from sulfonylureas and biguanides, the oral hypoglycemic drugs that are currently in clinical use. Chronic administration of these agents to insulin-resistant diabetic animals leads to enhanced insulin sensitivity *in vivo* [6–9]. They potentiate insulin-mediated glucose uptake and utilization in muscle by stimulating its receptor tyrosine kinase activity [8, 10]. *In vivo* administration of these agents leads to suppression of gluconeogenesis and stimulation of glycolysis in liver at the level of expression of key regulatory enzymes [12, 13]. Englitazone decreases glucagon-induced glycogenolysis in hepatocytes and troglitazone enhances glycogen synthase activation in Hep G2 cells [14, 15]. However, there are no studies to examine if pioglitazone can correct the abnormalities of glycogen metabolism in diabetic liver.

Diabetic liver is characterized by defective glycogen synthesis during fasted to refed transition [16]. The expression of glycogen synthase is not altered in the insulin-deficient

Address for offprints: R.L. Khandelwal, Department of Biochemistry, University of Saskatchewan, 107 Wiggins Road, Saskatoon, SK, S7N 5E5, Canada

diabetes [17]. However, the dephosphorylated active-form (G-6-P independent) is significantly decreased as a result of decrease in protein phosphatase-1 and increase in glycogen synthase kinase-3 activities [18–20]. Recent studies have demonstrated that the mitogenic and the metabolic actions of insulin result from two distinct signal transduction pathways [21, 22]. Insulin-mediated activation of glycogen synthase proceeds through the ras-independent, wortmannin-sensitive, phosphatidylinositol (PI)3-kinase pathway. Pioglitazone has been shown to potentiate insulin action at the level of PI3-kinase in Chinese hamster ovary cells expressing human insulin receptor and 3T3-L1 adipocytes [23, 24]. Hence, this action of pioglitazone is likely to stimulate the metabolic pathways downstream of PI3-kinase. The objective of the present study is, therefore, to examine if pioglitazone can potentiate insulin action on the activation of glycogen synthase and protein phosphatase-1 (PP-1) in cultured diabetic hepatocytes.

Materials and methods

Materials

The reagents for hepatocyte culture were purchased from GIBCO (Burlington, Canada). UDP-[^{14}C]glucose and [γ-^{32}P]ATP were products of NEN. Streptozotocin, UDP-glucose and glucose-6-phosphate were obtained from Sigma Chemical Company (Missouri, USA). Collagenase and okadaic acid were from Worthington Biochemical Corp. (Freehold, USA) and L.C Services (MA, USA), respectively. Phosphorylase b and phosphorylase kinase were purified from rabbit skeletal muscle by the procedures described earlier [18]. [^{32}P]phosphorylase a was prepared from phosphorylase b in presence of [γ-^{32}P]ATP, Mg^{2+} and phosphorylase kinase by the method of Krebs *et al*. [25]. Pioglitazone was a generous gift from the Upjohn Company (Michigan, USA).

Diabetic rats and hepatocyte culture

Male Sprague Dawley rats (150–200 g) were made diabetic by the intraperitoneal injection of streptozotocin (50 mg/kg body weight) dissolved in 100 mM citrate (pH 4.5). Normal control rats received the same volume of citrate buffer. The animals were fed *ad libitum* and kept under a constant 12-h light-dark cycle. Diabetic rats with a non-fasting plasma glucose concentration of 17–20 mM were used one week after STZ injection. In one set of experiment, varying degree of insulin deficiency and hyperglycemia were produced by using 40, 50 and 60 mg/kg of STZ. These

diabetic animals showed blood glucose values ranging from 12–24 mM. In all cases, hepatocytes were isolated from normal and diabetic rats in fed condition by collagenase perfusion as described earlier [26, 27]. The isolated hepatocytes with a cell viability of more than 90% were suspended in serum-free DMEM containing 0.1% bovine serum albumin. Three millions cells were placed in collagen coated plastic dishes (15 × 60 mm) and cultured in a humidified atmosphere (5% CO$_2$) at 37°C. After 4 h, the unattached cells were removed and the culture was continued in the absence and presence of hormones and pioglitazone for 18 h. Glycogen synthase was stimulated at this stage with glucose (30 mM) for different time intervals.

Enzyme assays

For the assay of glycogen synthase, liver cells were homogenized in 20 mM Tris-HCl (pH 7.4) containing 0.25 M sucrose, 0.05 mM dithiothreitol and 50 mM NaF. The homogenates were centrifuged at 2000 × g for 30 min and the supernatants were used for the assay. The active and total enzyme activities were determined by measuring the incorporation of [^{14}C]glucose from UDP-[^{14}C]glucose into glycogen in the absence and presence of glucose-6-P, respectively [18]. The radioactive glycogen was precipitated on Whatman No. 31 ET paper, washed in 66% (v/v) ethanol and counted for radioactivity.

For the assay of PP-1, the cells were homogenized in 50 mM HEPES (7.5) containing 2 mM EDTA, 1 mM DTE, 10 µg/ml aprotinin, 20 µg/ml leupeptin, 1 mM benzamidine and 1 mM phenylmethyl sulfonyl fluoride. After removal of the cytosol by centrifugation at 100,000 × g for 1 h, the pellet was solubilized with 1% Triton X-100 for 1 h at 4°C. The supernatant from centrifugation of this sample was used as the particulate fraction. [^{32}P] labeled phosphorylase was used as the substrate for PP-1 assay [28]. The assay was carried out in presence of okadaic acid (3 nM) to inhibit the activity of protein phosphatase 2A [29]. The released radioactive phosphate was measured in the supernatant after precipitation of proteins with trichloroacetic acid.

Analytical and statistical methods

Blood glucose and plasma insulin levels were measured by the methods described earlier [30]. Protein assay was carried out by the Bradford method using bovine serum albumin as the standard [31]. Statistical analysis was done by the student's *t*-test.

Results

Degree of hyperglycemia vs defective glycogen synthase activation

The defective glycogen synthase activation was first examined in STZ-treated rats with varying degree of diabetes. STZ injection at the doses of 40, 50 and 60 mg/kg in rats produced diabetes with blood glucose values of 12–24 mM. Hepatocytes were isolated from each one of these rats and cultured. The correlation between glycogen synthase activation in hepatocytes and blood glucose levels is shown in Fig. 1. The active glycogen synthase at 1 h after incubation of hepatocytes with 30 mM glucose had significant ($r = 0.93$, $p < 0.001$) inverse correlation with the blood glucose level of the corresponding rat. This finding suggests that the defective glycogen synthase activation in diabetic liver could be one of the factors in the development of hyperglycemia. The degree of insulin deficiency was established by measuring plasma insulin levels in diabetic rats from different groups. STZ-treated rats showed values ranging from 45–75 pM depending on the dose of STZ as compared to 387.2 ± 14.0 pM insulin levels in nondiabetic rats. For further studies, diabetic animals produced with a STZ dose of 50 mg/kg were used.

Reversal of defective glycogen synthase activation by hormones and pioglitazone

The glucose-induced glycogen synthase activation in normal and diabetic hepatocytes as a function of incubation time is shown in Fig. 2A. Glucose increased the active-form of glycogen synthase from 16–57% of the total activity at 1 h in nondiabetic hepatocytes. However, in diabetic hepatocytes the active glycogen synthase increased only to 27% at 1 h. Glucose-6-phosphate-dependent total glycogen synthase activity did not undergo any change in nondiabetic as well as diabetic hepatocytes during this 3 h incubation period (results not shown). Effects of preincubation of diabetic hepatocytes with insulin, dexamethasone and/or pioglitazone on the reversal of defective activation of glycogen synthase is shown in Fig. 2B. Insulin alone significantly ($p < 0.01$) increased the level of active-form of glycogen synthase in diabetic hepatocytes and its action was further potentiated in the presence of dexamethasone. Similarly, the action of insulin was also potentiated ($p < 0.05$–0.01) by pioglitazone. The combined effect of all three compounds showed the maximum activation of glycogen synthase. Pioglitazone or dexamethasone alone (i.e. in the absence of insulin) did not have any effect in the restoration of glycogen synthase activity (results not shown).

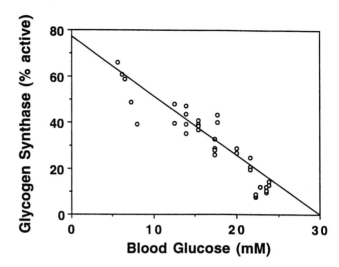

Fig. 1. Effect of varying degree of diabetes on the glucose mediated activation of glycogen synthase. Hepatocytes were isolated from non-diabetic and diabetic rats with varying degrees of hyperglycemia. Cultured hepatocytes were incubated in the medium containing 30 mM glucose for 1 h. Glycogen synthase activity in extracts of these cells was measured as described in the methods section. Correlation of active glycogen synthase (% of total activity) with blood glucose level: $r = 0.93$, $p < 0.001$.

Acute effects of insulin and pioglitazone on PP-1 activation

The defect in activation of glycogen synthase in diabetic hepatocytes could be due to a defect (or decrease in activity) in PP-1. The acute effects of insulin on the particulate PP-1 activity (the form responsible for the dephosphorylation and activation of glycogen synthase) were initially examined in nondiabetic and diabetic hepatocytes. As shown in Fig. 3A, insulin stimulated PP-1 activity significantly in nondiabetic hepatocytes in a time-dependent manner. For example, at 20 min insulin increased PP-1 activity by 60% ($p < 0.001$). In contrast, insulin did not exert any acute effect on PP-1 activity in diabetic hepatocytes. The acute effects of insulin on hepatocytes preincubated with pioglitazone for 18 h is shown in Fig. 3B. As expected, preincubation of nondiabetic hepatocytes with pioglitazone had no effect on PP-1 activity. However, pioglitazone-preincubated nondiabetic hepatocytes showed further increase in insulin-mediated stimulation of PP-1 activity by 28% ($p < 0.01$). Again, insulin did not show any acute stimulatory effect on PP-1 activity in diabetic hepatocytes preincubated without and with pioglitazone.

Long-term effects of hormones and pioglitazone in restoration of PP-1 activity

The long-term effect of insulin and other agents on the basal PP-1 activity in diabetic hepatocytes was further determined

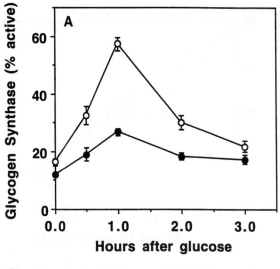

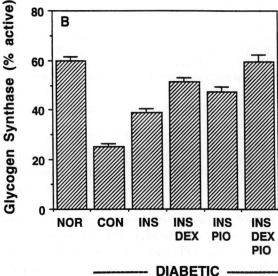

Fig. 2. Effect of diabetes on the activation of glycogen synthase and amelioration by insulin, pioglitazone and dexamethasone. (A) Cultured hepatocytes from nondiabetic (-O-) and diabetic (50 mg/kg streptozotocin;-●- rats were incubated in the medium containing 30 mM glucose for different time intervals. Glycogen synthase activity in extracts of these cells was measured as described in the methods section. The total activity of glycogen synthase measured in presence of glucose-6-phosphate did not alter under these conditions. Each point is the mean ± S.E.M. of 4 independent experiments. (B) Cultured hepatocytes from diabetic (50 mg/kg streptozotocin) rats were preincubated for 18 h in the absence (CON) and presence of 100 nM insulin (INS), 5 μM pioglitazone (PIO) and 100 nM dexamethasone (DEX) in different combinations. Nondiabetic hepatocytes (NOR) were also preincubated in the absence of drugs for comparison. These liver cells were then incubated in the medium containing 30 mM glucose for 1 h. Glycogen synthase activity in extracts of these cells was measured as described in the methods section. The total activity of glycogen synthase measured in presence of glucose-6-phosphate did not alter under these conditions. Each point is the mean ± S.E.M. of 4 independent experiments. Statistical significance. Nondiabetic (NOR) vs diabetic control (CON) hepatocytes as well as the effects of different agents in diabetic hepatocytes: p < 0.001.

and results are shown in Fig. 4. As previously stated, diabetes caused a 52% decrease in the particulate PP-1 activity. Long-term incubation of diabetic hepatocytes with insulin increased PP-1 activity by 36% (p < 0.05). The addition of pioglitazone and insulin together increased the PP-1 activity by 83% (p < 0.001). However, pioglitazone alone in the absence of insulin did not have any effect on the activity of PP-1 (results not shown). The addition of dexamethasone did not significantly enhance the insulin effect. However, the combination of all three agents i.e. insulin, pioglitazone and dexamethasone almost restored PP-1 activity to the non-diabetic level.

Discussion

The antidiabetic agent pioglitazone is known to potentiate the metabolic actions of insulin in various tissues. We have shown in this study that the reversal of defective glycogen synthase activation by insulin in diabetic hepatocytes is potentiated by this agent. Further we have demonstrated that pioglitazone enhances insulin action at the level of PP-1, the enzyme that dephosphorylates and activates glycogen synthase.

The regulation of glycogen turnover is impaired at several steps in diabetic liver and these defects contribute to the development of hyperglycemia. Decreased glycogen synthase activation in response to glucose is one of the defects in diabetic hepatocytes. The elevated blood glucose levels in diabetic rats were found to have an inverse relationship with the glucose-induced activation of glycogen synthase in cultured diabetic hepatocytes (Fig. 1). The defective glycogen synthase activation in diabetic hepatocytes was reversed by incubation with the hormones, insulin and dexamethasone (Fig. 2B). Miller *et al.* [32] earlier made similar observation on the in vitro reversal of defective glycogen synthase activation in diabetic hepatocytes with insulin and steroids. We have demonstrated in this investigation that the effects of these hormones in restoring glycogen synthase activation are significantly potentiated by the antidiabetic agent pioglitazone (Fig. 3B). To further understand the reversal of defect in glycogen synthase activation we examined the acute and chronic effect of insulin on PP-1 activation.

Glycogen synthase undergoes complex multisite phosphorylation in presence of several kinases leading to its inactivation [33]. Dephosphorylation and activation of this enzyme occurs in presence of serine/threonine protein phosphatases [34–36]. These phosphatases are classified into four major types based on their substrate specificity, inhibition by inhibitors 1 and 2 and requirement for divalent cations. Among these, both types 1 and 2A can dephosphorylate glycogen synthase *in vitro*. However, type 1 (PP-1) has been shown to play a significant role *in vivo* [34]. We had provided further evidence in support of the role of PP-1 by using the differential inhibitory effects of calyculin A and okadaic acid in cultured hepatocytes [26]. Four isoforms (α, γ; $\gamma 2$ and δ) of the catalytic subunit of PP-1 have been identified in

several tissues including liver [37]. Takizawa *et al.* [38] have demonstrated that the levels of hepatic catalytic subunit PP-1α have an inverse relation to the blood glucose levels in non-obese diabetic mice.

The native PP-1 is a 1:1 complex of the catalytic subunit and several regulatory subunits which act as targeting subunits [36]. The particulate fraction contains the active form consisting of the catalytic subunit and the glycogen binding subunit. Hepatic protein phosphatase has been shown to be stimulated by acute administration of insulin in perfused liver and inhibited by similar treatment with glucagon [39]. Both insulin and glucocorticoids are also known to induce the synthesis of PP-1 [34]. In the present study, insulin was able to activate PP-1 by short-term incubation with normal hepatocytes (Fig. 3A). This acute stimulation of PP-1 activity by insulin was further enhanced by pioglitazone (Fig. 3B). The decreased PP-1 activity in diabetic hepatocytes were significantly restored by long-term incubation with insulin and this effect was also enhanced by pioglitazone (Fig. 4).

In addition to potentiation of insulin action, thiazo-lidinediones have been shown to have direct insulin-mimetic effects *in vitro*. Troglitazone increases glucose transport in

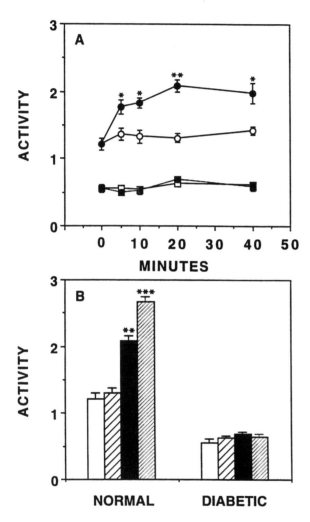

Fig. 3. Acute activation of protein phosphatase-1 by insulin in nondiabetic and diabetic hepatocytes. (A) Cultured hepatocytes from nondiabetic (-O-, -●-) and diabetic (50 mg/kg streptozotocin, -□-, -■-) animals were incubated in the presence (-●-, -■-) and absence (-O-, -□-) of 100 nM insulin for different time intervals. The particulate fractions from the hepatocytes were isolated and the activity of protein phosphatase-1 was measured as described in the methods section. Each point is the mean ± S.E.M. of 3 independent experiments. Statistical significance. Insulin-treated hepatocytes vs. control hepatocytes at respective time intervals: *$p < 0.05$; **$p < 0.01$. (B) Cultured nondiabetic and diabetic hepatocytes were preincubated in the absence (□, ■) and presence of 5 μM pioglitazone (▨, ▨) for 18 h and then stimulated without (□, ▨) and with (■, ▨) 100 nM insulin for 20 min. The particulate fractions from the hepatocytes were isolated and the activity of protein phosphatase-1 was measured as described in Materials and methods section. Values are the mean ± S.E.M. of 3 independent experiments. Statistical significance. Effect of insulin and pioglitazone. **$p < 0.01$, ***$p < 0.001$.

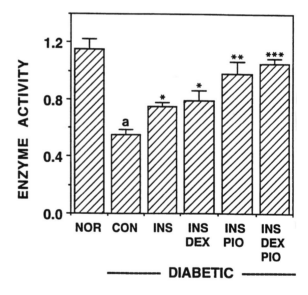

Fig. 4. Restoration of protein phosphatase-1 activity in diabetic hepatocytes by longterm incubation with insulin, pioglitazone and dexamethasone. Cultured hepatocytes from diabetic (50 mg/kg streptozotocin) rats were preincubated for 18 h in the absence (CON) and presence of 100 nM insulin (INS), 5 μM pioglitazone (PIO) and 100 nM dexamethasone (DEX) in different combinations. Nondiabetic hepatocytes (NOR) were also preincubated in the absence of drugs for comparison. The particulate fractions from the hepatocytes were isolated and the activity of protein phosphatase-1 was measured as described in the methods section. Values are the mean ± S.E.M. of 4 independent experiments. Statistical significance. Nondiabetic (NOR) vs diabetic control (CON) hepatocytes: [a]$p < 0.001$. Effects of different agents in diabetic hepatocytes: *$p < 0.05$; **$p < 0.01$; ***$p < 0.001$.

L6 muscle cells and activates glycogen synthase in Hep G2 cells in the absence of insulin [14, 40]. Englitazone has been shown to stimulate glucose uptake in 3T3-L1 adipocytes [8]. However, we did not observe any direct stimulating effect of pioglitazone on glycogen synthase or PP-1 in the absence of insulin (result not shown). Hofmann *et al.* [11] also demonstrated that pioglitazone did not have a direct beneficial effect on the expression of hepatic phosphoenolpyruvate carboxykinase and glucokinase in insulin-deficient diabetic rats whereas it potentiated the insulin action. It is possible that these antidiabetic agents could possess both insulin-sensitizing as well as insulin-mimetic actions depending on the metabolic pathway and the type of cell.

Several mechanisms have been proposed to explain the insulin-sensitizing effect of pioglitazone. This agent stimulates insulin receptor tyrosine kinase activity in the muscle of insulin resistant Wistar fatty rats [11]. It reverses high glucose-induced insulin resistance by normalizing protein tyrosine phosphatase activities in Rat 1 fibroblasts expressing human insulin receptor [41]. The intracellular free magnesium concentration has been shown to be elevated by pioglitazone and magnesium is known to potentiate insulin action at the post-receptor level [42]. Zhang *et al.* [23] have demonstrated that pioglitazone specifically potentiates insulin mediated activation of PI3-kinase in Chinese hamster ovary cell expressing human insulin receptor. It also prevents cAMP-induced antagonistic effect on this enzyme in 3T3-L1 adipocytes, and 3T3-L1 fibroblasts [24]. In a recent study, Bonini *et al.* [43] examined abnormalities in the expression of downstream components of insulin-signaling pathway in KKAY mice and observed partial correction after treatment with pioglitazone. Recent studies have clearly demonstrated that there are two distinct pathways of insulin mediated signal transduction [21, 22]. The mitogenic actions of insulin proceed through the ras-dependent pathway via raf-1 kinase, MEK and MAP kinase. The metabolic actions of insulin such as the activation of PP-1 proceed through ras-independent, wortmannin-sensitive, PI3-kinase-mediated pathway. Our finding that the insulin sensitizing effect of pioglitazone is at a level downstream of PI3-kinase agrees with the previous observation that pioglitazone enhances insulin mediated activation of this enzyme.

Orally effective insulin sensitizers have significant therapeutic value in reversing insulin resistance in diabetic patients. The new class of thiazolidinedione derivatives have been shown to potentiate insulin action by *in vivo* as well as *in vitro* studies. Recent clinical studies have demonstrated the blood glucose lowering effect of troglitazone in NIDDM patients [44, 45]. Pioglitazone has been shown to correct several metabolic aberrations such as enhanced hepatic gluconeogenesis, glucose transporter deficiency and hyper-insulinemia in insulin resistant diabetic animal models such as obese Zucker rats, ob/ob mice, KKAY obese mice and db/db mice [8, 9, 12, 13]. In the present investigation, we have observed that the insulin-mediated reversal of glycogen synthase activation in diabetic hepatocytes is potentiated by pioglitazone. This normalization seems to proceed through activation of PP-1, a crucial step in the complex regulation of glycogen synthase activity. Further studies are needed to examine the effect of pioglitazone on glycogen synthase kinase-3, the enzyme which phosphorylates and inactivates glycogen synthase.

Acknowledgements

This study was supported by an operating grant from the Medical Research Council of Canada.

References

1. Olefsky JM, Garvey WT, Henry RR, Britton D, Matthaei S, Freidenberg GR: Cellular mechanisms of insulin resistance in non-insulin-dependent (type II) diabetes. Am J Med 85: (Suppl 5A) 86–105, 1988
2. DeFronzo RA, Hendler R, Simonson D: Insulin resistance is a prominent feature of insulin-dependent diabetes. Diabetes 31: 795–801, 1982
3. Momose Y, Meguro K, Ikeda H, Hatanaka C, Oi S, Sohda T: Studies on antidiabetic agents. X. Synthesis and biological activities of Pioglitazone and related compounds. Chem Pharm Bull 39: 1440–1445, 1991
4. Sohda T, Mizuno K, Momose Y, Ikeda H, Fujita T, Meguro K: Studies on antidiabetic agents. 11. Novel thiazolidinedione derivatives as potent hypoglycemic and hypolipidemic agents. J Med Chem 35: 2617–2626, 1992
5. Hofmann CA, Colca JR: New oral thiazolidinedione antidiabetic agents act as insulin sensitizers. Diabetes Care 15: 1075–1078, 1992
6. Hofmann C, Lorenz K, Colca JR: Glucose transport deficiency in diabetic animals is corrected by treatment with the oral antihyperglycemic agent pioglitazone. Endocrinology 129: 1915–1925, 1991
7. Kemnitz JW, Elson DF, Roecker EB, Baum ST, Bergman RN, Meglasson MD: Pioglitazone increases insulin sensitivity, reduces blood glucose, insulin, and lipid levels, and lowers blood pressure in obese, insulin-resistant rhesus monkeys. Diabetes 43: 204–211, 1994
8. Stevenson RW, Hutson NJ, Krupp MN, Volkmann RA, Holland GF, Eggler JF, Clark DA, McPherson RK, Hall KL, Danbury BH, Gibbs EM, Kreutter DK: Actions of novel antidiabetic agent englitazone in hyperglycemic hyperinsulinemic ob/ob mice. Diabetes 39: 1218–1227, 1990
9. Bowen L, Stein PP, Stevenson R, Shulman GI: The effect of CP 68,722, a thiozolidinedione derivative, on insulin sensitivity in lean and obese Zucker rats. Metabolism 40: 1025–1030, 1991
10. Kobayashi M, Iwanishi M, Egawa Katsuya, Shigeta Y: Pioglitazone increases insulin sensitivity by activating insulin receptor kinase. Diabetes 41: 476–483, 1992
11. Hofmann C, Lorenz K, Williams D, Palazuk BJ, Colca JR: Insulin senitization in diabetic rat liver by an antihyperglycemic agent. Metabolism 44: 384–389, 1995
12. Hofmann CA, Edwards CW III, Hillman RM, Colca JR: Treatment of insulin-resistant mice with the oral antidiabetic agent Pioglitazone: Evaluation of liver GLUT2 and phosphoenolpyruvate carboxykinase expression. Endocrinology 130: 735–740, 1992

13. Fujiwara T, Okuno A, Yoshioka S, Horikoshi H: Suppression of hepatic gluconeogenesis in long-term troglitazone treated diabetic KK and C57BL/]Ksj-db/db mice. Metabolism 44: 486–490, 1995

14. Ciaraldi TP, Gilmore A, Olefsky JM, Goldberg M, Heidenreich KA: *In vitro* studies on the action of CS-045, a new antidiabetic agent. Metabolism 39: 1056–1062, 1990

15. Blackmore PF, McPherson K, Stevenson RW: Actions of the novel antidiabetic agent Englitazone in rat hepatocytes. Metabolism 42: 1583–1587, 1993

16. Van de Werve G, Sestoft L, Folke M, Kristensen LO: The onset of liver glycogen synthesis in fasted-refed rats. Effects of streptozotocin diabetes and of peripheral insulin replacement. Diabetes 33: 944–949, 1984

17. Rao PV, Pugazhenthi S, Khandelwal RL: The effects of streptozotocin-induced diabetes and insulin supplementation on expression of the glycogen phosphorylase gene in rat liver. J Biol Chem 270: 24955–24960, 1995

18. Pugazhenthi S, Khandelwal RL: Insulin-like effects of vanadate on hepatic glycogen metabolism in nondiabetic and streptozotocin-induced diabetic rats. Diabetes 39: 821–827, 1990

19. Foulkes JG, Jefferson LS: Protein phosphatase-1 and -2A activities in heart, liver, and skeletal muscle extracts from control and diabetic rats. Diabetes 33: 576–579, 1984

20. Rao PV, Pugazhenthi S, Khandelwal RL: Insulin decreases the glycogen synthase kinase-3α mRNA levels by altering its stability in streptozotocin-induced diabetic rat liver. Biochem Biophys Res Commun 217: 250–256, 1995

21. Lazar DF, Wiese RJ, Brady MJ, Mastick CC, Waters SB, Yamauchi K, Pessin JE, Cuatrecasas P, Saltiel AR: Mitogen-activated protein kinase kinase inhibition does not block the stimulation of glucose utilization by insulin. J Biol Chem 270: 20801–20807, 1995

22. Sakaue H, Hara K, Noguchi T, Matozaki T, Kotani K, Ogawa W, Yonezawa K, Waterfield MD, Kasuga M: Ras-independent and wortmannin-sensitive activation of glycogen synthase by insulin in Chinese hamster ovary cells. J Biol Chem 270: 11304–11309, 1995

23. Zhang B, Szalkowski D, Diaz E, Hayes N, Smith R, Berger L: Potentiation of instimulation of phosphatidylinositol 3-kinase by thiazolidine-derived antidiabetic agents in Chinese hamster ovary cell expressing human insulin receptors and L6 myotubes. J Biol Chem 269: 25735–25741, 1994

24. Sizer KM, Smith CL, Jacob CS, Swanson ML, Bleasdale JE: Pioglitazone promotes insulin-induced activation of phosphoinositide 3-kinase in 3T3-L1 adipocytes, by inhibiting a negative control mechanism. Mol Cell Endocrinol 102: 119–129, 1994

25. Krebs EG, Kent AB, Fischer EH: The muscle phosphorylase b kinase reaction. J Biol Chem 231: 73–78, 1958

26. Pugazhenthi S, Yu B, Gali RR, Khandelwal RL: Differential effects of calyculin A and okadaic acid on the glucose-induced regulation of glycogen synthase and phosphorylase activities in cultured hepatocytes. Biochim Biophys Acta 1179: 271–276, 1993

27. Yu B, Pugazhenthi S, Khandelwal RL: Effects of metformin on glucose and glucagon regulated gluconeogenesis in cultured normal and diabetic hepatocytes. Biochem Pharmacol 48: 949–954, 1994

28. Khandelwal RL, Vandenheede JR, Krebs EG: Purification, properties and substrate specificities of phosphoprotein phosphatases(s) from rabbit liver. J Biol Chem 251: 4850–4858, 1976

29. Srinivasan M, Begum N: Regulation of protein phosphatase 1 and 2A activities by insulin during myogenesis, in rat skeletal muscle cells in culture. J Biol Chem 269: 12514–12520, 1994

30. Pugazhenthi S, Angel JF, Khandelwal RL: Long-term effects of vanadate treatment on glycogen metabolizing and lipogenic enzymes of liver in genetically diabetic (db/db) mice. Metabolism 40: 941–946, 1991

31. Bradford MM: A rapid and sensitive method for the quantitation of protein utilizing the principle of protein dye binding. Anal Biochem 72: 246–254, 1976

32. Miller TB Jr, Garnache AK, Cruz J, McPherson K, Wolleben C: Regulation of glycogen metabolism in primary cultures of rat hepatocytes. J Biol Chem 261: 785–790, 1986

33. Pugazhenthi S, Khandelwal RL: Regulation of glycogen synthase activation in isolated hepatocytes. Mol Cell Biochem 149/150: 95–101, 1995

34. Cohen P: The structure and regulation of protein phosphatases. Ann Rev Biochem 58: 453–508, 1989

35. Shenolikar S, Nairn AC: Protein phosphatases: Recent progress. Adv second messenger phosphoprotein Res 23: 1–121, 1991

36. Wera S, Hemmings BA: Serine/threonine protein phosphatases. Biochem J 311: 17–29, 1995

37. Sasaki K, Shima H, Kitagawa Y, Irino S, Sugimura T, Nagao M: Identification of members of the protein phosphatase 1 gene family in the rat and enhanced expression of protein phosphatase 1a gene in rat hepatocellular carcinomas. Jpn J Cancer Res 81: 1272–1280, 1990

38. Takizawa N, Mizuno Y, Ito Y, Kikuchi K: Tissue distribution of isoforms, of type-1 protein phosphatase PP-1 in mouse tissues and its diabetic alterations. J Biochem 116: 411–415, 1994

39. Toth B, Bollen M, Stalmans W: Acute regulation of hepatic protein phosphatases by glucagon, insulin and glucose. J Biol Chem 263: 14061–14066, 1988

40. Ciaraldi TP, Huber-Knudsen K, Hickman M, Olefsky JM: Regulation of glucose transport in cultured muscle cells by novel hypoglycemic agents. Metabolism 44: 976–982, 1995

41. Maegawa H, Ide R, Hasegawa M, Ugi S, Egawa K, Iwanishi M, Kikkawa R, Shigeta Y, Kashiwagi A, Thiazolidine derivatives ameliorate high glucose-induced insulin resistance via the normalization of protein-tyrosine phosphatase activities. J Biol Chem 270: 7724–7730, 1995

42. Nadler J, Scott S: Evidence that pioglitazone increases intracellular free magnesium concentration in freshly isolated rat adipocyte. Biochem Biophys Res Commun 202: 416–421, 1994

43. Bonini JA, Colca JR, Dailey C, White M, Hofmann C: Compensatory alterations for insulin signal transduction and glucose transport in insulin-resistant diabetes. Am J Physiol 269: E759–E765, 1995

44. Kuzuya T, Iwamoto Y, Kosaka K, Takebe K, Ymanouchi T, Kasuga M, Kajinuma H, Akanuma Y, Yoshida S, Shigeta Y, Baba S: A pilot clinical trial of a new oral hypoglycemic agent, CS-045 in patients with non-insulin dependent diabetes mellitus. Diabetes Res Clin Pract 11: 147–154, 1991

45. Suter SL, Nolan JJ, Wallace P, Gumbiner B, Olefsky JM: Metabolic effects of new oral hypoglycemic agent, CS-045 in NIDDM subjects. Diabetes Care 15: 193–203, 1992

Molecular and Cellular Biochemistry **182:** 193–194, 1998.

Index to Volume 182